I0060877

Frontiers in Clinical Drug Research - Dementia

(Volume 1)

Edited by
Prof. Atta-ur-Rahman, *FRS*
Kings College, University of Cambridge, Cambridge, UK

Frontiers in Clinical Drug Research - Dementia

Volume # 1

Editor: Atta-ur-Rahman, *FRS*

ISBN (Print): 978-981-14-1093-2

© 2020, Bentham Books imprint.

Published by Bentham Science Publishers Pte. Ltd. Singapore. All Rights Reserved.

BENTHAM SCIENCE PUBLISHERS LTD.
End User License Agreement (for non-institutional, personal use)

This is an agreement between you and Bentham Science Publishers Ltd. Please read this License Agreement carefully before using the book/echapter/ejournal (**"Work"**). Your use of the Work constitutes your agreement to the terms and conditions set forth in this License Agreement. If you do not agree to these terms and conditions then you should not use the Work.

Bentham Science Publishers agrees to grant you a non-exclusive, non-transferable limited license to use the Work subject to and in accordance with the following terms and conditions. This License Agreement is for non-library, personal use only. For a library / institutional / multi user license in respect of the Work, please contact: permission@benthamscience.net.

Usage Rules:

1. All rights reserved: The Work is the subject of copyright and Bentham Science Publishers either owns the Work (and the copyright in it) or is licensed to distribute the Work. You shall not copy, reproduce, modify, remove, delete, augment, add to, publish, transmit, sell, resell, create derivative works from, or in any way exploit the Work or make the Work available for others to do any of the same, in any form or by any means, in whole or in part, in each case without the prior written permission of Bentham Science Publishers, unless stated otherwise in this License Agreement.
2. You may download a copy of the Work on one occasion to one personal computer (including tablet, laptop, desktop, or other such devices). You may make one back-up copy of the Work to avoid losing it.
3. The unauthorised use or distribution of copyrighted or other proprietary content is illegal and could subject you to liability for substantial money damages. You will be liable for any damage resulting from your misuse of the Work or any violation of this License Agreement, including any infringement by you of copyrights or proprietary rights.

Disclaimer:

Bentham Science Publishers does not guarantee that the information in the Work is error-free, or warrant that it will meet your requirements or that access to the Work will be uninterrupted or error-free. The Work is provided "as is" without warranty of any kind, either express or implied or statutory, including, without limitation, implied warranties of merchantability and fitness for a particular purpose. The entire risk as to the results and performance of the Work is assumed by you. No responsibility is assumed by Bentham Science Publishers, its staff, editors and/or authors for any injury and/or damage to persons or property as a matter of products liability, negligence or otherwise, or from any use or operation of any methods, products instruction, advertisements or ideas contained in the Work.

Limitation of Liability:

In no event will Bentham Science Publishers, its staff, editors and/or authors, be liable for any damages, including, without limitation, special, incidental and/or consequential damages and/or damages for lost data and/or profits arising out of (whether directly or indirectly) the use or inability to use the Work. The entire liability of Bentham Science Publishers shall be limited to the amount actually paid by you for the Work.

General:

1. Any dispute or claim arising out of or in connection with this License Agreement or the Work (including non-contractual disputes or claims) will be governed by and construed in accordance with the laws of Singapore. Each party agrees that the courts of the state of Singapore shall have exclusive jurisdiction to settle any dispute or claim arising out of or in connection with this License Agreement or the Work (including non-contractual disputes or claims).
2. Your rights under this License Agreement will automatically terminate without notice and without the

need for a court order if at any point you breach any terms of this License Agreement. In no event will any delay or failure by Bentham Science Publishers in enforcing your compliance with this License Agreement constitute a waiver of any of its rights.

3. You acknowledge that you have read this License Agreement, and agree to be bound by its terms and conditions. To the extent that any other terms and conditions presented on any website of Bentham Science Publishers conflict with, or are inconsistent with, the terms and conditions set out in this License Agreement, you acknowledge that the terms and conditions set out in this License Agreement shall prevail.

Bentham Science Publishers Pte. Ltd.
80 Robinson Road #02-00
Singapore 068898
Singapore
Email: subscriptions@benthamscience.net

**BENTHAM
SCIENCE**

CONTENTS

PREFACE

The book series *Frontiers in Clinical Drug Research-Dementia* presents cutting edge reviews written by the specialists in the field. The chapters in the 1st volume focus on drug research with special emphasis on clinical trials, and research on drugs in advanced stages of development for dementia and related disorders.

Khow & Yong, in chapter 1, discuss the challenges of bone and hip fractures in patients of Alzheimer's disease. Djordjevic *et al.*, in chapter 2, discuss the role of cholesterol in brain health and pathologies with special emphasis on Alzheimer's disease. Chapter 3 by Lake *et al.* reviews the advances in the treatment of Mild Cognitive Impairment (MCI) and dementia.

Chapter 4 by Uslu *et al.* gives an overview of analytical methods for the drugs used in the treatment of this disease. Kaushik and Jha, in chapter 5, present the nanotechnological advancements for the treatment of Alzheimer's disease. Chapter 6 by Mushtaq *et al.* presents the details of current challenges in Alzheimer's disease. The last chapter by Hani Nasser Abdelhamid, summarizes recent studies on the links between metals and Alzheimer's disease.

I am grateful to all the eminent scientists for their excellent contributions. The efforts of Ms. Fariya Zulfiqar (Manager Publications) and the leadership of Mr. Mahmood Alam (Director Publications) are greatly appreciated.

Prof. Atta-ur-Rahman, *FRS*
Kings College
University of Cambridge
Cambridge
UK

List of Contributors

Aleksandra Mladenovic Djordjevic	Institute for Biological Research "Sinisa Stankovic", University of Belgrade, Belgrade, Serbia
Amrina Shafi	Department of Biotechnology, University of Kashmir, Srinagar, Jammu and Kashmir-190006, India
Bengi Uslu	Department of Analytical Chemistry, Faculty of Pharmacy, Ankara University, Ankara, Turkey
Burcin Bozal-Palabiyik	Department of Analytical Chemistry, Faculty of Pharmacy, Ankara University, Ankara, Turkey
Cem Erkmen	Department of Analytical Chemistry, Faculty of Pharmacy, Ankara University, Ankara, Turkey
Dennis Chang	NICM Health Research Institute, Western Sydney University, Westmead, New South Wales, Australia
Dhriti Jha	Bhaskaracharya College of Applied Sciences, University of Delhi, Delhi, India
Firdous A. Khanday	Department of Biotechnology, University of Kashmir, Srinagar, Jammu and Kashmir-190006, India
Hani Nasser Abdelhamid	Advanced Multifunctional Materials Laboratory, Department of Chemistry, Assiut University, Assiut 71516, Egypt
James Lake	NICM Health Research Institute, Western Sydney University, Westmead, New South Wales, Australia
Kareeann Khow	Aged and Extended Care Services, The Queen Elizabeth Hospital, Woodville Road, Woodville South, SA 5011, Australia Adelaide Geriatric Training Research and Aged Care Centre, University of Adelaide, Adelaide SA 5005, Australia
Kosara Smiljanic	Institute for Biological Research "Sinisa Stankovic", University of Belgrade, Belgrade, Serbia
Mahima Kaushik	Cluster Innovation Centre, University of Delhi, Delhi, India
Natasa Loncarevic-Vasiljkovic	Institute for Biological Research "Sinisa Stankovic", University of Belgrade, Belgrade, Serbia
Selma Kanazir	Institute for Biological Research "Sinisa Stankovic", University of Belgrade, Belgrade, Serbia
Tuck Yean Yong	Internal Medicine, Flinders Private Hospital, Adelaide, South Australia, Australia
Umar Mushtaq	Department of Biotechnology, School of life Science, Central University of Kashmir, Ganderbal, Jammu and Kashmir-191201, India
Zahra Ayati	Department of Traditional Pharmacy, School of Pharmacy, Mashhad University of Medical Sciences, Mashhad, Iran

Frontiers in Clinical Drug Research - Dementia, 2020, Vol. 1, 1-25

Meeting the Challenges of Falls and Hip Fractures in People with Alzheimer's Disease

Kareeann Sok Fun Khow[1,2,*] and **Tuck Yean Yong**[3]

[1] *Aged and Extended Care Services, The Queen Elizabeth Hospital, Woodville Road, Woodville South, SA 5011, Australia*

[2] *Adelaide Geriatric Training Research and Aged Care Centre, University of Adelaide, Adelaide SA 5005, Australia*

[3] *Internal Medicine, Flinders Private Hospital, Adelaide, South Australia, Australia*

Abstract: Falls and hip fractures are common conditions among older people with Alzheimer's disease (AD) and are associated with high risk of morbidity and mortality. People with AD have up to an 8-fold increased risk of falling and 3-fold higher risk of hip fractures, compared with those who are cognitively intact. The increased risk of hip fractures among people with dementia may occur through a few pathways, including (a) risk factors that are common to both conditions, (b) the presence of dementia increasing incidence of hip fracture through intermediate risk factors such as falls and osteoporosis, and (c) side effects of treatment used in AD increasing hip fracture risk. A better understanding of these mechanisms and their effects on outcomes after hip fracture will assist in developing effective interventions and improving preventive strategies. Population aging heightens the need to recognize the interactions of these conditions in order to improve efforts to prevent hip fractures, improve outcomes through high-quality acute care and rehabilitation that returns patients to premorbid level of functioning, and provide evidence-based secondary prevention of falls or fragility fractures. Acute care of hip fractures focusing on orthogeriatric co-management has been shown to reduce length of hospital stay, perioperative complications including delirium, readmission rate and premature mortality. Secondary prevention of falls and further fractures is essential by ensuring risk factors for falling are addressed and osteoporosis is treated. New experimental approaches are being investigated to manage osteoporosis through surgical approaches in people with extremely high risk of recurrent hip fractures.

Keywords: Alzheimer's disease, Cognitive impairment, Dementia, Falls, Hip fracture, Orthogeriatric, Prevention, Rehabilitation, Risk factors, Treatment.

---

* **Corresponding author Dr Kareeann Khow:** Aged and Extended Care Services, The Queen Elizabeth Hospital, Woodville Road, Woodville South, SA 5011, Australia; Tel: +61 8 8222 8178, Fax: +61 8 8222 8593; E-mail: kareeann.khow@adelaide.edu.au

Atta-ur-Rahman (Ed.)
All rights reserved-© 2020 Bentham Science Publishers

INTRODUCTION

Alzheimer's disease (AD) is the most frequent form of dementia in many countries. Falls and hip fractures are common interrelated problems among older people with AD. The prevalence of AD increases with age, rising from 3.0% among those aged 65 to 74 years to 18.7% among the 75 to 84 years age group and even higher at 47.2% in people over 85 years [1]. Globally, the prevalence of dementia has been doubling every 20 years and is projected to affect 81 million by 2040, with many being affected by AD [2]. AD is a progressive condition that can interfere with self-care and social functioning.

Older people are more likely to fall resulting in catastrophic and life-threatening events. In addition, falls also have adverse consequences on their quality of life and that of their families or carers. It is estimated that 28-35% of community dwelling individuals older than 65 years fall each year and more than 50% of those living in residential care facilities [3]. In the United States, the absolute number of fall-related deaths has almost tripled from 8613 in 2000 to 25819 in 2016, according to the National Vital Statistics System [4]. The same study also reported that the age-adjusted fall-related mortality rate has doubled, rising from 46.3 to 105.9 per 100,000 women and from 60.7 to 116.4 per 100,000 men [4]. Falls also impose a heavy financial cost on the healthcare system. In 2015, the total medical costs attributable to fatal and non-fatal falls in the United States was nearly $50 billion [5].

About 85% of hip fractures happen in people over the age of 65 years [6]. With an aging population, the number of hip fractures worldwide is expected to rise from 1.7 million in 1990 to 6.3 million by 2050 [7]. In the first year after a hip fracture, only 50% regain their pre-fracture ambulation level and about 25% of people living in the community before the fracture would require care in a residential facility [8, 9]. Within three months after hip fracture, mortality is five to eight-fold higher than in age- and sex-matched controls and this excess mortality persists even after 10 years [10, 11].

This chapter will review the risk factors for falls and hip fractures in people with AD. This review will also describe the evidence for effective acute and post-acute care for hip fractures. In addition, effective strategies for primary and secondary prevention of falls and hip fractures in AD will also be examined.

RISK FACTORS FOR FALLS IN ALZHEIMER'S DISEASE

Patients with dementia including those with AD experience an eight-fold higher risk of falls compared to those without dementia, and especially for sustaining recurrent falls [12]. In one case-control study of four-year duration, 36% of patients with AD experienced a serious fall compared to 11% among those without the condition [13]. Several risk factors for falls are more prominent among people with AD (Table **1**).

Table 1. Risk factors for falls in people with Alzheimer's disease.

Changes in gait
• Shorter stride length
• Reduced gait speed
• Increased step-to-step variability
• Lower stepping frequency
• Increased double support ratio
• Increased sway path
Neurovascular instability
Delirium, especially acute hospital setting
Behavioural and psychological symptoms
Use of psychotropic medications

The ability to walk without falling depends on an intact motor, sensory, balance, postural reflexes, and vision to maintain a steady upright position and to move safely. Moreover, cognitive abilities including attention, reaction time, executive function, visuospatial skills, and navigation are important for walking safely. Studies have found that people with AD have a shorter stride length, slower gait speed, greater step-to-step variability, lower stepping frequency, increased double support ratio (more time spent in stance phase), and larger sway path than those without the condition [14 - 16]. Stride length and gait speed have been found to be associated with the risk of falls [15]. About 40% of people with AD have a gait apraxia [17]. Impaired central integration of signals for maintaining gait and balance is thought to contribute to apraxia in AD [18]. People with AD can also be impulsive and take unnecessary risks due to impaired insight and lack of perception in relation to environmental hazards [19].

AD is associated with a high prevalence of autonomic dysfunction, leading to postural hypotension [20, 21]. This may lead to falls and hip fractures. In addition, a study of participants with AD found a prevalence of carotid sinus hypersensitivity in 28% [22].

In the acute hospital setting, delirium is common among people with AD and is a major risk factor for falls [23]. Delirium persisting after hospital admission is associated with a six-fold increase in the risk of falls [24].

Behavioural and psychological symptoms are common among people with AD. Behavioural symptoms, such as wandering, restlessness, and attention seeking, are associated with higher risk of falls [25, 26]. Psychological symptoms, namely depression, psychosis, and paranoia are associated with higher risk of falls [25, 26]. In addition, medications used to manage these symptoms, such as antidepressants and antipsychotics, are also risk factors for increasing falls [27]. Tricyclic antidepressants (TCAs) poses a high risk for falls and should be avoided in older people [28]. Anticholinergic side effects of TCAs include cognitive impairment and delirium [28]. All have some alpha-blocking activity and can cause orthostatic hypotension [28]. The antihistamine properties of this group of drugs can also cause sedation, impair balance, and slow reaction times. All these effects can potentiate the risk of falling in older people. Selective serotonin reuptake inhibitors (SSRIs) can cause impaired psychomotor function, ataxia, and syncope, resulting in increased fall risk [28]. Serotonin norepinephrine reuptake inhibitor (SNRI) antidepressants are similar adverse effects associated with SSRIs including orthostatic hypotension [28].

Older people including those with AD, are commonly exposed to drugs with anticholinergic effects. Exposure to drugs with anticholinergic effect as a class was associated with increased odds of cognitive impairment (odds ratio, OR 1.45; 95% Confidence interval, CI 1.16-1.73) [29]. Meta-analyses results for falls showed a 116% and 79% increase in the odds of a fall occurring with the use of olanzapine and trazodone, respectively [29].

The relationship between the severity of AD and risk of falls is uncertain. In a study of 97 nursing home residents with AD, those with mild disease had a 20% incidence of falls, and 38% among those with moderate disease [30]. Another study of community-dwelling older people with cognitive impairment, the risk of falls increased by 70% for every 10-point decrease in Mini-Mental State Examination (MMSE) score [31]. However, other studies have not found similar association between falls and duration of dementia [13, 21].

One study evaluated the risk of falls and fractures in a cohort with AD (mean MMSE score of 17.6 ± 7.5) [12]. In this study, axial rigidity, poor tandem gait, agitation and wandering were associated with lower MMSE scores, and these characteristics could elevate the risk of falls [12].

COMMON RISK FACTORS FOR ALZHEIMER'S DISEASE AND HIP FRACTURES

Both AD and hip fracture share several common risk factors. These include increasing age, vitamin D insufficiency, reduced bone mineral density (BMD), weight loss and depression (Table **2**).

Table 2. Common risk factors shared between Alzheimer's disease and hip fractures.

Advanced age
Vitamin D insufficiency
Reduced bone mineral density
Weight loss
Depression

Advanced age is a risk factor commonly present in both AD and hip fractures [32, 33]. Older people with cognitive impairment are at significantly higher risk of sustaining a hip fracture than those who are cognitively intact [34]. In a study by Melton and colleagues, hip fracture risk over 10 years following the diagnosis of AD was 2.7-fold higher than in age- and sex-matched controls [34]. In addition, the incidence slope is relatively constant over time [34].

Vitamin D is an important intermediate risk factor between dementia and hip fracture. The prevalence of vitamin D among people with AD is high and increases as the disease progresses in severity [35 - 37]. In one study of patients with AD living in residential aged care facilities, 54% had severe 25–hydroxy–vitamin D deficiency with concentration of less than 5 ng/mL [37]. There is also some data suggesting that low vitamin D may be associated with cognitive decline [38]. People with both AD and hip fractures have a lower dietary intake of calcium and vitamin D than AD without fractures [36]. Similarly, sunlight exposure is lowest in people with AD and fractures [37].

People with AD have been found to have lower bone mineral density (BMD) but other studies have not supported this finding [36, 39, 40]. In a study of 140 participants, people with early AD were found to have lower mean BMD compared to those without, after adjustment of age, sex, smoking, physical activity, oestrogen replacement, depression, and apoE4 carrier status [41].

Patients with AD also tend to have lower body mass index (BMI) compared with those who are cognitively intact [42]. In addition, weight loss is relatively common among people with AD, especially during the late stages [43]. In turn, low BMI is associated with increased risk of hip fracture, as reported in several but not all studies [44 - 47]. Evaluation of community-dwelling older women and men have found that a weight loss of 10% or more was associated with an adjusted relative risk for hip fracture of 1.8 and 2.9, respectively [45, 46]. On the other hand, weight gain of 10% or greater was associated with a relative risk reduction of 0.4 and 0.7, respectively [45, 46].

Another common risk factor shared by both AD and hip fracture is depression.

Depression is prevalent among community-dwelling people with possible or probable AD, estimated to be 20% and 13%, respectively [48]. A meta-analysis of 14 prospective studies have found that depression was associated with a hazard ratio of 1.17 for sustaining a fracture [49, 50]. In addition, depression is frequently associated with risk factors for hip fractures, namely higher risk of falls, reduced bone mineral density, higher prevalence of smoking, more sedentary lifestyle and lower calcium intake [51 - 53]. Furthermore, people with depression also have an increased risk of developing dementia [54]. Medications used for the treatment of depression can also influence the risk for hip fractures. Antidepressants can contribute to bone loss, increase falls and fracture risk [55 - 57].

Patients with AD are more likely to take medications that are associated with increased risk for falls and hip fractures [27, 58]. Antipsychotics and anxiolytics, which are used in many patients with AD, can increase the risk of hip fracture. In particular, anxiolytics with a longer half-life are associated with an OR of 1.8 for sustaining a hip fracture [59]. However, anxiolytics have a limited role in the management of AD.

HIP FRACTURE AND COGNITIVE DECLINE

There is also evidence that sustaining a hip fracture can lead to cognitive decline. In a small study, 25 out of 26 patients with both AD and hip fracture had the onset of dementia after the fracture [34]. This observation may have arisen because patients had further cognitive decline after experiencing a hip fracture or the fracture event brought the diagnosis of AD to clinical attention. The latter postulation is probably more likely.

Delirium is a common complication after hip fracture surgery, affecting 35-65% of patients [60]. In some patients, delirium can persist for months after hip fracture surgery. In a study of patients who had hip fracture surgery, delirium persisted for a month in 32% and for 6 months in 6% [61]. Therefore, delirium may also be an important contributing factor for cognitive decline after hip fracture [62].

PREVENTING AND MANAGING HIP FRACTURES IN ALZHEIMER'S DISEASE

Hip fracture prevention through falls prevention and maintaining good bone health is paramount among people with AD (Table **3**). In the event that hip fractures has occurred, patients with AD should be managed according to evidence-based acute care and secondary preventive measures.

Table 3. Strategies for preventing and managing hip fractures in people with Alzheimer's disease.

Primary prevention of falls and hip fractures
• Multifactorial intervention addressing falls risk factors
• Management of bone health
Acute care
• Orthogeriatric co-management
• Optimal pain control
• Prevention and management of delirium
• Early mobilization
• Education of family and carers
Post-acute care
• Rehabilitation
• Secondary prevention of falls
• Management of osteoporosis
• Management of depression

PRIMARY PREVENTION OF FALLS AND HIP FRACTURES

Interventions to reduce falls are largely based on research among older people without cognitive impairment. There is limited evidence for effective falls intervention among people with dementia. In a Cochrane review of falls prevention, participants with cognitive impairment were specifically excluded from 66 of 111 studies among community-dwelling older people [63]. In addition, cognitive impairment was not specified as an inclusion or exclusion criterion in 44 studies. Only one of 41 studies conducted in residential facilities included participants with dementia [64]. This study involved 274 participants with a mean MMSE score of 13 who received multifactorial intervention based on risk factors for falls [64]. There was no difference in the proportion of participants who experienced falls in between those who received the intervention and those who did not during follow-up of a year.

Falls Prevention

Falls prevention interventional studies are urgently needed to determine which risk factors are modifiable and how the interventions can be best delivered among people with AD. In the absence of evidence, general measures such as discontinuation of inappropriate medications, minimise postural hypotension and modifications of environmental hazards are vital for preventing falls in this group of older people.

Falls prevention should begin with asking patients or their caregivers about falls. Older people with AD who have had a fall or fear falling should have their balance and gait evaluated [65]. If balance or gait is impaired, a thorough evaluation of fall risk factors should include assessing visual impairment, postural blood pressure change, footwear, and medications [65]. Older people with AD and impaired balance or gait should be referred for supervised exercises [65, 66]. Inappropriate medication use is a common modifiable risk factor for falls in older people including those with AD [66]. Evidence support the rationalisation of, either by reducing or discontinuing, psychotropics and cardiometabolic drugs as an effective strategy of reducing the risk of falls [66].

Managing and Preventing Osteoporosis

Preventing hip fracture is a high priority in older people because more than 80% reported that they would prefer to die rather than suffer a hip fracture resulting in requiring long-term residential care [67]. Management of osteoporosis and vitamin D deficiency is one of the cornerstones of reducing hip fractures in older people including among those with AD. Osteoporosis remains a commonly underdiagnosed and undertreated condition in many older people, contrary to evidence-based recommendations. Misunderstandings among clinicians about the treatment of osteoporosis among people with dementia represent a missed opportunity for treating this condition and preventing fragility fractures [68]. Although current recommendations suggest using 10-year fracture risk estimation to guide pharmacological treatment, such approach do not address decision-making required among patients with life expectancies less than 10 years [69]. Currently used hip fracture risk estimator do not include many comorbidities or frailty features common among older people which would be important in the consideration of preventive pharmacological treatment for osteoporosis.

Similar to the general population, fracture risk can be estimated with a number of tools in people with Alzheimer's disorder [69]. Bone mineral density (BMD) can be determined by dual-energy x-ray absorptiometry (DXA) and T-scores generated are a strong predictor of future fractures, which are also used in some fracture risk predictor tools. The FRAX® tool (University of Sheffield) is the most validated and commonly used calculator for fracture risk assessment [70]. A 10-year estimated risk between 10% and 20% would indicate a moderate fracture risk while an estimated risk higher than 20% indicates a high risk. Although the FRAX® tool has many strengths for fracture risk estimation, this prediction model does not take into account risk factors in older people including cognitive impairment, falls, urinary incontinence, neurological disorders and polypharmacy. Therefore, some of these additional factors need to be taken into consideration when a person with AD is near the threshold for pharmacological intervention.

Among people with AD, treatment of osteoporosis has been found to be effective for prevention of hip fractures in a randomized controlled trial [71]. In this study of 500 participants with AD with a mean MMSE of 16.5 who were randomized to receive risedronate with calcium and vitamin D or calcium and vitamin D only, a relative risk reduction of 74% was observed in the bisphosphonate group. The rate of serious side effects was low and comparable between the two study groups.

People with AD may have reduced adherence oral bisphosphonate therapy in the long-term. Therefore, consideration can be given to using once a year intravenous zoledronic acid or subcutaneous denosumab once every six months [72, 73]. Although not specifically studied in people with cognitive impairment, zoledronic acid and denosumab has been demonstrated to be effective in reducing the risk of hip fractures [72, 73].

Medical practitioners can be conflicted between deprescribing and commencing anti-osteoporosis medications to reduce fracture risk. Nevertheless, this concurrent approach of rationalising medications that cause falls and starting a medication of osteoporosis is appropriate for older people with multiple coexisting medical conditions requiring prevention of fragility fractures.

ACUTE TREATMENT OF HIP FRACTURES

The treatment of choice for almost all hip fractures involves surgical repair because this approach enables faster pain control and functional preservation. AD is not a contraindication for hip fracture surgery except for patients who have other coexisting medical illness that can lead to the risk of surgery exceeding its benefits. Sheehan *et al* have found that cognitive impairment is one of the more consistent factors to adversely affect the functional outcome of older people after hip fracture [74].

Once hip fracture surgery is intended for an individual, the goals of acute hospital care in people with AD is no different from standard care. These goals include short time to surgery, minimize perioperative complications, adequate pain control, early mobilization to restore gait and function.

Orthogeriatric Model of Care

An orthogeriatric model of care involving co-management by both the orthopaedic and geriatric team has consistently been found to have improved outcomes for older people with hip fractures [75, 76]. This model is based on early preoperative assessment, daily co-management, protocol-based geriatric care and early initiation of discharge planning. Patients managed under this model have shorter-than-expected length of hospital stay, shorter time to surgery, low complications, and reduced mortality and readmission rates [76, 77]. In addition, geriatric comanagement has also been demonstrated to reduce postoperative complication rates for pressure ulcers, urinary tract infection, undernutrition, disturbance of sleep, recurrent falls and the need for institutionalization [78]. However, there is still considerable heterogeneity of studies which have assessed the efficacy of orthogeriatric units in managing patients with hip fracture and cognitive impairment resulting in limitations in conclusive findings.

Timing of Surgery

There is no reason for patients with AD to have a longer time to surgery unless pre-existing medical conditions require such a delay. Whenever possible, early surgical treatment is beneficial for patients because it improves pain and this leads to lower incidence of delirium. Delay to surgery of greater than two days for stable patients with hip fractures has been found to be predictive of longer length of stay and higher one-year mortality [79]. A meta-analysis including 35 studies with a total of 191,873 participants, have reported that early surgery within 24-48 hours of fracture was associated with reduced risk of mortality (pooled OR 0.74; 95%CI 0.67-0.81) [80]. Although this finding was impressive, confounding effects from failure of observational studies to identify patients whose surgery were delayed for valid medical reasons have not been fully accounted for and this group is predisposed to adverse outcomes.

Prevention of Delirium

AD is an independent risk factor for perioperative delirium [60]. In an observational study of 541 participants with hip fractures, cognitive impairment was an independent risk factor for delirium (adjusted relative risk of 3.6; 95% confidence interval 1.8-7.2) [81]. Other risk factors included advanced age, more severe cognitive impairment before fracture, pre-existing stroke or neurodegenerative disorders, communication difficulties and social isolation [82]. A systematic review and meta-analysis of hip fractures in the setting of impaired cognition have suggested that the NEECHAM Confusion Scale can be useful for assessment of delirium [83]. Such evaluation is most useful when clinicians also took into consideration identified risk factors for delirium.

Several interventions have been found to effectively reduce the incidence of delirium after hip fractures but these are not specifically in people with AD. The first intervention involving proactive geriatric consultation has shown an absolute risk reduction of 18%, in comparison with the usual care group [60]. The number needed to treat to prevent one case of delirium was 5.6. Although such an approach was found to reduce the incidence of delirium, the duration or severity of this complication was not reduced once it has occurred. The second approach is a program adopting daily co-management involving both a geriatrician and an orthopaedic surgeon [84]. One study has reported that this approach lowers the incidence of delirium by 24% [84]. Evidence supporting the use of pharmacological intervention to reduce the incidence of postoperative delirium is lacking among people with AD.

Adequate Analgesia

Optimal pain control is essential after hip fracture surgery including for people with AD. Adequate analgesia leads to faster functional recovery, reduces the risk of delirium and improves quality of life. One study has shown that lower pain score in people with dementia after hip fracture surgery have a shorter hospital length of stay and higher likelihood of walking by 3 days post-operation [85].

Early Mobilization

Early mobilization after hip fracture is vital for preventing venous thromboembolism, pressure sores and pneumonia [86]. Cognitive impairment has not been found to be an impediment to immediate weight bearing and early mobilization [87]. Therefore, people with AD should be encouraged to mobilize under supervision after having hip fracture fixation, as permitted by their surgeon.

Surgery in Severe Alzheimer's Disease

People with AD is a heterogeneous group in relation to severity of cognitive impairment, functional capacity and burden of coexisting diseases. Therefore, an individualized approach has to be taken to managing this group of patients when they sustain a hip fracture. Surgical repair of hip fracture among people with severe Alzheimer's and limited life expectancy should reflect patients' advance directives, family preferences as wells clinicians' assessment of risks and benefits of treatment. However, there is a paucity of data about non-operative care for patients with hip fracture in the setting of severe cognitive impairment.

POST-ACUTE CARE

The goal of post-acute care after hip fracture is to ensure optimal recovery after hip fracture surgery and prevent future recurrent fractures. After the first hip fracture, the incidence of a second hip fracture rises by six-fold from 3.6 to 20 per 1000 person-years in women and by nine-fold from 1.6 to 15 per 1000 person-years in men [88]. This risk is even higher than in people with dementia compared to those without [89].

Rehabilitation

In the past, rehabilitation has been thought to be of doubtful benefit for people with AD who sustain hip fractures. However, evidence support benefits of rehabilitation among participants with mild-to-moderate dementia who sustain hip fractures [90, 91]. The loss of function that occurs after hip fracture among people with mild dementia can be recovered through post-surgery rehabilitation program. In one RCT, 91% of participants with mild dementia (MMSE: 18-23) in the interventional group who received rehabilitation were living independently at 3 months compared to 67% in the control (P = 0.009) [76]. For those with moderately severe dementia (MMSE: 12-17), 63% were living independently among those in the interventional group, compared to 17% among the control (P = 0.009). For participants with mild dementia, the median hospital length of stay was significantly lower in the interventional group compared with the control (29 *vs* 46 days). Among those with moderate dementia, the median length of hospital stay was also shorter in the interventional group compared to the control (47 *vs* 147 days). Another prospective longitudinal study in a geriatric rehabilitation centre has reported similar results [92].

In a systematic review by Muir and Yohannes, intensive inpatient rehabilitation among patients with cognitive impairment after hip fracture surgical repair was beneficial for recovery physical function [93]. However, severe cognitive impairment still predicted a less favourable return to living independently.

Preventing Recurrent Falls

Hip fractures are usually related to minimal trauma arising from a fall. After sustaining a hip fracture, many patients are at risk of another fall. Therefore, secondary prevention of hip fracture must incorporate evaluation of falls risk factors.

One study has demonstrated the effectiveness of a multidisciplinary interventional program to prevent inpatient falls and injuries during rehabilitation, including among patients with dementia [94, 95]. The intervention involved management of pain and delirium, removal of urinary catheter within 24 hours after surgery, prevention of urinary retention, supplementation of calcium and vitamin D, and protein drinks daily. However, this intervention has not been shown to sustain the effectiveness after discharge [96]. Further work is needed to identify ways of preventing falls beyond the hospital stay for acute care and rehabilitation following hip fracture, especially among people with AD.

Treatment of Osteoporosis

Studies have found that the treatment rate of osteoporosis was low even after patients have experienced hip fractures [97 - 100]. In a multi-centre study in the US involving 51,386 patients who had hip fracture repair, 6.6% received calcium and vitamin D after the surgery, 7.3% was treated with a bone antiresorptive therapy, and only 2% received both [101]. Similar low rates of treatment have been observed in other studies [99, 100]. Concern is heightened among patients with dementia who sustain hip fractures because they are even less likely to receive treatment for osteoporosis compared with those without cognitive impairment [102].

In the United States, pharmacological treatment is recommended in people with (a) hip or vertebral fracture, (b) BMD T-scores of -2.5 or less, or (c) BMD T-scores between -1.0 and -2.5 and a 10-year risk of hip fracture of ≥3% or 10-year risk of major osteoporotic fracture of ≥20% [69].

In the management of osteoporosis in older people with AD, life expectancy should be taken into consideration in the selection of pharmacological treatment (Fig. **1**). Among older people, life expectancy is substantially heterogeneous. For example, the median life expectancy for an 80-year-old woman in a developed country is about 10 years. However, the life expectancy can range from less than 4 years for the "least healthy" quartile to more than 14 years for the "healthiest" quartile [103]. Clinicians should be aware that the benefits of pharmacological intervention for osteoporosis may be evident within 6 to 12 months and benefits may be derived immediately from effective fall prevention interventions. With increasing age, the number needed to treat to prevent one hip fracture decreases until after the age of 80 years [104]. Although a 90-year-old woman has a shorter life expectancy, she still has a significantly higher life-time fracture risk and lower number needed to treat to prevent one hip fracture, in comparison with a woman who is 70-year-old. Economic modelling suggest that treating older women for fracture reduction is cost effective when life expectancies are of at least two years [103]. One might expect that such findings would still apply to older women with mild-to-moderate AD with a life expectancy of two years or more. However, no similar study has been performed among male subjects.

After addressing any calcium and vitamin D deficiency, bisphosphonate therapy is still the recommended first line secondary prevention of fractures (Fig. **1**). The number of older people (average age of 85 years) needed to be treated with oral bisphosphonate to prevent one hip fracture is about 200 [105]. Even though oral bisphosphonate is considered the most cost-effective treatment for preventing hip fractures, therapy selection among people with AD should consider factors such as pill burden and comorbidities. Older patients already taking multiple oral medications for other coexisting conditions may prefer a once or twice a year formulation. For people with AD who have dysphagia or medication adherence difficulties, intravenous bisphosphonate such as zolendronic acid, or subcutaneous formulation such as denosumab, may be preferred. As stage IV and V chronic kidney disease is common in older people, and denosumab may be more appropriate in patients with this condition [69]. There are often barriers to initiating appropriate osteoporosis treatment including risk of interaction with other medications, lack of adherence and perceived shorter life expectancy [106]. Management of osteoporosis among older people should involve shared decision-making with patients and their family or caregivers, especially those with AD. Shared decision-making based on a clear understanding of the risks and benefits can encourage better treatment adherence.

Older people with AD at high risk of hip or other fractures	
• Previous hip or vertebral fracture • T-score ≤-2.5 at any site • FRAX 10-year fracture risk ≥3% for hip or ≥20% for major osteoporotic fractures • Intermediate-risk FRAX score with previous fall or additional risk factors	
MEDICATION REVIEW	
Prescribe medications for bone protection	**Rationalise medications predisposing to falls**
• **Ensure vitamin D adequate and dietary/supplemental calcium intake ≥1200 mg/day** • **Determine life expectancy ≥2 years?**	**Consider ceasing or reducing dose to the lowest effective of any of the following medications**: Antipsychotics Antidepressants Anticholinergics Benzodiazepines Sedative-hypnotics Opioids Antihypertensives if blood pressure is below target or postural hypotension Hypoglycaemics if glycohaemoglobin <53 mmol/mol (or 7.0%)
Current first-line treatment Oral bisphosphonates	
Alternate first-line treatment	
Condition | **Medication**	

Condition	Medication
Chronic kidney disease, stage IV-V*	Denosumab
Severe osteoporosis	Romosozumab or abaloparatide
Dysphagia	Zoledronic acid or denosumab

Fig. (1). Managing older people with Alzheimer's disease at risk of hip fracture.
*Chronic kidney disease, stage IV: creatinine clearance 15-29 mL/min; Chronic kidney disease, stage V:

Management of Depression

The presence of AD and depression in older people have an additive effect. In a cohort study of patients with hip fracture, the coexistence of depression and dementia was associated with increased 12-month mortality (adjusted hazard ratio [HR] of 8.7 compared with 3.4 when only dementia was present) [107]. Therefore, recognising depressive symptoms in patients with hip fracture and AD is essential. Appropriate psychotherapy, family therapy and judicious use of antidepressant can be useful approaches to managing the depression in this setting

[108]. If pharmacological treatment is required, SSRI is the preferred treatment for depression in AD and the initial dose should be low [108]. The duration of treatment with antidepressants need to be reviewed regularly. Appropriate management of depression can often help the patient to better engage in rehabilitation and also prevent future falls or recurrent fractures.

GAPS IN CURRENT EVIDENCE AND FUTURE RESEARCH

Despite recent advances, there is still a gap in the understanding of the interaction between AD, falls and hip fractures. In particular, there is a need to better understand the psychomotor aspect of AD which will be important for the prevention of falls.

There is currently a paucity of evidence on effective intervention to prevent falls in people with AD or other forms of dementia. Residency in a nursing home and cognitive impairment were the main reasons for exclusions from participation in clinical trials on rehabilitation after hip fracture surgery [109]. To achieve better outcomes, this barrier needs to be addressed in future clinical trials, which will help translation to effective clinical practice. The Developing an Intervention for Fall-Related Injuries in Dementia (DIFRID) study is currently in progress to evaluate the feasibility of developing and implementing a new intervention to improve outcomes for people with dementia who sustain fall-related injuries [110]. Broadly, this approach is intended to enhance short-term recovery and reduce the likelihood of future falls. Key components of this new intervention cover three areas: (a) optimizing the circumstances of rehabilitation for people with dementia, (b) compensating strategies for the reduced ability to self-manage, and (c) equipping health providers with the necessary skills and information to manage this patient group. The results of this study will be awaited with interest.

The ability to use gerontechnology tools to help prevent falls is another area that requires further investigation. A fall detection and fall risk assessment (FDFRA) system was developed using a sensor system [111]. Such intervention has been used to address fall risk factors and subsequently reduce risk of future falls. However further work is needed to evaluate the effectiveness of FDFRA system in preventing falls and related injuries. Sensor-based fall risk assessments of postural sway, functional mobility, stepping and walking can discriminate between fallers and non-fallers [112].

The most cost-effective way of rehabilitation after hip fracture among people with AD has not been evaluated fully. For example, the effectiveness of community or home-based rehabilitation rather than inpatient for people with mild AD needs further research.

Although current treatments are effective in reducing the risk of hip fracture, the residual risk in 'real-world' clinical practice is amplified by poor adherence [113]. In addition, the relative risk of recurrent hip fracture rises significantly after the initial fracture and is maximal within the first year. The risk of a second fracture persists for 10 years after the first one and secondary prevention should continue beyond the initial post-fracture period. To address this unmet gap in preventive measures, a subgroup of patients may require additional treatment such as surgical bone enhancement. Therefore, procedures that immobilise fracture or fill up local bone defect such as bone cysts, hemangioma or avascular necrosis, have been suggested to strengthen bone fragility of the proximal femur before the occurrence of fracture [114].

Proposed strategies include prophylactic nailing, bone grafting with osteoinductive or osteoconductive materials and femoroplasty with cement [114 - 116]. Prophylactic nailing has been recommended for incomplete atypical femoral fractures but this approach is currently not performed in clinical practice for severe osteoporotic hips [117]. Femoroplasty involves injecting surgical cement, usually polymethylmethacrylate (PMMA), into an osteoporotic proximal femur. This procedure has been evaluated in animals and showed to improve bone strength by 30-80% [118, 119]. However, there is still no translation of this technique into humans. Some bone graft substitutes, such as calcium phosphate bone cements, resorbable allograft or polymers, can be used as an intra-osseous scaffold for delivery of bone anabolic agents for enhancing bone formation and consolidating bone mass [120]. As an example of osteoanabolic substance use, synthetic bone grafts can be carriers for growth factors to stimulate osteoblast proliferation, such as bisphosphonates or strontium [121 - 123]. The main disadvantage of bone graft substitutes is that these materials have been used to fill a defect such as a cyst or space left after tumour extraction. However, in osteoporosis, injecting osteoinductive or osteoconductive materials require fluidity to avoid administration under high pressure into the trabecular bone of the proximal femur.

CONCLUSION

AD is the commonest cause of dementia but remains a challenging problem for people at risk of hip fractures and for their recovery after such fracture. Risk factors for falls that are more specific for people with AD need to be clarified so that effective interventions can be designed. Evidence indicate that acute management of hip fractures involving orthogeriatric co-management offers a better clinical outcome including for patients with AD. Among people with mild-to-moderate dementia, intensive geriatric rehabilitation after hip fracture surgery is beneficial in returning this group of patients to their premorbid residence and therefore avoid long-term residential facility. Effective secondary prevention of hip fractures among people with AD requires further research. New materials such as osteoconductive and osteoinductive materials are currently under development and are promising ways to strengthen osteoporotic bones.

CONSENT FOR PUBLICATION

Not applicable.

CONFLICT OF INTEREST

The authors declare no conflict of interest, financial or otherwise.

ACKNOWLEDGEMENTS

Declared none.

REFERENCES

[1]   Livingston G, Sommerlad A, Orgeta V, *et al.* Dementia prevention, intervention, and care. Lancet 2017; 390(10113): 2673-734.
        [http://dx.doi.org/10.1016/S0140-6736(17)31363-6] [PMID: 28735855]

[2]   Scheltens P, Blennow K, Breteler MM, *et al.* Alzheimer's disease. Lancet 2016; 388(10043): 505-17.
        [http://dx.doi.org/10.1016/S0140-6736(15)01124-1] [PMID: 26921134]

[3]   Campbell AJ, Borrie MJ, Spears GF, Jackson SL, Brown JS, Fitzgerald JL. Circumstances and consequences of falls experienced by a community population 70 years and over during a prospective study. Age Ageing 1990; 19(2): 136-41.
        [http://dx.doi.org/10.1093/ageing/19.2.136] [PMID: 2337010]

[4]   Hartholt KA, Lee R, Burns ER, van Beeck EF. Mortality From Falls Among US Adults Aged 75 Years or Older, 2000-2016. JAMA 2019; 321(21): 2131-3.
        [http://dx.doi.org/10.1001/jama.2019.4185] [PMID: 31162561]

[5]   Florence CS, Bergen G, Atherly A, Burns E, Stevens J, Drake C. Medical Costs of Fatal and Nonfatal Falls in Older Adults. J Am Geriatr Soc 2018; 66(4): 693-8.
        [http://dx.doi.org/10.1111/jgs.15304] [PMID: 29512120]

[6]   Braithwaite RS, Col NF, Wong JB. Estimating hip fracture morbidity, mortality and costs. J Am Geriatr Soc 2003; 51(3): 364-70.
        [http://dx.doi.org/10.1046/j.1532-5415.2003.51110.x] [PMID: 12588580]

[7]   Marks R. Hip fracture epidemiological trends, outcomes, and risk factors, 1970-2009. Int J Gen Med 2010; 3: 1-17.
        [PMID: 20463818]

[8]   Vochteloo AJ, Moerman S, Tuinebreijer WE, *et al.* More than half of hip fracture patients do not regain mobility in the first postoperative year. Geriatr Gerontol Int 2013; 13(2): 334-41.
        [http://dx.doi.org/10.1111/j.1447-0594.2012.00904.x] [PMID: 22726959]

[9]   Cumming RG, Klineberg R, Katelaris A. Cohort study of risk of institutionalisation after hip fracture. Aust N Z J Public Health 1996; 20(6): 579-82.
        [http://dx.doi.org/10.1111/j.1467-842X.1996.tb01069.x] [PMID: 9117962]

[10]   Abrahamsen B, van Staa T, Ariely R, Olson M, Cooper C. Excess mortality following hip fracture: A systematic epidemiological review. Osteoporos Int 2009; 20(10): 1633-50.

[11]   Haentjens P, Magaziner J, Colón-Emeric CS, *et al.* Meta-analysis: excess mortality after hip fracture among older women and men. Ann Intern Med 2010; 152(6): 380-90.
       [http://dx.doi.org/10.7326/0003-4819-152-6-201003160-00008] [PMID: 20231569]

[12]   Buchner DM, Larson EB. Falls and fractures in patients with Alzheimer-type dementia. JAMA 1987; 257(11): 1492-5.
       [http://dx.doi.org/10.1001/jama.1987.03390110068028] [PMID: 3820464]

[13]   Morris JC, Rubin EH, Morris EJ, Mandel SA. Senile dementia of the Alzheimer's type: an important risk factor for serious falls. J Gerontol 1987; 42(4): 412-7.
       [http://dx.doi.org/10.1093/geronj/42.4.412] [PMID: 3598089]

[14]   Visser H. Gait and balance in senile dementia of Alzheimer's type. Age Ageing 1983; 12(4): 296-301.
       [http://dx.doi.org/10.1093/ageing/12.4.296] [PMID: 6660138]

[15]   Tanaka A, Okuzumi H, Kobayashi I, Murai N, Meguro K, Nakamura T. Gait disturbance of patients with vascular and Alzheimer-type dementias. Percept Mot Skills 1995; 80(3 Pt 1): 735-8.
       [http://dx.doi.org/10.2466/pms.1995.80.3.735] [PMID: 7567389]

[16]   Nakamura T, Meguro K, Yamazaki H, *et al.* Postural and gait disturbance correlated with decreased frontal cerebral blood flow in Alzheimer disease. Alzheimer Dis Assoc Disord 1997; 11(3): 132-9.
       [http://dx.doi.org/10.1097/00002093-199709000-00005] [PMID: 9305498]

[17]   Della Sala S, Spinnler H, Venneri A. Walking difficulties in patients with Alzheimer's disease might originate from gait apraxia. J Neurol Neurosurg Psychiatry 2004; 75(2): 196-201.
       [PMID: 14742586]

[18]   Shaw FE. Falls in cognitive impairment and dementia. Clin Geriatr Med 2002; 18(2): 159-73.
       [http://dx.doi.org/10.1016/S0749-0690(02)00003-4] [PMID: 12180241]

[19]   Whitney J, Close JC, Jackson SH, Lord SR. Understanding risk of falls in people with cognitive impairment living in residential care. J Am Med Dir Assoc 2012; 13(6): 535-40.
       [http://dx.doi.org/10.1016/j.jamda.2012.03.009] [PMID: 22561138]

[20]   Femminella GD, Rengo G, Komici K, *et al.* Autonomic dysfunction in Alzheimer's disease: tools for assessment and review of the literature. J Alzheimers Dis 2014; 42(2): 369-77.
       [http://dx.doi.org/10.3233/JAD-140513] [PMID: 24898649]

[21]   Allan LM, Ballard CG, Rowan EN, Kenny RA. Incidence and prediction of falls in dementia: a prospective study in older people. PLoS One 2009; 4(5)e5521
       [http://dx.doi.org/10.1371/journal.pone.0005521] [PMID: 19436724]

[22]   Ballard C, Shaw F, McKeith I, Kenny R. High prevalence of neurovascular instability in neurodegenerative dementias. Neurology 1998; 51(6): 1760-2.
       [http://dx.doi.org/10.1212/WNL.51.6.1760] [PMID: 9855544]

[23]   Brand CA, Sundararajan V. A 10-year cohort study of the burden and risk of in-hospital falls and fractures using routinely collected hospital data. Qual Saf Health Care 2010; 19(6)e51
       [PMID: 20558479]

[24]   Flaherty JH, Morley JE. Delirium in the nursing home. J Am Med Dir Assoc 2013; 14(9): 632-4.
       [http://dx.doi.org/10.1016/j.jamda.2013.06.009] [PMID: 23911952]

[25]   Eriksson S, Gustafson Y, Lundin-Olsson L. Risk factors for falls in people with and without a diagnose of dementia living in residential care facilities: a prospective study. Arch Gerontol Geriatr 2008; 46(3): 293-306.
       [http://dx.doi.org/10.1016/j.archger.2007.05.002] [PMID: 17602762]

[26]   Eriksson S, Strandberg S, Gustafson Y, Lundin-Olsson L. Circumstances surrounding falls in patients with dementia in a psychogeriatric ward. Arch Gerontol Geriatr 2009; 49(1): 80-7.
       [http://dx.doi.org/10.1016/j.archger.2008.05.005] [PMID: 18635273]

[27]   Leipzig RM, Cumming RG, Tinetti ME. Drugs and falls in older people: a systematic review and

meta-analysis: I. Psychotropic drugs. J Am Geriatr Soc 1999; 47(1): 30-9.
[http://dx.doi.org/10.1111/j.1532-5415.1999.tb01898.x] [PMID: 9920227]

[28] Darowski A, Chambers SA, Chambers DJ. Antidepressants and falls in the elderly. Drugs Aging 2009; 26(5): 381-94.
[http://dx.doi.org/10.2165/00002512-200926050-00002] [PMID: 19552490]

[29] Ruxton K, Woodman RJ, Mangoni AA. Drugs with anticholinergic effects and cognitive impairment, falls and all-cause mortality in older adults: A systematic review and meta-analysis. Br J Clin Pharmacol 2015; 80(2): 209-20.
[http://dx.doi.org/10.1111/bcp.12617] [PMID: 25735839]

[30] Nakamura T, Meguro K, Sasaki H. Relationship between falls and stride length variability in senile dementia of the Alzheimer type. Gerontology 1996; 42(2): 108-13.
[http://dx.doi.org/10.1159/000213780] [PMID: 9138973]

[31] Buchner DM, Larson EB. Transfer bias and the association of cognitive impairment with falls. J Gen Intern Med 1988; 3(3): 254-9.
[http://dx.doi.org/10.1007/BF02596341] [PMID: 3259981]

[32] Hebert LE, Scherr PA, Beckett LA, *et al.* Age-specific incidence of Alzheimer's disease in a community population. JAMA 1995; 273(17): 1354-9.
[http://dx.doi.org/10.1001/jama.1995.03520410048025] [PMID: 7715060]

[33] Farmer ME, Harris T, Madans JH, Wallace RB, Cornoni-Huntley J, White LR. Anthropometric indicators and hip fracture. The NHANES I epidemiologic follow-up study. J Am Geriatr Soc 1989; 37(1): 9-16.
[http://dx.doi.org/10.1111/j.1532-5415.1989.tb01562.x] [PMID: 2909610]

[34] Melton LJ III, Beard CM, Kokmen E, Atkinson EJ, O'Fallon WM. Fracture risk in patients with Alzheimer's disease. J Am Geriatr Soc 1994; 42(6): 614-9.
[http://dx.doi.org/10.1111/j.1532-5415.1994.tb06859.x] [PMID: 8201146]

[35] Buell JS, Dawson-Hughes B. Vitamin D and neurocognitive dysfunction: preventing "D"ecline? Mol Aspects Med 2008; 29(6): 415-22.
[http://dx.doi.org/10.1016/j.mam.2008.05.001] [PMID: 18579197]

[36] Kipen E, Helme RD, Wark JD, Flicker L. Bone density, vitamin D nutrition, and parathyroid hormone levels in women with dementia. J Am Geriatr Soc 1995; 43(10): 1088-91.
[http://dx.doi.org/10.1111/j.1532-5415.1995.tb07005.x] [PMID: 7560696]

[37] Sato Y, Asoh T, Oizumi K. High prevalence of vitamin D deficiency and reduced bone mass in elderly women with Alzheimer's disease. Bone 1998; 23(6): 555-7.
[http://dx.doi.org/10.1016/S8756-3282(98)00134-3] [PMID: 9855465]

[38] Dhesi JK, Bearne LM, Moniz C, *et al.* Neuromuscular and psychomotor function in elderly subjects who fall and the relationship with vitamin D status. J Bone Mineral Res 2002; 17(5): 891-7.

[39] Weller I, Schatzker J. Hip fractures and Alzheimer's disease in elderly institutionalized Canadians. Ann Epidemiol 2004; 14(5): 319-24.
[http://dx.doi.org/10.1016/j.annepidem.2003.08.005] [PMID: 15177270]

[40] Johansson C, Skoog I. A population-based study on the association between dementia and hip fractures in 85-year olds. Aging (Milano) 1996; 8(3): 189-96.
[http://dx.doi.org/10.1007/BF03339676] [PMID: 8862194]

[41] Loskutova N, Honea RA, Vidoni ED, Brooks WM, Burns JM. Bone density and brain atrophy in early Alzheimer's disease. J Alzheimers Dis 2009; 18(4): 777-85.
[http://dx.doi.org/10.3233/JAD-2009-1185] [PMID: 19661621]

[42] Sato Y, Kanoko T, Satoh K, Iwamoto J. Risk factors for hip fracture among elderly patients with Alzheimer's disease. J Neurol Sci 2004; 223(2): 107-12.
[http://dx.doi.org/10.1016/j.jns.2004.03.033] [PMID: 15337610]

[43]   Poehlman ET, Dvorak RV. Energy expenditure, energy intake, and weight loss in Alzheimer disease. Am J Clin Nutr 2000; 71(2): 650S-5S.
[http://dx.doi.org/10.1093/ajcn/71.2.650s] [PMID: 10681274]

[44]   Langlois JA, Harris T, Looker AC, Madans J. Weight change between age 50 years and old age is associated with risk of hip fracture in white women aged 67 years and older. Arch Intern Med 1996; 156(9): 989-94.
[http://dx.doi.org/10.1001/archinte.1996.00440090089009] [PMID: 8624179]

[45]   Langlois JA, Mussolino ME, Visser M, Looker AC, Harris T, Madans J. Weight loss from maximum body weight among middle-aged and older white women and the risk of hip fracture: the NHANES I epidemiologic follow-up study. Osteoporos Int 2001; 12(9): 763-8.

[46]   Langlois JA, Visser M, Davidovic LS, Maggi S, Li G, Harris TB. Hip fracture risk in older white men is associated with change in body weight from age 50 years to old age. Arch Intern Med 1998; 158(9): 990-6.
[http://dx.doi.org/10.1001/archinte.158.9.990] [PMID: 9588432]

[47]   Greenspan SL, Myers ER, Kiel DP, Parker RA, Hayes WC, Resnick NM. Fall direction, bone mineral density, and function: risk factors for hip fracture in frail nursing home elderly. Am J Med 1998; 104(6): 539-45.
[http://dx.doi.org/10.1016/S0002-9343(98)00115-6] [PMID: 9674716]

[48]   Landes AM, Sperry SD, Strauss ME. Prevalence of apathy, dysphoria, and depression in relation to dementia severity in Alzheimer's disease. J Neuropsychiatry Clin Neurosci 2005; 17(3): 342-9.
[http://dx.doi.org/10.1176/jnp.17.3.342] [PMID: 16179656]

[49]   Wu Q, Liu J, Gallegos-Orozco JF, Hentz JG. Depression, fracture risk, and bone loss: a meta-analysis of cohort studies. Osteoporosis int 2010; 21(10): 1627-35.

[50]   Wu Q, Liu B, Tonmoy S. Depression and risk of fracture and bone loss: an updated meta-analysis of prospective studies. Osteoporos Int 2018; 20(6): 1303-12.

[51]   van Doorn C, Gruber-Baldini AL, Zimmerman S, et al. Epidemiology of Dementia in Nursing Homes Research Group. Dementia as a risk factor for falls and fall injuries among nursing home residents. J Am Geriatr Soc 2003; 51(9): 1213-8.
[http://dx.doi.org/10.1046/j.1532-5415.2003.51404.x] [PMID: 12919232]

[52]   Diem SJ, Blackwell TL, Stone KL, et al. Study of Osteoporotic Fractures. Depressive symptoms and rates of bone loss at the hip in older women. J Am Geriatr Soc 2007; 55(6): 824-31.
[http://dx.doi.org/10.1111/j.1532-5415.2007.01194.x] [PMID: 17537081]

[53]   Diem SJ, Harrison SL, Haney E, et al. Depressive symptoms and rates of bone loss at the hip in older men. Osteoporos Int 2013; 24(1): 111-9.

[54]   Saczynski JS, Beiser A, Seshadri S, Auerbach S, Wolf PA, Au R. Depressive symptoms and risk of dementia: the Framingham Heart Study. Neurology 2010; 75(1): 35-41.
[http://dx.doi.org/10.1212/WNL.0b013e3181e62138] [PMID: 20603483]

[55]   Diem SJ, Blackwell TL, Stone KL, et al. Use of antidepressants and rates of hip bone loss in older women: the study of osteoporotic fractures. Arch Intern Med 2007; 167(12): 1240-5.
[http://dx.doi.org/10.1001/archinte.167.12.1240] [PMID: 17592096]

[56]   Diem SJ, Blackwell TL, Stone KL, et al. Study of Osteoporotic Fractures Research Group. Use of antidepressant medications and risk of fracture in older women. Calcif Tissue Int 2011; 88(6): 476-84.
[http://dx.doi.org/10.1007/s00223-011-9481-5] [PMID: 21455735]

[57]   Diem SJ, Ruppert K, Cauley JA, et al. Rates of bone loss among women initiating antidepressant medication use in midlife. J Clin Endocrinol Metab 2013; 98(11): 4355-63.
[http://dx.doi.org/10.1210/jc.2013-1971] [PMID: 24001746]

[58]   Ray WA, Griffin MR, Schaffner W, Baugh DK, Melton LJ III. Psychotropic drug use and the risk of

hip fracture. N Engl J Med 1987; 316(7): 363-9.
[http://dx.doi.org/10.1056/NEJM198702123160702] [PMID: 2880292]

[59]   Ray WA, Griffin MR, Downey W. Benzodiazepines of long and short elimination half-life and the risk of hip fracture. JAMA 1989; 262(23): 3303-7.
[http://dx.doi.org/10.1001/jama.1989.03430230088031] [PMID: 2573741]

[60]   Marcantonio ER, Flacker JM, Wright RJ, Resnick NM. Reducing delirium after hip fracture: a randomized trial. J Am Geriatr Soc 2001; 49(5): 516-22.
[http://dx.doi.org/10.1046/j.1532-5415.2001.49108.x] [PMID: 11380742]

[61]   Marcantonio ER, Flacker JM, Michaels M, Resnick NM. Delirium is independently associated with poor functional recovery after hip fracture. J Am Geriatr Soc 2000; 48(6): 618-24.
[http://dx.doi.org/10.1111/j.1532-5415.2000.tb04718.x] [PMID: 10855596]

[62]   Olofsson B, Persson M, Bellelli G, Morandi A, Gustafson Y, Stenvall M. Development of dementia in patients with femoral neck fracture who experience postoperative delirium-A three-year follow-up study. Int J Geriatr Psychiatry 2018; 33(4): 623-32.
[http://dx.doi.org/10.1002/gps.4832] [PMID: 29292537]

[63]   Gillespie LD, Robertson MC, Gillespie WJ, *et al.* Interventions for preventing falls in older people living in the community. Cochrane Database Syst Rev 2012; (9): CD007146
[http://dx.doi.org/10.1002/14651858.CD007146.pub3] [PMID: 22972103]

[64]   Shaw FE, Bond J, Richardson DA, *et al.* Multifactorial intervention after a fall in older people with cognitive impairment and dementia presenting to the accident and emergency department: randomised controlled trial. BMJ 2003; 326(7380): 73.
[http://dx.doi.org/10.1136/bmj.326.7380.73] [PMID: 12521968]

[65]   Stevens JA. The STEADI Tool Kit: A Fall Prevention Resource for Health Care Providers. IHS Prim Care Provid 2013; 39(9): 162-6.
[PMID: 26766893]

[66]   Hopewell S, Adedire O, Copsey BJ, *et al.* Multifactorial and multiple component interventions for preventing falls in older people living in the community. Cochrane Database Syst Rev 2018; 7CD012221
[http://dx.doi.org/10.1002/14651858.CD012221.pub2] [PMID: 30035305]

[67]   Salkeld G, Cameron ID, Cumming RG, *et al.* Quality of life related to fear of falling and hip fracture in older women: a time trade off study. BMJ 2000; 320(7231): 341-6.
[http://dx.doi.org/10.1136/bmj.320.7231.341] [PMID: 10657327]

[68]   Vestergaard P, Rejnmark L, Mosekilde L. Osteoporosis is markedly underdiagnosed: a nationwide study from Denmark. Osteoporos Int 2005; 16(2): 134-41.2005;

[69]   Viswanathan M, Reddy S, Berkman N, *et al.* Screening to Prevent Osteoporotic Fractures: Updated Evidence Report and Systematic Review for the US Preventive Services Task Force. JAMA 2018; 319(24): 2532-51.
[http://dx.doi.org/10.1001/jama.2018.6537] [PMID: 29946734]

[70]   Watts NB, Lewiecki EM, Miller PD, Baim S. National Osteoporosis Foundation 2008 Clinician's Guide to Prevention and Treatment of Osteoporosis and the World Health Organization Fracture Risk Assessment Tool (FRAX): what they mean to the bone densitometrist and bone technologist. J Clin Densitom 2008; 11(4): 473-7.
[http://dx.doi.org/10.1016/j.jocd.2008.04.003] [PMID: 25936482]

[71]   Sato Y, Honda Y, Umeno K, *et al.* The prevention of hip fracture with menatetrenone and risedronate plus calcium supplementation in elderly patients with Alzheimer disease: a randomized controlled trial. Kurume Med J 2011; 57(4): 117-24.
[http://dx.doi.org/10.2739/kurumemedj.57.117] [PMID: 21778673]

[72]   Black DM, Delmas PD, Eastell R, *et al.* HORIZON Pivotal Fracture Trial. Once-yearly zoledronic

acid for treatment of postmenopausal osteoporosis. N Engl J Med 2007; 356(18): 1809-22.
[http://dx.doi.org/10.1056/NEJMoa067312] [PMID: 17476007]

[73]   Cummings SR, San Martin J, McClung MR, *et al.* FREEDOM Trial. Denosumab for prevention of fractures in postmenopausal women with osteoporosis. N Engl J Med 2009; 361(8): 756-65.
[http://dx.doi.org/10.1056/NEJMoa0809493] [PMID: 19671655]

[74]   Sheehan KJ, Williamson L, Alexander J, *et al.* Prognostic factors of functional outcome after hip fracture surgery: a systematic review. Age Ageing 2018; 47(5): 661-70.
[http://dx.doi.org/10.1093/ageing/afy057] [PMID: 29668839]

[75]   Friedman SM, Mendelson DA, Bingham KW, Kates SL. Impact of a comanaged Geriatric Fracture Center on short-term hip fracture outcomes. Arch Intern Med 2009; 169(18): 1712-7.
[http://dx.doi.org/10.1001/archinternmed.2009.321] [PMID: 19822829]

[76]   Mendelson DA, Friedman SM. Principles of comanagement and the geriatric fracture center. Clin Geriatr Med 2014; 30(2): 183-9.
[http://dx.doi.org/10.1016/j.cger.2014.01.016] [PMID: 24721359]

[77]   Menzies IB, Mendelson DA, Kates SL, Friedman SM. Prevention and clinical management of hip fractures in patients with dementia. Geriatr Orthop Surg Rehabil 2010; 1(2): 63-72.
[http://dx.doi.org/10.1177/2151458510389465] [PMID: 23569664]

[78]   White JJ, Khan WS, Smitham PJ. Perioperative implications of surgery in elderly patients with hip fractures: an evidence-based review. J Perioper Pract 2011; 21(6): 192-7.
[http://dx.doi.org/10.1177/175045891102100601] [PMID: 21823308]

[79]   Shiga T, Wajima Z, Ohe Y. Is operative delay associated with increased mortality of hip fracture patients? Systematic review, meta-analysis, and meta-regression. Can J Anaesth 2008; 55(3): 146-54.

[80]   Moja L, Piatti A, Pecoraro V, *et al.* Timing matters in hip fracture surgery: patients operated within 48 hours have better outcomes. A meta-analysis and meta-regression of over 190,000 patients. PLoS One 2012; 7(10)e46175
[http://dx.doi.org/10.1371/journal.pone.0046175] [PMID: 23056256]

[81]   Morrison RS, Magaziner J, Gilbert M, *et al.* Relationship between pain and opioid analgesics on the development of delirium following hip fracture. J Gerontol A Biol Sci Med Sci 2003; 58(1): 76-81.
[http://dx.doi.org/10.1093/gerona/58.1.M76] [PMID: 12560416]

[82]   Juliebø V, Bjøro K, Krogseth M, Skovlund E, Ranhoff AH, Wyller TB. Risk factors for preoperative and postoperative delirium in elderly patients with hip fracture. J Am Geriatr Soc 2009; 57(8): 1354-61.
[http://dx.doi.org/10.1111/j.1532-5415.2009.02377.x] [PMID: 19573218]

[83]   Smith T, Hameed Y, Cross J, Sahota O, Fox C. Assessment of people with cognitive impairment and hip fracture: a systematic review and meta-analysis. Arch Gerontol Geriatr 2013; 57(2): 117-26.
[http://dx.doi.org/10.1016/j.archger.2013.04.009] [PMID: 23680535]

[84]   Morrison RS, Chassin MR, Siu AL. The medical consultant's role in caring for patients with hip fracture. Ann Intern Med 1998; 128(12 Pt 1): 1010-20.
[http://dx.doi.org/10.7326/0003-4819-128-12_Part_1-199806150-00010] [PMID: 9625664]

[85]   Morrison RS, Magaziner J, McLaughlin MA, *et al.* The impact of post-operative pain on outcomes following hip fracture. Pain 2003; 103(3): 303-11.
[http://dx.doi.org/10.1016/S0304-3959(02)00458-X] [PMID: 12791436]

[86]   Siu AL, Penrod JD, Boockvar KS, Koval K, Strauss E, Morrison RS. Early ambulation after hip fracture: effects on function and mortality. Arch Intern Med 2006; 166(7): 766-71.
[http://dx.doi.org/10.1001/archinte.166.7.766] [PMID: 16606814]

[87]   Hung WW, Egol KA, Zuckerman JD, Siu AL. Hip fracture management: tailoring care for the older patient. JAMA 2012; 307(20): 2185-94.
[http://dx.doi.org/10.1001/jama.2012.4842] [PMID: 22618926]

[88]   Schrøder HM, Petersen KK, Erlandsen M. Occurrence and incidence of the second hip fracture. Clin Orthop Relat Res 1993; (289): 166-9.
[PMID: 8472408]

[89]   Yamanashi A, Yamazaki K, Kanamori M, *et al.* Assessment of risk factors for second hip fractures in Japanese elderly. Osteoporos Int 2005; 16(10): 1239-46.

[90]   Huusko TM, Karppi P, Avikainen V, Kautiainen H, Sulkava R. Randomised, clinically controlled trial of intensive geriatric rehabilitation in patients with hip fracture: subgroup analysis of patients with dementia. BMJ 2000; 321(7269): 1107-11.
[http://dx.doi.org/10.1136/bmj.321.7269.1107] [PMID: 11061730]

[91]   Huusko TM, Karppi P, Avikainen V, Kautiainen H, Sulkava R. Intensive geriatric rehabilitation of hip fracture patients: a randomized, controlled trial. Acta Orthop Scand 2002; 73(4): 425-31.
[http://dx.doi.org/10.1080/00016470216324] [PMID: 12358116]

[92]   Goldstein FC, Strasser DC, Woodard JL, Roberts VJ. Functional outcome of cognitively impaired hip fracture patients on a geriatric rehabilitation unit. J Am Geriatr Soc 1997; 45(1): 35-42.
[http://dx.doi.org/10.1111/j.1532-5415.1997.tb00975.x] [PMID: 8994485]

[93]   Muir SW, Yohannes AM. The impact of cognitive impairment on rehabilitation outcomes in elderly patients admitted with a femoral neck fracture: a systematic review. J Geriatr Phys Therapy 2009; 32(1): 24-32.

[94]   Stenvall M, Olofsson B, Lundstrom M, *et al.* A multidisciplinary, multifactorial intervention program reduces postoperative falls and injuries after femoral neck fracture. Osteoporos Int 2007; 18(2): 167-75.

[95]   Stenvall M, Berggren M, Lundström M, Gustafson Y, Olofsson B. A multidisciplinary intervention program improved the outcome after hip fracture for people with dementia--subgroup analyses of a randomized controlled trial. Arch Gerontol Geriatr 2012; 54(3): e284-9.
[http://dx.doi.org/10.1016/j.archger.2011.08.013] [PMID: 21930310]

[96]   Berggren M, Stenvall M, Olofsson B, Gustafson Y. Evaluation of a fall-prevention program in older people after femoral neck fracture: a one-year follow-up. Osteoporos Int 2008; 19(6): 801-9.

[97]   Davis JC, Ashe MC, Guy P, Khan KM. Undertreatment after hip fracture: a retrospective study of osteoporosis overlooked. J Am Geriatr Soc 2006; 54(6): 1019-20.
[http://dx.doi.org/10.1111/j.1532-5415.2006.00761.x] [PMID: 16776813]

[98]   Lüthje P, Nurmi-Lüthje I, Kaukonen JP, Kuurne S, Naboulsi H, Kataja M. Undertreatment of osteoporosis following hip fracture in the elderly. Arch Gerontol Geriatr 2009; 49(1): 153-7.
[http://dx.doi.org/10.1016/j.archger.2008.06.007] [PMID: 18706704]

[99]   Lai MM, Ang WM, McGuiness M, Larke AB. Undertreatment of osteoporosis in regional Western Australia. Australas J Ageing 2012; 31(2): 110-4.
[http://dx.doi.org/10.1111/j.1741-6612.2011.00544.x] [PMID: 22676170]

[100] Kim SR, Park YG, Kang SY, Nam KW, Park YG, Ha YC. Undertreatment of osteoporosis following hip fractures in jeju cohort study. J Bone Metab 2014; 21(4): 263-8.
[http://dx.doi.org/10.11005/jbm.2014.21.4.263] [PMID: 25489575]

[101] Jennings LA, Auerbach AD, Maselli J, Pekow PS, Lindenauer PK, Lee SJ. Missed opportunities for osteoporosis treatment in patients hospitalized for hip fracture. J Am Geriatr Soc 2010; 58(4): 650-7.
[http://dx.doi.org/10.1111/j.1532-5415.2010.02769.x] [PMID: 20398147]

[102] Kamel HK, Hussain MS, Tariq S, Perry HM, Morley JE. Failure to diagnose and treat osteoporosis in elderly patients hospitalized with hip fracture. Am J Med 2000; 109(4): 326-8.
[http://dx.doi.org/10.1016/S0002-9343(00)00457-5] [PMID: 10996585]

[103] Pham AN, Datta SK, Weber TJ, Walter LC, Colón-Emeric CS. Cost-effectiveness of oral bisphosphonates for osteoporosis at different ages and levels of life expectancy. J Am Geriatr Soc

2011; 59(9): 1642-9.
[http://dx.doi.org/10.1111/j.1532-5415.2011.03571.x] [PMID: 21883116]

[104]   Wells GA, Cranney A, Peterson J, *et al.* Alendronate for the primary and secondary prevention of osteoporotic fractures in postmenopausal women. Cochrane Database Syst Rev 2008; (1): CD001155
[http://dx.doi.org/10.1002/14651858.CD001155.pub2] [PMID: 18253985]

[105]   Zullo AR, Zhang T, Lee Y, *et al.* Effect of Bisphosphonates on Fracture Outcomes Among Frail Older Adults. J Am Geriatr Soc 2019; 67(4): 768-76.
[http://dx.doi.org/10.1111/jgs.15725] [PMID: 30575958]

[106]   Brauner DJ, Muir JC, Sachs GA. Treating nondementia illnesses in patients with dementia. JAMA 2000; 283(24): 3230-5.
[http://dx.doi.org/10.1001/jama.283.24.3230] [PMID: 10866871]

[107]   Bellelli G, Frisoni GB, Turco R, Trabucchi M. Depressive symptoms combined with dementia affect 12-months survival in elderly patients after rehabilitation post-hip fracture surgery. Int J Geriatr Psychiatry 2008; 23(10): 1073-7.
[http://dx.doi.org/10.1002/gps.2035] [PMID: 18489008]

[108]   Thompson S, Herrmann N, Rapoport MJ, Lanctôt KL. Efficacy and safety of antidepressants for treatment of depression in Alzheimer's disease: a metaanalysis. Can J Psychiatry 2007; 52(4): 248-55.
[http://dx.doi.org/10.1177/070674370705200407] [PMID: 17500306]

[109]   Sheehan KJ, Fitzgerald L, Hatherley S, *et al.* Inequity in rehabilitation interventions after hip fracture: a systematic review. Age Ageing 2019; 48(4): 489-97.
[http://dx.doi.org/10.1093/ageing/afz031] [PMID: 31220202]

[110]   Wheatley A, Bamford C, Shaw C, *et al.* Developing an Intervention for Fall-Related Injuries in Dementia (DIFRID): an integrated, mixed-methods approach. BMC Geriatr 2019; 19(1): 57.
[http://dx.doi.org/10.1186/s12877-019-1066-6] [PMID: 30819097]

[111]   Rantz M, Skubic M, Abbott C, *et al.* Automated In-Home Fall Risk Assessment and Detection Sensor System for Elders. Gerontologist 2015; 55 (Suppl. 1): S78-87.
[http://dx.doi.org/10.1093/geront/gnv044] [PMID: 26055784]

[112]   Ejupi A, Lord SR, Delbaere K. New methods for fall risk prediction. J Bone Mineral Res 2014; 17(5): 407-11.
[http://dx.doi.org/10.1097/MCO.0000000000000081] [PMID: 24992225]

[113]   Kanis JA, Svedbom A, Harvey N, McCloskey EV. The osteoporosis treatment gap. J Bone Mineral Res 2004; 29(9): 1926-8.

[114]   Chiarello E, Tedesco G, Cadossi M, *et al.* Surgical prevention of femoral neck fractures in elderly osteoporotic patients. A literature review. Clin Cases Miner Bone Metab 2016; 13(1): 42-5.
[http://dx.doi.org/10.11138/ccmbm/2016.13.1.042] [PMID: 27252744]

[115]   Moore WR, Graves SE, Bain GI. Synthetic bone graft substitutes. ANZ J Surg 2001; 71(6): 354-61.
[http://dx.doi.org/10.1046/j.1440-1622.2001.02128.x] [PMID: 11409021]

[116]   Hak DJ. The use of osteoconductive bone graft substitutes in orthopaedic trauma. J Am Acad Orthop Surg 2007; 15(9): 525-36.
[http://dx.doi.org/10.5435/00124635-200709000-00003] [PMID: 17761609]

[117]   Blum L, Cummings K, Goulet JA, Perdue AM, Mauffrey C, Hake ME. Atypical femur fractures in patients receiving bisphosphonate therapy: etiology and management. Eur J Orthop Surg Traumatol 2016; 26(4): 371-7.
[http://dx.doi.org/10.1007/s00590-016-1742-6] [PMID: 26943872]

[118]   Basafa E, Armiger RS, Kutzer MD, Belkoff SM, Mears SC, Armand M. Patient-specific finite element modeling for femoral bone augmentation. Med Eng Phys 2013; 35(6): 860-5.
[http://dx.doi.org/10.1016/j.medengphy.2013.01.003] [PMID: 23375663]

[119] Basafa E, Murphy RJ, Kutzer MD, Otake Y, Armand M. A Particle Model for Prediction of Cement Infiltration of Cancellous Bone in Osteoporotic Bone Augmentation. PLoS One 2013; 8(6)e67958
[http://dx.doi.org/10.1371/journal.pone.0067958] [PMID: 23840794]

[120] Sterling JA, Guelcher SA. Biomaterial scaffolds for treating osteoporotic bone. Curr Osteoporos Rep 2014; 12(1): 48-54.
[http://dx.doi.org/10.1007/s11914-014-0187-2] [PMID: 24458428]

[121] Mavrogenis AF, Dimitriou R, Parvizi J, Babis GC. Biology of implant osseointegration. J Musculoskelet Neuronal Interact 2009; 9(2): 61-71.
[PMID: 19516081]

[122] Dimitriou R, Jones E, McGonagle D, Giannoudis PV. Bone regeneration: current concepts and future directions. BMC Med 2011; 9: 66.
[http://dx.doi.org/10.1186/1741-7015-9-66] [PMID: 21627784]

[123] Dinopoulos H, Dimitriou R, Giannoudis PV. Bone graft substitutes: What are the options? Surgeon 2012; 10(4): 230-9.

CHAPTER 2

Cholesterol in Brain Health and Pathologies: The Role in Alzheimer's Disease

Kosara Smiljanic, Natasa Loncarevic-Vasiljkovic, Selma Kanazir and **Aleksandra Mladenovic Djordjevic**[*]

Institute for Biological Research "Sinisa Stankovic", University of Belgrade, Belgrade, Serbia

Abstract: Cholesterol is the molecule essential for life, but also with a possible detrimental role. Apart from being a vital structural constituent of the cells, cholesterol is a factor involved in many important cell processes. However, it has been known that high blood cholesterol is associated with many pathological conditions. An elevated level of cholesterol is linked with cardiovascular disease, diabetes and neurodegenerative disorders.

Almost quarter of the total cholesterol in the body resides in the brain. This vast pool is synthesized *in situ* and it is almost completely isolated and independent from the periphery due to the presence of blood-brain barrier. In the central nervous system, cholesterol plays important role in neural cells structure and functions, including synaptic transmission. Due to this, its content must be precisely maintained in order to keep brain function well. However, cholesterol is critically challenged in the aging brain and disturbed in several of pathological conditions, like Huntington's disease (HD), Parkinson's disease (PD), Niemann-Pick type C (NPC) disease and Smith-Lemli Opitz syndrome (SLOS), traumatic brain injury, multiple sclerosis (MS) and in Alzheimer's disease (AD).

Altered cholesterol metabolism has been extensively implicated in the pathogenesis of AD. A growing amount of evidence underscores the link between disturbed intracellular trafficking of cholesterol in the brain and the formation of amyloid plaques. The inheritance of the epsilon4 allele of the Apolipoprotein E (ApoE), the main transport protein for cholesterol in the brain represents the main risk factor for late onset form of Alzheimer's disease. Other genetic polymorphisms associated with critical points in cholesterol metabolism may also contribute to the AD pathogenesis. Hypercholesterolemia has been considered nowadays also as a risk factor, and all of these players are thought to promote the production of beta-amyloid and development of AD. Additional proof towards cholesterol involvement in the pathogenesis of AD gave epidemiological data of the cholesterol-lowering drugs, statins that have been shown to decrease the risk for AD.

---

[*] **Corresponding author Aleksandra Mladenovic Djordjevic:** Institute for Biological Research "Sinisa Stankovic", University of Belgrade, Belgrade, Serbia; Tel: +381 11 2078 338; Fax: +381 11 2761 433; E-mail: anamikos@ibiss.bg.ac.rs

Atta-ur-Rahman (Ed.)
All rights reserved-© 2020 Bentham Science Publishers

This chapter is aimed to summarize existing knowledge about the brain cholesterol metabolism, how the homeostasis is changed during aging and in various neuro-degenerative diseases, with special emphasis on Alzheimer's disease. As a final point, we will try to give a full insight into the environmental influences (including dietary restriction and statins therapy) on brain cholesterol homeostasis.

Keywords: Aging, Alzheimer's disease, Brain injury, Central nervous system, Cholesterol homeostasis, Cholesterol metabolism, Dietary restriction, Neurodegenerative diseases, Oxysterols, Statins.

INTRODUCTION

Nowadays, there is almost no molecule like cholesterol, surrounded by such a big debate. It has triggered so much attention since its discovery in the second half of the 18th century, and does not cease to provoke controversy among scientists, but also in the general population. It is indispensable for life in so many ways, but it can still be harmful and fatal for health. The fact that so far 13 Nobel Prizes were awarded for the achievements connected to the structure, function and metabolism of cholesterol, tells us a story about its significance. Cholesterol is often called the secret killer, but still, no cell membrane would exist without it. The cholesterol is "the one" that builds the membranes, keeps their integrity, permeability and cell viability. In contrast to a rigid cell wall that plants possess, animal cell's membranes rich in cholesterol are fluid. Its role does not stop here- cholesterol is a precursor in the biosynthesis of vitamin D, bile acids and all steroid hormones.

Representing double-edged sword cholesterol is still in the focus of a large number of studies that attempt to shed light on its role in physiological and pathological processes. French physician-chemist François Poulletier first isolated cholesterol from gallstones in 1784. It was named cholesterine by French chemist Michel E. Chevreul (Greek: chole for bile and stereos for solid). It took an entire century to get to exact molecular formula of cholesterol established by Austrian botanist Friedrich Reinitzer in 1888. Elucidating its structure was extremely difficult. Marvelous work of Heinrich O. Wieland and Adolf Windaus brought them the Nobel Prize in Chemistry in 1927 and 1928, respectively, for revealing tetracyclic cholesterol skeleton [for a review see 1]. Konrad Bloch (1964) won the Nobel Prize for the contribution to the clarification of cholesterol biosynthesis. Michael Brown and Joseph Goldstein, who received the Nobel Prize in 1985 for their work on cholesterol regulation, named cholesterol Janus, a molecule with two faces.

Every animal cell is capable of synthesizing cholesterol by a complex 37-step process and most of our cholesterol is indeed made in our own organism. However, cholesterol can be also absorbed directly from the food. The content of cholesterol in our body is a result of a fine balance between the process of synthesis and the process of absorption. If cholesterol homeostasis is preserved, then the body compensates for any absorption of additional cholesterol by reducing cholesterol synthesis and on the contrary - it will synthetize additional cholesterol when we need it, in cases like pregnancy, or recovery from illnesses and injuries. To make things more complicated, our body recycles cholesterol whenever it is possible, rather than synthetizing it. Typically, about 50% of the cholesterol excreted by the liver is reabsorbed by the small intestine back into the bloodstream.

Cholesterol is essential for cell functioning. It is like the brick in the house wall, a structural component of each and every cell in our bodies, required for building and maintaining the integrity of animal cell membranes, thus providing a protective barrier. However, cholesterol is not only **mechanical constituent** of plasma membranes; it also **regulates the permeability and fluidity** of cell membrane over the range of temperatures. Tetracyclic ring structure (Fig. **1**) provides cholesterol capability to decrease membrane fluidity and makes membrane rigid.

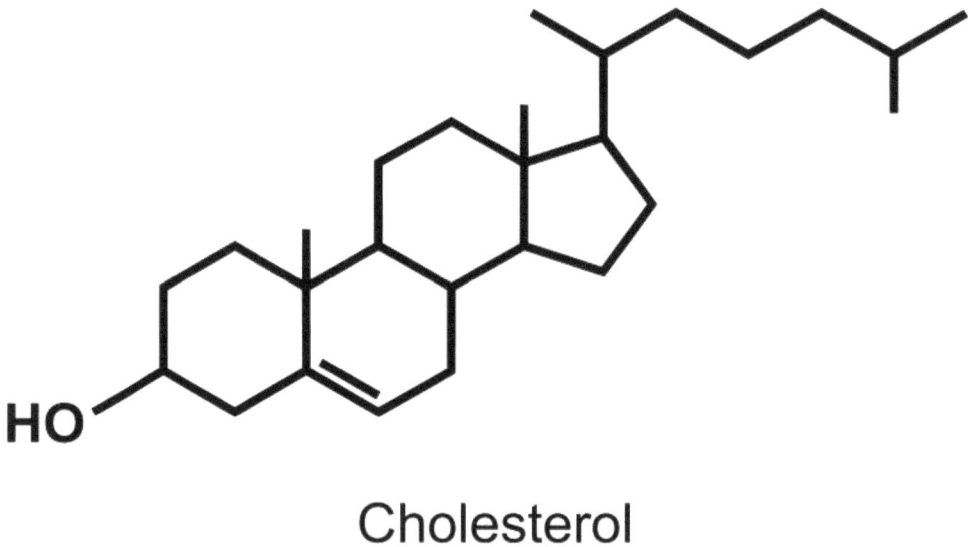

Fig. (1). Tetracyclic ring structure of cholesterol.

Cholesterol is a compelling modulator of the membrane proteins and also participates in numerous membrane trafficking and transmembrane signaling processes [2].

One of the most important roles played by cholesterol is in the **hormone production**. Cholesterol is stored in the adrenal glands, ovaries and the testes and is converted to various steroid hormones like cortisol and aldosterone, and also the sex hormones estrogens, progesterone, and testosterone. It is also an important precursor molecule essential for the biosynthesis of Vitamin D.

Cholesterol plays an important role in our body's digestion. In the liver, cholesterol is converted to bile and then stored in the gallbladder; **bile acids and bile salts** are essential for the absorption and digestion of fat molecules and fat soluble Vitamins – Vitamin A, D, E and K.

Some recent research suggest that cholesterol may act as an **antioxidant** [3].

By being essential constituent of the brain cells and by regulating ion permeability, cell shape, cell–cell interaction, and transmembrane signaling, cholesterol is necessity for neuronal physiology, both during development and in the adulthood. However, in order to play its function well, cholesterol homeostasis needs to be tightly regulated. Mutations in cholesterol-related genes result in impaired brain functions during early life and hereditary diseases. Later in life, defects in brain cholesterol metabolism can occur due to variety of reasons, and may contribute to neurological disorders, such as Parkinson's, Alzheimer's or Huntington's disease.

The aim of this chapter is to describe and summarize the role of cholesterol in physiological and pathological processes in the brain, especially its role in AD pathogenesis. In the first part we will focus on the cholesterol's role in the brain, especially during aging, while in the second part emphasis will be put on its contribution to the pathological processes in the brain. The third part will be dedicated to the environmental interventions that can be used in order to maintain the homeostasis of brain cholesterol.

CHOLESTEROL IN THE BRAIN

Brain weight to total body weight ratio is 1 to 50, but brain cholesterol content to total body cholesterol content ratio is 1 to 4. This fact makes the brain being the organ richest in the cholesterol and cholesterol being the major sterol in the adult brain. The majority of cholesterol (70-80%) inhabits in myelin sheaths, while the other pool resides in plasmalemmal and subcellular membranes of neural cells [4]. Cholesterol is not uniformly distributed within membranes but is enriched in the cytosolic (inner) leaflet of the plasma membrane [5] (Fig. **2**). Further, cholesterol is concentrated in nano/micro-domains within plasma membrane, so-called 'lipid rafts'. These rafts represent highly dynamic arrangements dispersed throughout the membrane of cells that recruit downstream signaling molecules upon activation by external or internal signals [6]. Rafts also contain a high amount of sphingomyelin, which is augmented in the outer leaflet of the plasma membrane, indicating that some trans-bilayer translocation must occur to form and stabilize these domains [7]. In neurons, rafts have been perceived at synapses, suggesting their role in synaptic transmission [8 - 10]. Lipid rafts are also fundamental in neuronal cell adhesion, axonal guidance and in growth-factor mediated signal transduction, vesicular trafficking and membrane associated proteolysis and help the neurons to compartmentalize and regulate these processes [11, 12].

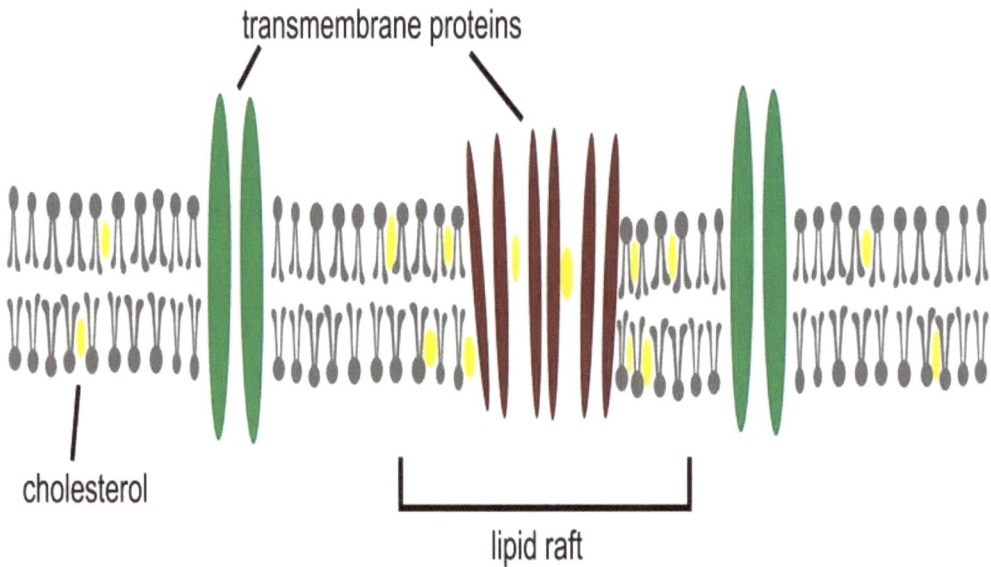

Fig. (2). Cell membrane organization with standard lipid bilayer and lipid raft.

As a multifaced molecule, cholesterol fulfills a variety of roles in the central nervous system (CNS). In addition to its well-known functions in all cells of an organism, in the brain, cholesterol meets a number of specific roles.

It is prerequisite component for synapse and dendrite formation [13, 14], as well as for axonal guidance [15], making it essential for neuronal physiology through-out entire lifespan. Cholesterol is important component of synaptic vesicles and also represents a universal regulator in multiple forms of vesicle endocytosis at mammalian central synapses. Cholesterol disruption impairs three different forms of endocytosis, including slow endocytosis, rapid endocytosis, and endocytosis of the retrievable membrane that exists at the surface before stimulation [16]. It supports intrinsic negative curvature of membranes [17, 18] and facilitates the fusion process [19]. In addition, cholesterol may favor membrane fusion through its interaction with synaptophysin, an integral membrane protein enriched in synaptic vesicles [20]. Cholesterol depletion in neurons impairs synaptic vesicle exocytosis, neuronal activity and neurotransmission, leads to dendritic spine and synapse degeneration [21].

The lipid arrangement of the subsynaptic membranes participates in the regulation of the number and composition of the postsynaptic glutamate receptors which contribute to the induction and consolidation of memory formation [22 - 24]. Cholesterol is also a protector of cholinergic processes [11]. In addition to cholesterol, its biologically active oxidized products oxysterols can influence cell function as well [25, 26].

As mentioned above, myelin sheaths are particularly rich in cholesterol. Additionally, the lipid-protein composition of myelin differs extremely from the other cell membranes. Namely, in comparison with standard membrane composition, where distribution is about of 30% lipids and 70% of proteins, myelin sheets have the opposite ratio, consisting of about 70% of lipids and 30% of proteins. Of those lipids, cholesterol, phospholipids, and glycosphingolipids stand in a molar ratio of about 4: 4: 2. This specific composition of myelin sheath, in addition to a reduced permeability to ions that cholesterol enrichment gives to the myelin, is exactly what is needed for myelin to fulfill its essential role to enable facilitated signal propagation throughout the axons.

When the brain undergoes a massive cellular expansion, like during embryonic development or during recovery, cholesterol plays an essential role, enabling proper cell division and growth.

The importance of cholesterol in learning and memory is also well studied although with some controversial results. There are studies pointing to detrimental influence of high cholesterol levels to human learning and memory, mainly as it is considered that elevated serum cholesterol is a risk factor for mild cognitive impairment [27 - 31] and dementia [32, 33]. According to some studies, lanosterol and lathosterol, cholesterol precursors levels were correlated with low memory performance [34], while cholesterol levels correlated with measures of intelligence in some cases [35, 36, 31], but not in the elderly [37, 38]. On the other hand, it has been shown that low HDL cholesterol can be correlated with deficits and declines in memory in midlife [39] and that increased cholesterol, on contrary, can improve learning and memory [40, 41, 38]. Moreover, there are cases when high cholesterol was correlated with high cognitive functioning [42, 41].

Animal studies revealed that elevating cholesterol can improve learning and memory in mice [43, 44] and rats [45], whereas animals that are either deficient in cholesterol or have cholesterol synthesis blocked experienced wide range of deficits in learning and memory [46 - 51].

Accordingly, disturbing the cholesterol synthesis or lipoprotein transport reduces synaptic plasticity and, consequently, cognitive functions [52, 53].

Cholesterol Metabolism in the Brain

In general, cholesterol metabolism shares the same characteristics in the entire organism. However, due to the specific requirements and features of the central nervous system and the presence of blood-brain barrier (BBB), cholesterol metabolism in the brain exhibits some specifics and we will pay special attention to those in this chapter.

As mentioned before, cholesterol is prerequisite for proper bodily functioning. Outside the CNS, tissues meet their cholesterol requirement in two ways: from endogenous synthesis and from exogenous lipoproteins delivered from the circulation and originating from the food. Due to the presence of the BBB, which, when intact is impermeable to plasma lipoproteins, almost all the cholesterol in the CNS is produced *in situ* [4], although there are evidences that certain lipoproteins carrying cholesterol can pass BBB [54]. Another specificity of the brain cholesterol is that its half-life in the adult organism is much longer than the half-life of plasma cholesterol. Namely, brain cholesterol half-life is between 6 months and 5 years in comparison to few days of half-life of peripheral cholesterol [55 - 57].

Cholesterol Synthesis

First evidence that brain cells synthesize cholesterol has been published in the first half of the 20th century by the group of Heinrich Waelsch. Their inventive experiments to study lipid synthesis in rats using heavy water implied that cholesterol is synthesized in the brain and that the rate of synthesis is higher during early postnatal development than in adult animals [58]. Nowadays it is widely known that majority of brain cholesterol accumulates between the perinatal period and adolescence, in the period when neurons get specialized myelin sheets. After this process, cholesterol synthesis in the brain becomes "silent" and the whole metabolism is characterized by a very low cholesterol turnover and minimal losses [59]. These findings were additionally confirmed in the study on the aging rats. With advanced age, assessed by the concentration of its precursors, cholesterol synthesis was increased in extracerebral tissues, while in the brain (cortex and hippocampus) it was diminished [60]. In addition to these findings, it is known that in adult brains, cholesterol synthesis can be induced, as it is changed after brain injury [61]. However, in physiological conditions, cholesterol synthesis is maintained at a low rate at adult state [62]. This is the result of a very effective recycling of cholesterol in the brain and a very long half-life of cholesterol [63].

However, there is a difference in the rate of cholesterol synthesis between different brain regions. In the cortex and hippocampus, a high transcript levels of the key enzymes involved in cholesterol synthesis can be found [64]. It seems that cholesterol content, the rate of its synthesis, as well as the expression levels of cholesterol-synthesizing enzymes and regulating factors like sterol-sensing proteins, intracellular transporters and shuttle proteins, vary significantly within different brain regions [65 - 67]. The levels of lipoprotein receptors responsible for uptake of cholesterol also vary greatly indicating diversity between brain regions not only in the synthesis rate, but also in the absorption.

Brain cholesterol synthesis is not only time- and region- specific, but it's also a cell specific. Namely, neural cells differ by their capacity to produce cholesterol. Oligodendrocytes have the highest capacity for cholesterol synthesis, while astrocytes synthesize 2 to 3 times more cholesterol in comparison to neurons [62]. Cholesterol synthesis in oligodendrocytes is highest during active myelination, but decreases by ~90% after myelination process has been completed and remains unaffected by gender or the cholesterol concentration in the blood [4]. In the adult brain, however, cholesterol synthesis occurs predominantly in the astrocytes, although cholesterol can be synthesized by neurons themselves [68, 69]. Both astrocytes and neurons *in vitro* would synthesize cholesterol at a rate inversely proportional to the cholesterol content in the growth environment. Neurons would however, rather import cholesterol through the ApoE–LDL receptor system, than synthesized it, while oligodendrocytes and Schwann cells, in contrast, would synthesize *de novo* cholesterol rather than importing it. Under normal circum-stances, cholesterol synthesis in the neurons is required only when intake of cholesterol produced by other CNS cells is insufficient to meet the cholesterol requirement of the neuronal cells. In general, cholesterol homeostasis in all the neural cells is thus maintained by the perfect balance between cholesterol influx and cholesterol synthesis and in accordance with cell type preference to one of those processes.

The reason why neurons after the birth reduce or stop the synthesis of cholesterol and become almost completely dependent on cholesterol synthesized by astrocytes [61] probably lies in the fact that synthesis of one molecule of cholesterol requires 37 reactions. Such a complicated and demanding synthesis, and a diverse array of enzymes distributed in different subcellular compartments, consume large amount of energy [70]. For those reasons, neurons may abort cholesterol synthesis and redirect energy to the generation of electrical activity.

However, recent analysis of *in situ* hybridization data from the Allen Mouse Brain Atlas (http://www.brain-map.org) showed that the neurons possess much higher transcript levels of several cholesterol biosynthetic enzymes in comparison to astrocytes [71]. Although those data do not necessarily mean that cholesterol synthesis is higher in neurons, they indicate that neurons can readily take a part in the cholesterol synthesis any time, if needed.

The concentration of cholesterol remains essentially steady under normal physical conditions. However, a fraction of the pool is constantly replaced. Mechanisms must be in place to constantly excrete or degrade cholesterol, and at the same time to constantly supply cell plasma membranes with an equivalent amount of new sterols. These two processes must be very tightly regulated [4].

Estimations of the metabolic stability of cholesterol throughout the brain are mainly based on the fact that the largest amount of cholesterol in the brain is found in myelin which is metabolically inert. However, since cholesterol turnover is proportional to the metabolic rate of different tissues and species, in individual neurons and astrocytes it could be very high and reaches to an estimated 20% per day, depending on the region of the brain and neuronal cell type [4]. Cholesterol metabolism in the different regions of the brain is not uniform, but it can be influenced by aging and gender [72].

Cellular cholesterol synthesis (Fig. **3**) is a complex and resource-intense process: it starts with the conversion of acetyl-CoA to 3-hydroxy-3-methylglutaryl-CoA by HMG-CoA synthase (HMGCS), which is then converted to mevalonate by HMG-CoA reductase (HMGCR). This enzyme constitutes the only known rate-limiting and irreversible step in cholesterol synthesis. This is followed by a long series of enzymatic reactions that convert mevalonate into 3-isopenenyl pyrophosphate, farnesyl pyrophosphate, squalene and lanosterol, first steroidal precursor of cholesterol.

From this point onwards, two alternative pathways lead to cholesterol formation: the Bloch pathway *via* the direct cholesterol precursor desmosterol, and the Kandutsch–Russell pathway *via* lathosterol as featuring intermediate [73]. Evidence for the alternative use of these pathways during aging has been reported. In young rodents, cerebral cholesterol is mainly synthesized by the Bloch pathway, while the Kandutsch–Russell pathway prevails in older rodents [74, 60]. This diversity is not only age-specific. There are also indications that astrocytes predominantly synthesize cholesterol through the Bloch pathway, while neurons use the Kandutsch-Russel pathway [75].

Acetyl-CoA

Mevalonate pathway

HMG- CoA

HMGCR (rate limiting step)

Mevalonate

Lanosterol

Kandutsch-Russel pathway

Bloch pathway

Desmosterol   Lathosterol

7-Dehydrocholesterol

Cholesterol

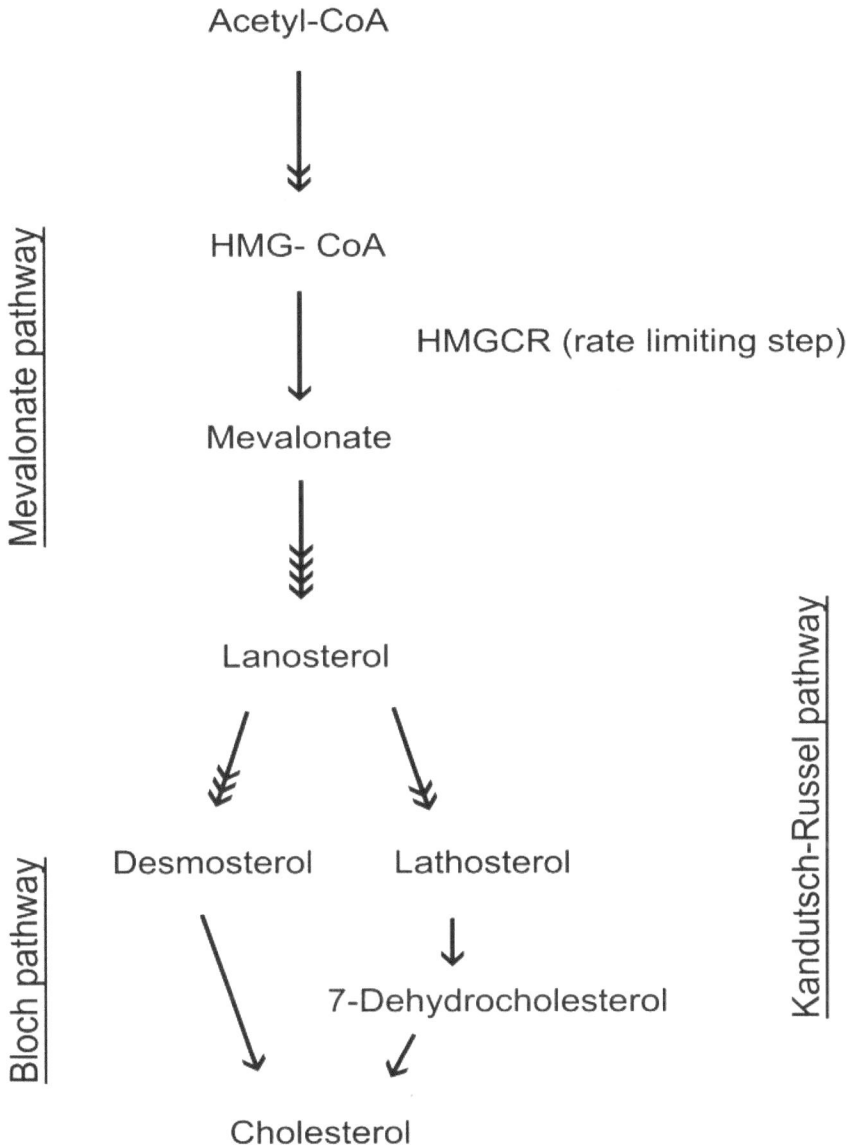

Fig. (3). Cholesterol synthesis.

Cholesterol Transport

Cholesterol produced by astrocytes is transported to neurons. Recycling and intercellular transport of cholesterol in the brain are efficiently maintained by specific lipoprotein transport system. For a long period of time, the question whether lipoproteins cross the BBB stays controversial. Brankatschk and Eaton showed that in Drosophila lipoprotein lipophorin is capable of crossing the BBB [54]. However, the opinion that brain lipoprotein system is completely independent of one in the peripheral circulation is still widely accepted. According to this estimation, lipoprotein particles are assembled *in situ* from the components both imported from the plasma and produced locally.

The main apolipoprotein present in the CNS is Apolipoprotein E (ApoE) [70]. Cholesterol and Apolipoprotein E, together with phospholipids generate lipoproteins that are similar in size to plasma high-density lipoproteins. The secretion of lipoprotein particles taking place *via* the process mediated by one or more of the ATP-binding cassette (ABC) transporters such as ABCA1, ABCG1 and/or ABCG4 (Fig. **4**).

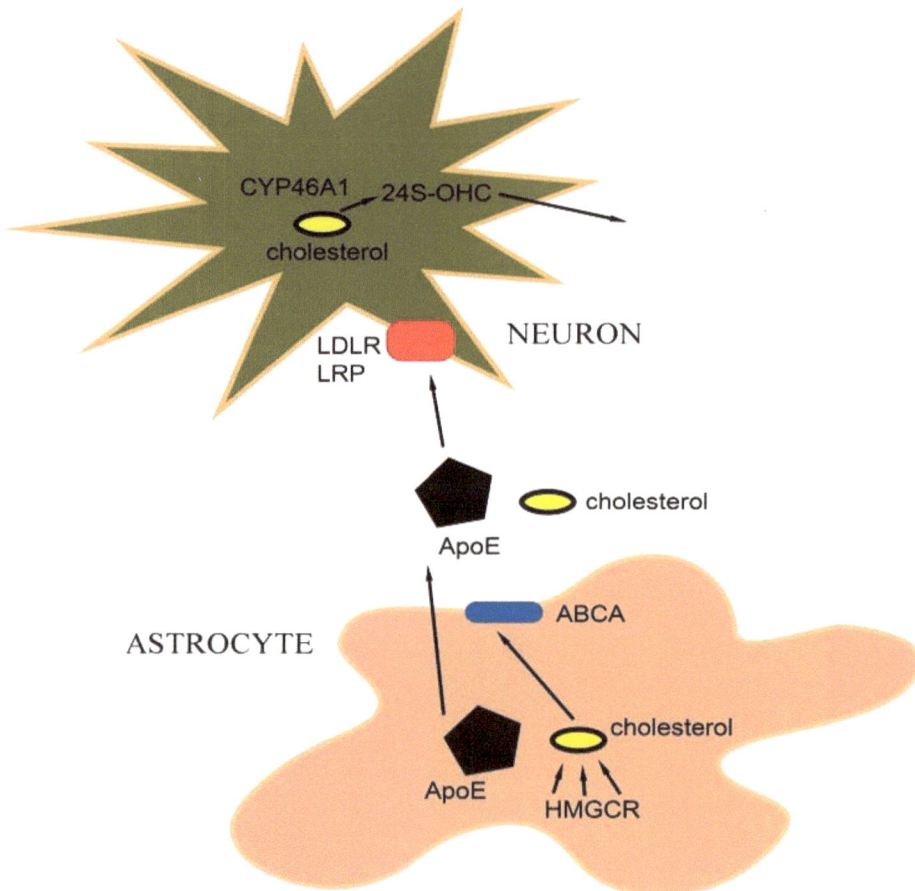

Fig. (4). The secretion and uptake of cholesterol.

The uptake of astrocyte-derived ApoE-containing lipoprotein particles by neurons is mediated by specific receptors and the process of endocytosis. Those receptors are LDL receptor, LDL-receptor-related protein (LRP) and ApoE receptor 2 (ApoER2) and they belong to the low-density lipoprotein (LDL) receptor family [76, 77]. The ApoER2 participates in reelin-mediated intracellular signaling while the LDL receptor and the LRP are those participating in cholesterol delivery to neurons and glia [78, 79]. Between those two, LDL receptor is predominately present in glial cells, and LDL receptor–related protein 1 receptors are more expressed in neurons.

In this manner, by using ABC transporters, Apolipoprotein E and its receptors, cholesterol is transferred from astrocytes to other neural cells [80, 81]. ApoE appears to be a key player in regulating cholesterol homeostasis and distribution among cells of the brain, and, consequently, it is essential for normal brain functioning. ApoE-containing lipoproteins stimulate synaptogenesis [80], enhance axonal growth [82] and prevent neuronal death [83]. Moreover, a role for ApoE in nerve repair has been indicated, because ApoE synthesis in glial cells increases by up to 150-fold after a nerve injury [84].

Cells regulate their cholesterol content by an exquisite feedback mechanism that balances biosynthesis and import. Cells sense their cholesterol level by membrane-bound transcription factors known as sterol regulatory element-binding proteins (SREBPs), which regulate the transcription of genes encoding enzymes for cholesterol and fatty acids metabolism. There are 3 different SREB proteins: SREBP1a and SREBP1c, for regulation of genes involved in biogenesis and homeostasis of both cholesterol and fatty acids, and SREBP2, regulating genes involved in cholesterol biosynthesis and metabolism solely. SREBP-cleavage-activating protein (SCAP) is bound to the full length immature SREBPs *via* its C-terminal domain. SCAP N-terminal domain is sterol-sensing and interacts with an insulin-induced gene-1 (Insig-1), another ER membrane-resident protein. *Via* SCAP–SREBP interaction, the whole protein complex stays bound to the ER in an inactive, immature form. When the cholesterol amount in the cell is low, Insig-1 dissociates from the SCAP–SREBP complex, the complex is then transported to the Golgi apparatus and SREBPs are processed sequentially by two proteases, site 1 protease and site 2 protease. The resulting active SREBPs homodimers enter the nucleus and regulate those genes containing so called sterol response element (SRE) [85]. SRE containing genes are mostly involved in lipid biogenesis and homeostasis [86].

Insig-1 is also responsible for the feedback inhibition of cholesterol biosynthesis [85]. When the cholesterol content in the cell is high Insig-1 stays connected to SCAP–SREBP complex in the ER, preventing the complex entering the nucleus and activating SRE. It can also bind the key enzyme in the biosynthesis of cholesterol, HMGCR and this action results in suppression of lipid biosynthesis and uptake [85, 87]. In the opposite cases of cell's increased need for cholesterol, Insig-1 is unable to bind SCAP or HMG-CoA and that results in proteolytic activation of SREB and in HMG-CoA stability. In those cases, Insig-1 becomes prone to degradation by proteasomal machinery [88].

The expression of SREBP genes is in addition controlled by retinoid X receptor (RXR)–liver X receptor (LXR) system and endogenous oxysterols such as 27-hydroxycholesterol and 24(S),25-epoxycholesterol, that are capable of inhibiting SREBP pathway by activating LXRs [89].

Cholesterol Elimination

The excess of cholesterol can be exported from the brain in two ways. One is to be associated with Apolipoprotein E and to be exported to cerebrospinal fluid (CSF). Astrocytes are in particular prone to this process, while neurons mostly use another way of excretion. The capacity of this pathway is very limited and in this manner only 1-2 mg cholesterol can be exported per day. On the other hand, neuronal cells possess a neuron specific enzyme, cholesterol 24S-hydroxylase (CYP46A1), responsible for the conversion of cholesterol into 24S-hydroxy-cholesterol (24S-OHC) [90]. 24(S)-hydroxycholesterol, as well as any other oxysterol that has a hydroxyl group in the side chain of cholesterol, can cross BBB more easily. Cholesterol, transformed in this way, passes the BBB, enters the circulation, and can be eliminated by the liver [91]. The amount of 24S-OHC removed from the brain is about 6-7 mg per day.

The enzyme itself is localized in endoplasmic reticulum [90, 92]. Immuno-histochemical staining revealed its presence in the pyramidal cells of the hippocampus, Purkinje cells of cerebellum, and in the cell bodies of the neurons of layers II / III, V and VI of the cortex. Also, the expression of this protein was confirmed in the interneurons of the hippocampus and cerebellum, in retinal ganglion cells, and in a subset of retinal cells localized in the inner nuclear layer [92]. Studies in CYP46A1-deficient mice show that at least 40% of the cholesterol that is excreted from the brain, is in the form of 24S-OHC. It is interesting that in those mice, although the elimination of cholesterol *via* 24-OHC pathway is disabled, cholesterol does not accumulate in the cells, as efficient compensatory mechanism reduces cholesterol synthesis by ~40% [93].

Brain is the major source of 24S-OHC and in comparison to the plasma and peripheral organs, where the ratio between cholesterol and oxysterols is about 1.000: 1 to 100.000: 1; in the brain that ratio amounts 500: 1 and 1000: 1. These facts imply the essential role of 24S-OHC in the CNS.

The specificity of the brain cholesterol metabolism goes beyond these quantitative differences. While the main oxysterol in the brain is 24S-OHC, in circulation another oxy-form of cholesterol is predominant. In the case of periphery, oxidation occurs at position 27 of cholesterol ring, and the enzyme responsible for this reaction is sterol 27-hydroxylase (CYP27A1). 27-hydroxycholesterol (27-OHC) produced this way is capable of passing the BBB and entering the brain. It is estimated that each day a 5 mg of 27-hydroxycholesterol arrives into the brain [94]. Of course, the exact amount depends on the integrity of BBB that can be disrupted in certain pathological conditions like AD. In that case, the amount of both oxysterols in the brain can be significantly changed.

Approximately the same amount of 24S-OHC that is produced and exported from the brain is taken up by the liver daily [63]. It is interesting to mention that in mice, liver is also capable of producing 24(S)-hydroxycholesterol, but while in 24S -hydroxylase knockout mice the synthesis of new cholesterol in the brain was reduced by approximately 40%, in the liver knocking out the gene had no effect on the level of its product [95]. This implies that although the liver is capable of the same oxy-conversion as brain, this process is not essential for the liver.

The fact that almost the entire 24S-OHC present in the circulation originates from the brain, makes it as a very convenient surrogate marker of cholesterol homeo-stasis in the brain.

24(S)-hydroxycholesterol activates liver X-receptors (LXR), LXRα and LXRβ, expressed in most tissues and organs and important regulators of genes involved in cholesterol homeostasis [96 - 98]. ABCA1, ABCG1 and ApoE are genes with LXR-regulating elements [99]. In this manner, a net export of cholesterol from the brain is regulated by a feedback mechanism *via* LXRβ, ABCA1 and CYP46.

It was previously mentioned that small portion of cholesterol is eliminated from the brain by the CSF and bounded with ApoE. Although the exact sites and mechanisms of this process are still unsolved, it is known that LDL receptors have some role in this excretion process. For instance, a LDL receptor overexpression induces a decrease, while deletion leads to an increase in ApoE level in CSF in mice [100].

Several studies showed that although the level of 24S-OHC can be increased in patients at the beginning of various neurodegenerative disorders and dementias in comparison to control subjects, it declines with the disease progression, and especially with longer duration of diseases and poorer memory performances. Thus, 24S-OHC level in plasma is generally lower in severe and advanced stages of AD, and it is also characterized by the lower global cognitive score [reviewed in 101].

In comparison to plasma 24S-OHC concentration, the level of 24S-OHC in CSF is considered to be a more reliable marker of neurodegeneration and dementias. It is shown that its level in AD and mild cognitive impairment does not vary as in plasma, but is steadily higher in comparison with control subjects [102 - 104]. Furthermore, the levels of 24-OHC in the CSF increased with the number of ApoE4 alleles [102]. However, it is not related to the cholesterol level itself [105] and it is not correlated with Aβ level in the CSF [104]. On contrary, it seems that levels of oxysterol in the plasma, but not in CSF, were correlated with amyloid beta Aβ42 and tau levels in CSF [106].

On the other hand, in brain tissue of AD cases only slightly lower concentrations of 24-OHC, and slightly elevated level of 27-OHC are detected, in comparison to control subjects.

The level of 24S-OHC in the circulation was shown to reflect the balance between brain production of this sterol and its metabolism in the liver [107]. Production in the brain is relatively constant, while its metabolism in the liver varies with age. As a result, the 24S-OHC level is relatively constant after the second decade of life, although there is a certain trend in increasing the amount of this sterol after the sixth decade of life. It has also been shown that various neuropathological conditions and diseases affect the level of 24S-OHC in the circulation [108 - 110].

CHOLESTEROL IN THE AGING BRAIN

Data indicating the fate of cholesterol in the aging brain are contradictory. Moderate cholesterol loss *in vitro* and *in vivo* was found in membranes of old hippocampal neurons [111, 112]. Synaptosomes isolated from the old mice are also characterized by decreased cholesterol content, as well as aged human hippocampus and cerebellum [112, 113], and the frontal and temporal cortex of people over 20 years of age [114]. The rearrangements in cholesterol localization can also occur during aging and increased amount of cholesterol in the outer layer of the membrane bilayer and consequential cholesterol asymmetry in the bipolar membrane compared to young mice [115] has been described.

However, not all the brain regions undergo changes in cholesterol level during the life span [113]. In addition, sometimes it is not the amount of cholesterol that is changing during aging, but the metabolism. For instance, cholesterol synthesis has been shown to decrease in the human hippocampus, while the amount of cholesterol did not change [116].

Cholesterol in Brain Pathologies

There are several brain conditions that might be related to the aging process, but strongly linked to defects in cholesterol metabolism. Alzheimer's disease is perhaps the most known neurodegenerative disorder linked to cholesterol, however numerous other conditions such as Niemann-Pick type C (NPC) disease and Smith-Lemli-Opitz syndrome, atherosclerosis, as well as Huntington's and Parkisnosn's diseases are linked to the changes related to cholesterol. In these cases, one or more of cholesterol metabolic pathways are affected, resulting in disturbed cholesterol homeostasis (Fig. **5**).

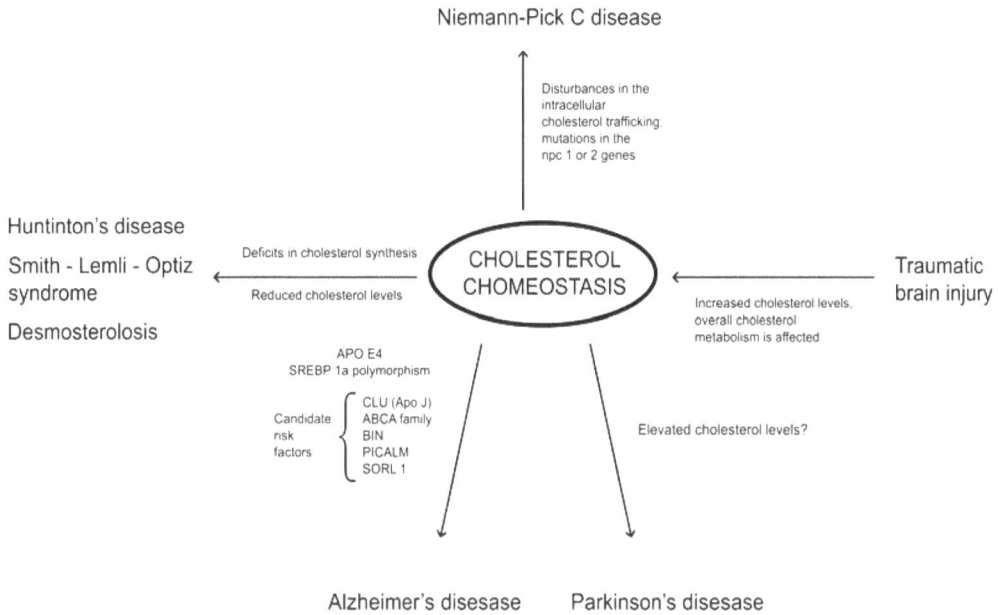

Niemann-Pick C disease

Disturbances in the intracellular cholesterol trafficking: mutations in the npc 1 or 2 genes

Huntinton's disease

Smith - Lemli - Optiz syndrome

Desmosterolosis

Deficits in cholesterol synthesis

Reduced cholesterol levels

CHOLESTEROL CHOMEOSTASIS

Traumatic brain injury

Increased cholesterol levels, overall cholesterol metabolism is affected

APO E4
SREBP 1a polymorphism

Candidate risk factors { CLU (Apo J) ABCA family BIN PICALM SORL 1

Elevated cholesterol levels?

Alzheimer's disesase   Parkinson's disesase

Smith-Lemli-Opitz syndrome (SLOS) is a rare autosomal recessive disorder caused by total or partial deficiency of enzyme 7-dehydrocholesterol reductase (Dhcr7) that leads to the altered composition of cell membranes. The absence of proper function of the enzyme that catalyzes the conversion of 7-dehydrocholesterol to cholesterol, the final step of cholesterol synthesis in the Kandutsch-Russell pathway, in this case leads to a markedly reduced level of tissue cholesterol and total sterol levels, and elevated 7-dehydrocholesterol level [117]. Clinical signs of the disease are developmental deformities, incomplete myelination, and mental retardation. This high 7-dehydrocholesterol concentration inhibits HMGCR, that further decreases internal synthesis and exacerbates cholesterol deficit in the cell [118]. As a consequence of changed cholesterol to 7-dehydrocholesterol ratio in the cell, the whole lipid composition of the membrane is changing. Disturbed cholesterol content in the membrane in the favor of increased 7-dehydro- cholesterol levels increases membrane fluidity [119] and thus induces functional changes in the membranes. People with severe form of this disorder have only 2% of the normal cholesterol plasma level, a significantly decreased level of choles- terol in all tissues, with the brain as the most severe affected tissue. In the case of moderate form of disease other tissues are capable to compensate cholesterol level to some extent due to the dietary import, but in the brain, due to the existence and efficacy of the BBB, the cholesterol level remains extremely low.

Huntington's disease (HD) is also characterized by certain abnormalities in cholesterol homeostasis [120]. Both in murine models of HD and in postmortem brains of HD patients decreased expression of genes involved in cholesterol biosynthetic pathways was observed [121]. The Huntingtin protein, a hallmark of Huntington's disease, in these cases is not properly palmitoylatiolated within neurons, which leads to defective subcellular trafficking and its accumulation and formation of cytoplasmic and nuclear inclusions [122, 123]. Accumulated Huntingtin further leads to neuronal dysfunction, cell loss and atrophy of putamen and neostriatum of diseased persons.

Another rare, autosomal recessive disease caused by the mutations in the *dhcr24* gene whose protein product catalyzes the reduction of desmosterol to cholesterol is **Desmosterolosis**. Deficits in cholesterol synthesis and its reduced level, or accumulation of desmosterol most probably lead to the severe brain defects including microcephalia, hydrocephalia, ventricular enlargement, defects in the corpus callosum, and thinning of white matter and seizures [124].

Niemann-Pick C disease (NPC) is an autosomal recessive disorder caused in 95% of the cases by mutations in the *npc1* or *npc2* genes, that leads to disturbances in the intracellular cholesterol trafficking. NPC proteins mediates the exit of cholesterol from the endosomal-lysosomal system and accumulation of unesterified cholesterol and in late endosomes and lysosomes. A consequential reduction of cholesterol content in plasma membranes and changes in the synaptic vesicle composition and morphology lead to various neurological symptoms. In the brain of NPC patients, a variety of disorders in cell extensions, like enlarged neurites, ectopic dendrites and axonal dystrophy can be seen. Neurofibrillary tangles, neuroinflammation, and neuronal death in the advanced cases are also present. Purkinje cells in cerebellum are affected in particular [125].

Taken together, these rare genetic diseases underline that the brain is highly sensitive to perturbations of cholesterol synthesis and transport. Both high and reduced levels of cholesterol and/or its precursors and products can lead to serious deficits in metabolism and consequently to serious neurodegeneration and these defects cannot be annulled by manipulations of peripheral cholesterol levels.

Parkinson's disease (PD) –Studies about cholesterol role in PD are controversial; while few of them demonstrated correlation between high plasma cholesterol and PD [126, 127] others did not [128 - 130].

Evidences for a causative link between high cholesterol and PD are not so straight forward as they are in AD, but rather indirect [for a review see 131]. For example, it has been shown that high fat diet exacerbated the parkinsonian pathology [132, 132, 133] while oxysterols can be significantly high in plasma, CSF or brain of PD and Lewy body demented patients [134 - 137]. High cholesterol can promote α-synuclein aggregation *in vitro*, while statin-induced decrease of cholesterol suppresses that process. There are additional beneficial effects of statins on PD pathology reported, starting from reduced incidence and slower progression of PD in statin users, to reduced aggregation of a-synuclein found *in vitro*, and *in vivo* and to up regulation of dopamine receptors [all reviewed in 131]. In MPTP model, one of the most widely used animal model of PD, cholesterol contributes to the loss of dopaminergic neurons [138]. Hypercholesterolemia induced dec- rease in the level of cortical and striatal biogenic amines, serotonin and dopamine, decrease in tyrosine hydroxylase immunoreactivity in striatum, motor behavioral abnormalities and depression, all of these being symptoms of PD [139].

Traumatic brain injury (TBI) Traumatic brain injury (TBI) is associated with significant neuropsychological deficits, and increases the risk, later in life, for neurodegenerative diseases such as AD [140] and PD [141]. The initial traumatic event is shearing, laceration, and/or contusion of brain tissue resulting from a physical impact. Secondary injury appears as a result of ischemia, alterations in ion levels, oxidative stress caused by reactive oxygen species (ROS), edema and axonal swelling [142, 143]. Thus, the neurodegeneration beyond the initial impact may be due to the inflammatory processes following TBI [144 - 147], since inflammation contributes to the chronic neurodegeneration seen in AD [148 - 152] and PD [153 - 156].

Due to the neuronal damage and proliferation of glial cells, cholesterol homeostasis is highly compromised [157, 158]. An increase in CSF cholesterol levels is observed in the TBI patients [159]. This excess cholesterol is likely due to injury-induced neuronal damage and the resulting membrane debris at the site of injury [160 - 162]. It has been proposed that hyperlipidemia may contribute to worsen proinflammatory condition in a later phase of TBI and an increased risk to new neurovascular events. Since inflammation after the TBI leads to cell death, lipid accumulation that leads to progression of inflammation, contributes to the pathophysiology of brain injury, and increased acute risk [148]. In the periphery, high cholesterol levels are known to trigger inflammatory mechanisms [163], and cholesterol clearance mechanisms are thought to be anti-inflammatory [164 -

167]. In the brain, cholesterol clearance *via* CYP46A1-mediated conversion could have anti-inflammatory role [165 - 167]. Namely, following TBI expression of CYP46A1 was shifted from neurons to glia [168, 169]. Since microglial cells play an important role in phagocytosis of the injured brain tissue [170] and have a role in the removal of lipid debris [161, 162] increased microglial CYP46A1 activity may play a role in removal of damaged cell membranes following injury. These data suggest that glial CYP46A1 expression could be part of a system for the post-injury removal of damaged cell membranes thus contributing to re-establishment of the brain cholesterol homeostasis. These findings suggest that while CYP46A1 activity in neurons plays a critical role in normal cholesterol homeostasis, increased CYP46A1 activity in glial cells may be a characteristic of the brain response to injury.

Both ABCA1 and ApoE have been shown to increase at the site of acute brain injury [171, 172]. ApoE accumulates at the injury site when peripheral nerves are injured, suggesting a role in the injury repair process [173]. Compared with wild-type ApoE-sufficient mice, ApoE-deficient mice show susceptibility to ischemic insult, with both neurological deficit and infarct size worsened [145]. Blood–nerve and blood–brain barriers are also impaired in ApoE-knockout mice [174]. ApoE modulates glial activation and the endogenous CNS inflammatory response. This is consistent with the known *in vitro* and *in vivo* immuno-modulatory properties of ApoE [175]. These findings indicate that ApoE is a key protein involved in maintaining the health of the central nervous system in response to injury.

CHOLESTEROL AND ALZHEIMER'S DISEASE

Alzheimer´s disease is the most notorious neurodegenerative disorder. Up to date aging represents the most important risk factor. However, link between cholesterol metabolism and susceptibility to AD must not be neglected. One of the major neuropathological features of AD, besides senile plaques and neuro-fibrillary tangles is accumulation of lipid droplets, pointing to the role of the disturbed lipid metabolism, especially dysregulated cholesterol metabolism in pathogenesis and progression of AD.

Experimental animals, starting from rabbits, to monkeys and transgenic (APP) mice, after being on a high-cholesterol and high-fat diet, showed an increased beta-amyloid deposition [175, 176]. In addition, considerable epidemiological evidences indicate that among many factors contributing to AD pathology, hypercholesterolemia and insulin resistance are among the most important risk factors [177 - 183].

High LDL and high total cholesterol levels are very often associated with cognitive impairment. Hypercholesterolemia accelerates intraneuronal accumu-lation of Aβ oligomers resulting in memory impairment in transgenic mouse model of AD [184]. Many studies indicated that cholesterol and LDL could be modifiable risk factors and that lowering them may represent a strategy for preventing cognitive impairment.

However, other studies point to controversial conclusions, especially those on late-life cholesterol levels and dementia risk. Some authors even reported that low, rather than high levels of cholesterol in late-life, are associated with increased dementia risk [185 - 187]. Although it seems that high cholesterol in late life may mean better brain health, high cholesterol in midlife could be associated with an increased risk of late-life dementia and cognitive decline [188]. Community-based longitudinal study of Chinese elderly demonstrated that higher blood cholesterol concentrations were associated with faster cognitive decline [189], while another, cross-sectional study of rural, elderly Chinese cohort indicated that there are factors, other than cholesterol, playing an intermediate, detrimental role between cholesterol and cognitive decline. One of those could be homocysteine. When homocysteine level was normal, both low and high cholesterol levels were associated with lower cognitive scores. On the other hand, when homocysteine levels were high, no significant association between cholesterol and cognition was found, indicating that determining factor here could be homocysteine, not cholesterol itself [190].

An interesting recent study by Paul and Borah in hypercholesterolemic mice showed, for example, that significantly elevated level of cholesterol in different brain regions occurs together with the depletion of acetyl cholinesterase activity. Authors consider that hypercholesterolemia caused BBB disruption and most probably led to increased cholesterol level in the brain [191]. Authors also found that global mitochondrial dysfunction occurs in the brains of hypercho-lesterolemic mice, which is in concordance with previously published study showing a significant relationship between hypercholesterolemia, cognitive impairment, and mitochondrial dysfunctional/oxidative stress [192]. Taken all together, data suggest that cholesterol associates with late-life cognitive function, but the association is strongly age-dependent.

The role of cholesterol in AD pathology is not only age-related, but also region specific. Namely, extensive efforts have been made in the last two decades to determine regional specificities in the brain cholesterol metabolism/ content in AD cases. Thus, there is a general attitude that the structures predominantly affected in AD, *i.e.* cortex and hippocampus, are exactly those characterized by a disturbed cholesterol homeostasis. On contrary, cerebellum by being a relatively "intact" in AD, showed no changes in cholesterol level [178, 193]. However, recently it has been found that cerebellum also suffer from the appearance of plaques [194 - 196]. As it has been shown that cholesterol homeostasis in cerebellum goes through different type of age-related changes in comparison to cortex and hippocampus [197], the question stays whether cholesterol metabolism is linked to Aβ pathology in this brain structure in a somewhat different manner.

Nowadays it is widely known that there are many links between different aspects of cholesterol metabolism and various pathways involved in AD pathogenesis, but it is still not fully elucidated what is the cause and what is the consequence. In the following subchapter these connections will be discussed in more detailed manner.

The Role of Apolipoprotein E in Alzheimer's Disease

In favor of cholesterol-AD causative link speaks the fact that the Apolipoprotein E gene is the most important genetic risk factor for the sporadic (or late-onset) form of AD (LOAD) [reviewed in 198 - 200]. Few other genes that can be marked as "candidate risk factors", like CLU (also named ApoJ), ABCA family members, BIN, PICALM, and SORL1, also belong to "lipid metabolism genes". One of the largest genome-wide association studies (GWASs) undertaken to date [201] revealed that those genes are among top 20 susceptibility loci for sporadic or late-onset AD For example, ApoJ/Clusterin is involved in lipid transport and has heat-shock-like chaperone activity. It seems that certain polymorphisms in this gene increase the LDL-cholesterol level [202]. ABCA1 mutations are characterized by reduced levels of circulating HDLs and the deposition of cholesterol esters in peripheral tissues, leading to the Tangier disease and familial HDL deficiency [203]. ABCA1 is also reported to be involved Aβ production and amyloid plaque formation [204, 205]. The amount of ABCA1 present in the cells influences the amount of Aβ levels, while the lack of ABCA1 leads to the increased Aβ and senile plaques levels [206] overexpression of ABCA1 decreases Aβ deposition [207]. On the other hand, Aβ is capable of inhibiting ABCA1 expression in cultured astrocytes [208]. However, none except ApoE was shown to be significantly associated with amyloid or tangle pathologies in AD [209], raising the issue of the clinical utility of these so called genetic risk variants [210].

There are three alleles of the ApoE gene in humans: E2, E3, and E4 and AD incidence is accelerated in ApoE4 individuals. The mean frequency in the general population is 8%, 78% and 14% for E2, E3 and E4, respectively [211]. ApoEE4 homozygous have a 50–90% higher chance for getting AD, in comparison to E2 and E3 carriers. The influence of the allele is additive. Those who are homozygous for the E4 ApoE have more chances of suffering from this disease than those who are heterozygous. In addition, possession of E4 allele lowers the age of AD onset. Frequency data of ApoE4 allele within AD female patients are interesting. Namely, it seems that there is a marked enrichment of the E4 frequency within AD group in women compared to man. Further, E4/3 phenotype is once again more prevalent in women than in men, while it is the opposite with the "protective" phenotype E3/3. This might be one of the causes why AD is more prevalent within females.

The exact mechanism is still not fully discovered, but until now, there are several possible pathways/ways described trough which ApoE4 exerts its detrimental effects. First of them is that E4 allele has a higher affinity for LDL receptor in comparison to ApoE3 proteins, while ApoE2 has the lowest affinity [reviewed in 212]. As a matter of fact, ApoE2 has a defective binding to the LDL-receptor and is responsible for type III hyperlipidemia.

This distinction is a result of a single amino acid difference between ApoE3 (Cys112) and ApoE4 (Arg112) [198] and points out that lipoproteins that contain E4 are cleared more efficiently from a circulation in comparison with those containing E3 and E2. In this manner, in ApoE4 carriers serum ApoE level is lower than in non-ApoE4 carriers. Interestingly, ApoE4 is not only associated with AD, but it also could be find in non-demented elderly subjects, where it is positively correlated to the degree of Aβ deposition in the cerebral cortex [213].

ApoE4 accelerates the cholesterol uptake and consequently increases cholesterol content. In addition, the ApoE4 genotype enhances Aβ production and Aβ fibril formation *in vitro* and *in vivo* [214 - 216] and synergizes with Aβ toxicity [217]. Furthermore, plaque density correlates to ApoE genotype in a clear ApoE allele 4 dose manners.

AD patients with ApoE4 show decreased ability of Aβ clearance, and in that way the presence of E4 allele further facilitates Aβ accumulation and amyloid plaque formation [218]. However, it seems that association of the APOE genotype and AD is dependent on the age, sex, and cholesterol distributions [219].

ApoE can be found within amyloid plaques. In the cortex of AD patients, ApoE3 and E 4 show affinity for binding to beta amyloid and may mediate beta amyloid cell internalization, by binding to the low density lipoprotein receptor-related protein [4]. Thus, it is possible that ApoE represents a kind of a pathological chaperon for beta amyloid by promoting the structure needed for the formation of nonsoluble amyloid [220].

Additionally, it is known that ApoE can be cleaved in the manner similar to APP, and ApoE fragments arising in this way can cause AD-like neurotoxicity in mouse models. In alignment with its already proven "pathological impact", ApoE4 is more susceptible to cleavage than ApoE3 [221, 222]. In this manner ApoE4 carriers are even more prone to AD in comparison to others. Another "bad influence" of ApoE4 can be seen by its connection to the BBB breakdown *via* activations of pro-inflammatory pathway(s) and induction of neuroinflammatory phenotype [223]. ApoE plays an essential role in maintenance of synaptic integrity and plasticity. Deteriorations in synaptic plasticity have been noticed in ApoE4 carriers within group of AD patients [224, 225] as well as in TBI cases, where E4 again has detrimental dose-dependent effect [226, 227].

It is interesting that ApoE4 variant is involved in AD pathogenesis in a cell specific manner. CRISPR/Cas9 gene editing technology showed that ApoE4 genotype induces changes in the expression of genes involved in synaptic function in neurons, while in astrocytes it changes the expression of genes involved in lipid metabolism [228].

Having in mind all above mentioned, it is evident that ApoE4 has a significant role in numerous steps of AD progression, including beta amyloid production and plaques accumulation, the extent of cellular loss and overall disease progression.

The Role of SREB Proteins in Alzheimer's Disease

The link between lipid metabolism perturbation and AD was also studied by Barbero-Camps and colleagues [229] who generated triple transgenic mice featuring SREBP2 overexpression in combination with APPswe/PS1ΔE9 mutations (APP/PS1). This model linking disturbed cholesterol homeostasis with pathological APP processing showed loss of synaptophysin and neuronal death, memory impairment, deficits in spatial memory and increased synaptotoxicity. It was obvious that SREBP2 and cholesterol play a role in pathogenesis of AD [229]. SREBP1a polymorphism has also been linked to the risk of AD in carriers of the ApoE4 allele [230]. The relationship is reciprocal- as APP fragments also regulate lipid homeostasis. Namely, APP co-localizes with SREBP1 in the Golgi apparatus, and in this manner decreases HMGCR level and HMGCR mediated cholesterol biosynthesis, and cholesterol 24S-hydroxylase and SREBP mRNA levels [231]. As APP co-localize with SREBP1 only in neurons, but not in astrocytes, it seems that APP controls cholesterol turnover needed for neuronal activity [231]. On the other hand, it was reported that link between cholesterol and APP exists also in astrocytes. Cholesterol exposure induced astrocyte activation, enrichment of ganglioside GM1-cholesterol patches in the astrocyte membrane, increased APP content and ROS production, and enhanced the APP-BACE1 interaction [232]. Furthermore, both Aβ42-induced inhibition of SREBP-2 transportation from the endoplasmic reticulum (ER) to the Golgi and cleavage have been reported [233, 234].

The Role of CYP46A1 in Alzheimer's Disease

Both 24S-OHC and CYP46A1 are implicated in the pathogenesis of AD. It has been shown that in patients with dementia, including AD, serum and CSF concentrations of 24S-OHC were found to be significantly higher in comparison to control subjects [102, 108]. A suppression of CYP46A1 role would lead to increased cholesterol content, disturbed cholesterol homeostasis, increased amount of APP present in cholesterol-enriched lipid rafts, increased amount of Aβ peptides and, as expected, to molecular, cellular and cognitive deteriorations and deficits [52, 235]. On the other hand, increased level and/or activity of cholesterol 24S-hydroxylase reduced the amyloid pathology in mouse models of AD [236, 237] and diminished Aβ production in cell culture models [238].

The Role of Cholesterol in Tau Pathology

The effects of high cholesterol in tau pathology in AD remain controversial. Neuronal cholesterol may play an important role in NFT formation by modulating tau phosphorylation *via* the isoprenoid cascade [239]. In addition, it has been showed that in THY-Tau22, animal model of tau pathology, the amounts of CYP46A1 and 24S-OHC in the hippocampus were lower than in control mice, indicating disturbed cholesterol homeostasis in this case. The vector-induced normalization of CYP46A1 and 24S-OHC content completely abolished cognitive deficits, impaired long-term depression and spine defects characteristic for this model [240]. On contrary, according to the recent study of Gratuze and colleagues, high-fat, high-cholesterol, and/or high-sugar diets had no effect on tau phosphorylation in a transgenic mouse model of AD-like tau pathology [241].

Cholesterol and APP Processing

One of the best-known direct molecular links between cholesterol and AD is for sure the hypothesis in which cholesterol is considered to be capable of changing membrane properties and functioning, by changes in its own quantity and distribution within membranes [242]. Increase in the cholesterol content and its accumulation in the parts of membrane so called lipid rafts, directly modulates the processing of APP [243] by promoting β-secretase activity and amyloidogenic, "pathological" pathway [244]. Namely, when elevated in the cell, cholesterol binds directly to C terminal transmembrane domain of APP [245, 246]. In addition, APP through its C-terminal domain is capable of interacting with flotillin-1, another structural protein highly expressed in lipid rafts. Those interactions result in APP clustering, subsequent endocytosis of APP, and APP insertion into lipid rafts close to the β- and γ-secretases [246, 247]. APP becomes prone to be processed by those two proteases, *i.e.* APP processing goes toward pathological, amyloidogenic pathway. It has been shown that β- and γ-secretase activities are stimulated by high level of cholesterol [244, 248]. Oppositely, when cholesterol content in the membrane is low, APP processing by α-secretase in non-raft domains, and non-amyloidogenic pathway considered to be "non-pathological" is promoted [249]. It is very clear that accessibility of APP to either α-secretase or β-secretase, and/or their ratio in lipid rafts represent a strong determinant to which direction APP processing will go (Fig. **6**).

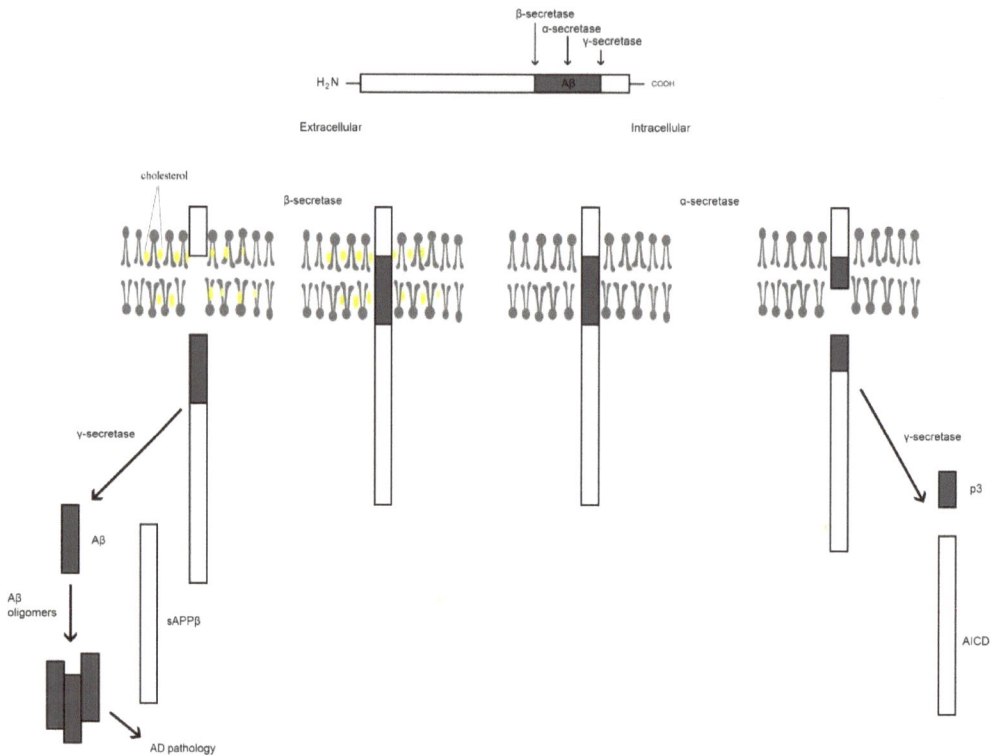

Fig. (6). The role of cholesterol in APP processing.

In addition to APP processing, evidence suggests that lipid rafts and cholesterol play a supplementary role in the AD pathogenesis, by influencing neurotrans-mitter and neurotransmitter receptors functioning. It seems that cholesterol level in the lipid rafts influences release, uptake and clearance of certain neurotrans-mitters and trafficking of their receptors and thus their function [described in details in 250]. Lipid rafts play additional role in AD pathogenesis *via* APP endocytosis and internalization to the acidic endosomes and generation of toxic Aβ species [243, 251]. Lipid rafts are parts of the membrane where accumulation of dimeric Aβ, Apolipoprotein E and phosphorylated tau occurs, and this coincided with the beginning of memory loss [252].

The Role of Cholesterol Esters in AD Pathogenesis

It has also been suggested that it is not only the level of the cholesterol plays an important role, but also the relationship between free and esterified cholesterol in the cell. It seems that non-esterified cholesterol even when high, might not affect APP processing, however, esterified cholesterol might up-regulate APP processing and Aβ generation [253]. There are several reports about elevated content of cholesterol esters in the brain regions most affected by AD in patients and in transgenic mouse models [254]. On contrary, a lack of acetyl-CoA cholesterol acyltransferase (ACAT1), an enzyme that esterifies cholesterol, in the case of application of pharmacological inhibitors results in a lower level of β-amyloid production or even an amelioration of AD pathology in animal models of AD [237, 254, 255].

The Role of Cholesterol-Lowering Drug Statins in Alzheimer's Disease

Since first experimental evidences were linking cholesterol to atherosclerosis and hearth disease attention was focused on searching for molecules that would block one of the steps in the cholesterol biosynthesis. Triparanol (MER/29), which was introduced into clinical use in the U.S. in 1959, was the first cholesterol-lowering agent that inhibited cholesterol synthesis. It inhibited the final step in the cholesterol synthetic pathway, resulting in the accumulation of other sterols [256, 257]. Due to serious side effects, including cataracts it was withdrawn from the market in the early 1960s. Citrinin, substance showing strong inhibition of HMGCR and, consequently, lower serum cholesterol levels in rats was isolated in 1972 from the culture broth of mold [258, 259]. However, the research was suspended because of its toxicity to the kidneys. In 1979 monacolin K which was later renamed into lovastatin was discovered. Lovastatin was given FDA approval to become the first commercial statin in 1987. Lovastatin was followed by a new statin, simvastatin. In 1989 pravastatin was developed. Today the most popular statin is Pfizer's atorvastatin (Lipitor®) [1].

Within the last decade, statins have attracted a lot of attention due to the potentially useful effect of AD prevention [260 - 262]. Although causative relations between cholesterol and AD exist, there are still many controversies regarding the possibility of statin usage as a potential therapy for AD. This idea came into focus with a large cohort study enrolled between 1994 and 2002, in which post mortem analysis showed less severe tangles in the brain of the patients who were subjected to statin therapy in comparison to statin non-users. This research was unfortunately so heavily transferred by TV and newspapers, and very often misinterpreted, that raised a lot of hope for the AD patients, and induced many miss-conclusions, especially in general population. The main idea was based on the capability of statins to inhibit HMGCR. By performing statins treatment in the neuronal culture, a reduction in cholesterol synthesis and content, and a consequential reduction of pathological APP processing have been observed [263]. Statin treatment also has been showed as effective in a reduction of the hyperphosphorylated tau and protection against glutamate excitotoxicity [239, 264, 265]. It is known that high cholesterol level leads to τ-hyper phosphorylation and Aβ production in rat brain and mice. At the same time, one of the widely used statins-Atorvastatin inhibited τ-hyperphosphorylation and decreased Aβ generation. Thus, it was expected statins to make a revolution in the treatment of AD.

However, in humans though a significant inhibition of brain cholesterol biosynthesis was noticed *via* the inhibition of HMGCR, AD biomarkers showed no changes [266] and there was no significant clinical benefit in patients [267, 268]. In addition, although the mechanism of action of statins on cholesterol peripherally is well described, it is not yet clear how statins exert their protective effect on AD pathology since they are not able passing the BBB. In the case of statins that are not lipophilic some authors believe that by lowering plasma cholesterol levels they indirectly influence the synthesis of cholesterol in the brain [269]. However, this is not in accordance with the basic postulate of completely autonomous homeostasis of cholesterol in the brain as it assumes cholesterol passage between circulation and brain. Some data indicate that cholesterol level in the plasma, as well as environmental manipulations like dietary restriction or high cholesterol diet can influence cholesterol metabolism in the brain, but those effects are rather indirect and they mainly do not change brain cholesterol level [197, 270]. There is also a hypothesis that the statin effects in the brain could be achieved at the level of small blood vessel walls within the blood-brain barrier [271].

In addition to possible use of statins in AD treatment, it has been suggested that statins could represent an ideal candidate therapy for acute brain injury. Statins influence multiple mechanisms of acute and secondary neuronal injury; they have endothelial and vasoactive properties, as well as anti-oxidant, anti-inflammatory, anti-excitotoxicity, and anti-thrombotic effects [272]. Study by He and colleagues [273] designed to elucidate the effect of a brain permeable statin (lovastatin) on cholesterol and oxysterol levels of the hippocampus after neuronal injury, revealed that lovastatin induced significantly lower levels of cholesterol, 24S-OHC, and 7-ketocholesterol, but also modulated hippocampal neuronal loss following injury. Since oxysterols, including 7-ketocholesterol [274, 275] and 24S-OHC [276] are toxic to cells in culture brain-permeable statins such as lovastatin could have a neuroprotective effect by limiting the levels of oxysterol in brain areas undergoing neurodegeneration. However, the statin activity most likely involves mechanisms other than inhibiting cholesterol synthesis, a prominent one being the drug's anti-inflammatory effects [272, 277, 278].

Nowadays there are dozens of papers about statin treatment and AD and it is known that statins are not so promising therapeutic possibility as it was thought, and that their effect on the brain are, at best controversial if not truly ineffective regarding AD. It seems that at best, as Geifman and colleagues claim "By focusing on subgroups of the AD population, it may be possible to detect endotypes responsive to statin therapy" with greater efficacy in patients homozygous for ApoE4 [279].

The Impact of Dietary Restriction on Cholesterol in the Brain

Together with potential beneficial effects of statins on the CNS functioning, their adverse effects are described as well. Several studies have shown that simvastatin causes sleep disorders and that the use of statins is associated with memory problems [280, 281] directly inhibiting long-term potentiation [52]. Also, high doses of simvastatin and pravastatin lead to a reduction in cholesterol in the cortex [282] which may have adverse effects on neuronal growth and signal transduction.

As a good alternative to statins treatment, environmental manipulations that could affect cholesterol metabolism should be taken into account. The most promising one is reduction of food intake. While it is well documented how dietary restriction (DR) modulates cholesterol metabolism peripherally [283, 284], little is known on how it affects cholesterol levels and metabolism in the aging brain. Up to date, only a few studies have addressed this issue. Study conducted by Pallottini and colleagues [285] revealed that although HMGCR activity was increased during aging both in the liver and the brain, the amount of protein was unchanged. Its activity was shown to be influenced by DR (both intermittent fasting- IF and 60% AL) solely in the liver. More recent study [270] showed that DR exerted significant impact on expression profile of the key enzymes involved in cholesterol metabolism. Alternative day feeding induced a significant decrease in HMGCR protein expression in the hippocampus of old rats. Changed ApoE level following IF in this study suggested that dietary regime could altered cholesterol transport and its reutilization in the brain. A more restrictive DR [197] promoted brain cholesterol catabolism by increasing the CYP46A1 protein level in the rat hippocampus.

Pallottini *et al.* [285] also followed the impact of both DR and every-other-day feeding on dolichol and cholesterol accumulation, while Mulas *et al.* [286] used the same dietary regime to analyze its effects on cholesterol and cholesterol ester levels in the whole brains of aging rats. Although cholesterol metabolism in the brain is mostly autonomous, recent studies have revealed comparable changes in the serum and brain lipid profiles [287, 288]. Studies on aging rats showed that IF is capable of provoking a switch to a less energy-consuming synthesis pathway that involves a desmosterol as a direct cholesterol in the old animals exposed to DR [270].

Dietary habits are capable of influencing the concentrations of intermediates in cholesterol biosynthetic pathway. In mice, 20 hours fasting did not affect the brain cholesterol concentration; however, major changes in post-lanosterol compounds were reported [289]. The brain responds to fasting by a lowering of sterol intermediates, resulting in decreased cholesterol synthesis. The cholesterol concentration was unaffected, probably due to a homeostatic mechanisms maintained in the brain such as efficient cholesterol reutilization and its long half-life in the brain [4]. Some studies have also reported that exposure of rats and mice to various types of diets had no effect on the overall brain cholesterol levels [286, 290]. Certain parallels can be drawn between statin treatment and long-term DR in regard to cholesterol precursors. Similarly to DR, short-term treatment with high dose of simvastatin as well as pravastatin, another HMGCR inhibitor, had no impact on cholesterol levels in whole-brain homogenates. However, such treatment significantly decreased brain lathosterol levels, indicating a reduction of cholesterol synthesis in the brain [116, 270]. The great advantage of the DR in comparison to statins is its capacity to increase desmosterol level in the hippocampus of old animals [270]. As mentioned, the Bloch pathway is prevailing in cholesterol biosynthesis in young animals [74], so this result suggests that long-term DR is capable of provoking a switch to a less energy-consuming synthesis pathway. However, when these results are interpreted, one should be very cautious and be aware that the influence of the DR on the brain cholesterol metabolism depends on the degree of restriction, as well as the duration of the applied dietary regime [270, 289, 291].

CONCLUDING REMARKS

Despite all the evidences showing that disturbed cholesterol homeostasis plays a role in AD pathology, there is still no definitive answer to the question does high level of cholesterol *per se* lead to AD. Epidemiological data are controversial and they open more questions than giving the answers. Some epidemiological studies showed that high midlife cholesterol is associated with an increased risk for dementia and AD [292, 32, 293]. However, other reports failed to show that association [294]. In some studies, cholesterol was followed over a very long period of time, and it was clearly shown that it is not related to AD [295]. The reason for this discrepancy could lie in diversity of population, ethnical groups and/or gender used for analysis [see in detail in 296]. Great diversity exists not only in general population, but among AD cases as well. ApoE represents the major susceptibility gene for AD [55]; however, ApoE4 is neither necessary, nor sufficient for AD and represents only a risk factor. Only 60% of AD patients are ApoE4 carriers and only 50% of ApoE4 homozygotes will develop AD [297].

Similarly, to epidemiological data, molecular studies are argumentative and indicate that all kind of dysregulation in lipids metabolism could be responsible for AD rather than high cholesterol itself. For example, effectiveness of cholesterol-lowering drugs-statins in treating or preventing AD is controversial [298]. In addition, then, it has been shown that deficiency of ABCA1 gene expression results in 80% loss of ApoE in the brain, while the rest of ApoE is poorly lipidated [299]. While this poorly lipidated ApoE increases amyloid burden in mouse models of AD, ApoE lipidation almost diminishes mature amyloid plaques [299, 300]. This indicates that lipid binding capacity of ApoE is one of the pathological mechanisms in AD.

In the same time, and in contrast to its putative deleterious role in AD, the essentiality of cholesterol in life processes stands together with the huge neural requirements for cholesterol. It is very likely that limiting factor in the synapse formation is the glial delivery of cholesterol to neurons. The very interesting experiment performed more than 20 years ago by Pfrieger and Barres, showed that pure neuronal culture is able to form synapses only and exclusively in the presence of astrocytes and cholesterol-rich lipoproteins originated from those astrocytes. Adding astrocytes to the neuronal culture increased the total number of synapses by 7-fold and synaptic activity by 70-fold. Furthermore, cholesterol solely was capable inducing formation of as many synapses as astrocytes [reviewed in 301]. It has been shown that improved memory functions are associated with higher total and LDL cholesterol levels in the very elderly or higher HDL levels [35, 38].

We do believe that impaired cholesterol metabolism in the brain, and dysregulated lipid metabolism in general, indeed represent factors involved in AD suscep-tibility and/or pathology. Many reports speak in the favor of that claim [302 - 305]. Hypercholesterolemia and elevated serum cholesterol level have been shown to increase susceptibility to AD [304 - 308] and hypercholesterolemia is a frequent among AD patients [179]. But having in mind other evidences, it is more probable that disturbed cholesterol homeostasis plays a detrimental role, rather than high cholesterol, while reasons for disturbed homeostasis could be numerous. As said at the beginning, cholesterol has at least two faces, the good and the bad one, and its role in AD pathogenesis is probably the same. Since its discovery, we have learned a lot about cholesterol. We discovered that it is its carrier that determines the susceptibility for late onset AD. We also know that there is other cholesterol related factors that increase AD susceptibility, as there are other diseases, apart from AD, caused by, or connected with deficits in cholesterol homeostasis. However, there are still many questions needed to be answered before giving a final conclusion about cholesterol pathological role. Although we know how to control cholesterol level, and which aspects of cholesterol homeostasis might be involved in AD pathogenesis, recent studies challenge the current opinion regarding risks that cholesterol brings for cardiovascular and Alzheimer's disease. Recent epidemiological and clinical data speak in favor of reconsideration current attitude that favors restriction of dietary cholesterol [309], and some countries decided not to have an upper limit for cholesterol intake in their dietary guidelines. The "bad reputation" cholesterol had for many years vanishes slowly and new perspectives in cholesterol researches are opening. According to all the data we collected for decades, only thing we can certainly say is that cholesterol is essential for all, but it can be detrimental for some.

CONSENT FOR PUBLICATION

Not applicable.

CONFLICT OF INTEREST

The authors confirm that this chapter contents have no conflict of interest.

ACKNOWLEDGEMENTS

This work was supported by the Ministry of Education, Science and Technological Development of the Republic of Serbia (grant N° ON173056).

REFERENCES

[1]   Endo A. A historical perspective on the discovery of statins. Proc Jpn Acad, Ser B, Phys Biol Sci 2010; 86(5): 484-93.
[http://dx.doi.org/10.2183/pjab.86.484] [PMID: 20467214]

[2]   Egawa J, Pearn ML, Lemkuil BP, Patel PM, Head BP. Membrane lipid rafts and neurobiology: age-related changes in membrane lipids and loss of neuronal function. J Physiol 2016; 594(16): 4565-79.
[http://dx.doi.org/10.1113/JP270590] [PMID: 26332795]

[3]   Smith LL. Another cholesterol hypothesis: cholesterol as antioxidant. Free Radic Biol Med 1991; 11(1): 47-61.
[http://dx.doi.org/10.1016/0891-5849(91)90187-8] [PMID: 1937129]

[4]   Dietschy JM, Turley SD. Thematic review series: brain Lipids. Cholesterol metabolism in the central nervous system during early development and in the mature animal. J Lipid Res 2004; 45(8): 1375-97.
[http://dx.doi.org/10.1194/jlr.R400004-JLR200] [PMID: 15254070]

[5]   Mondal M, Mesmin B, Mukherjee S, Maxfield FR. Sterols are mainly in the cytoplasmic leaflet of the plasma membrane and the endocytic recycling compartment in CHO cells. Mol Biol Cell 2009; 20(2): 581-8.
[http://dx.doi.org/10.1091/mbc.e08-07-0785] [PMID: 19019985]

[6]   Lingwood D, Simons K. Lipid rafts as a membrane-organizing principle. Science 2010; 327(5961):

46-50.
[http://dx.doi.org/10.1126/science.1174621] [PMID: 20044567]

[7] van Meer G. Dynamic transbilayer lipid asymmetry. Cold Spring Harb Perspect Biol 2011; 3(5): a004671.
[http://dx.doi.org/10.1101/cshperspect.a004671] [PMID: 21436058]

[8] Gil C, Cubí R, Blasi J, Aguilera J. Synaptic proteins associate with a sub-set of lipid rafts when isolated from nerve endings at physiological temperature. Biochem Biophys Res Commun 2006; 348(4): 1334-42.
[http://dx.doi.org/10.1016/j.bbrc.2006.07.201] [PMID: 16920068]

[9] Wasser CR, Kavalali ET. Leaky synapses: regulation of spontaneous neurotransmission in central synapses. Neuroscience 2009; 158(1): 177-88.
[http://dx.doi.org/10.1016/j.neuroscience.2008.03.028] [PMID: 18434032]

[10] Mailman T, Hariharan M, Karten B. Inhibition of neuronal cholesterol biosynthesis with lovastatin leads to impaired synaptic vesicle release even in the presence of lipoproteins or geranylgeraniol. J Neurochem 2011; 119(5): 1002-15.
[http://dx.doi.org/10.1111/j.1471-4159.2011.07474.x] [PMID: 21899539]

[11] Bohr IJ. Does cholesterol act as a protector of cholinergic projections in Alzheimer's disease? Lipids Health Dis 2005; 4: 13.
[http://dx.doi.org/10.1186/1476-511X-4-13] [PMID: 15949039]

[12] Tsui-Pierchala BA, Encinas M, Milbrandt J, Johnson EM Jr. Lipid rafts in neuronal signaling and function. Trends Neurosci 2002; 25(8): 412-7.
[http://dx.doi.org/10.1016/S0166-2236(02)02215-4] [PMID: 12127758]

[13] Goritz C, Mauch DH, Pfrieger FW. Multiple mechanisms mediate cholesterol-induced synaptogenesis in a CNS neuron. Mol Cell Neurosci 2005; 29(2): 190-201.
[http://dx.doi.org/10.1016/j.mcn.2005.02.006] [PMID: 15911344]

[14] de Chaves EI, Rusiñol AE, Vance DE, Campenot RB, Vance JE. Role of lipoproteins in the delivery of lipids to axons during axonal regeneration. J Biol Chem 1997; 272(49): 30766-73.
[http://dx.doi.org/10.1074/jbc.272.49.30766] [PMID: 9388216]

[15] Fester L, Zhou L, Bütow A, *et al.* Cholesterol-promoted synaptogenesis requires the conversion of cholesterol to estradiol in the hippocampus. Hippocampus 2009; 19(8): 692-705.
[http://dx.doi.org/10.1002/hipo.20548] [PMID: 19156851]

[16] Yue HY, Xu J. Cholesterol regulates multiple forms of vesicle endocytosis at a mammalian central synapse. J Neurochem 2015; 134(2): 247-60.
[http://dx.doi.org/10.1111/jnc.13129] [PMID: 25893258]

[17] Chen Z, Rand RP. The influence of cholesterol on phospholipid membrane curvature and bending elasticity. Biophys J 1997; 73(1): 267-76.
[http://dx.doi.org/10.1016/S0006-3495(97)78067-6] [PMID: 9199791]

[18] Coorssen JR, Rand RP. Effects of cholesterol on the structural transitions induced by diacylglycerol in phosphatidylcholine and phosphatidylethanolamine bilayer systems. Biochem Cell Biol 1990; 68(1): 65-9.
[http://dx.doi.org/10.1139/o90-008] [PMID: 2350502]

[19] Churchward MA, Rogasevskaia T, Höfgen J, Bau J, Coorssen JR. Cholesterol facilitates the native mechanism of Ca2+-triggered membrane fusion. J Cell Sci 2005; 118(Pt 20): 4833-48.
[http://dx.doi.org/10.1242/jcs.02601] [PMID: 16219690]

[20] Thiele C, Hannah MJ, Fahrenholz F, Huttner WB. Cholesterol binds to synaptophysin and is required for biogenesis of synaptic vesicles. Nat Cell Biol 2000; 2(1): 42-9.
[http://dx.doi.org/10.1038/71366] [PMID: 10620806]

[21] Linetti A, Fratangeli A, Taverna E, *et al.* Cholesterol reduction impairs exocytosis of synaptic vesicles.

J Cell Sci 2010; 123(Pt 4): 595-605.
[http://dx.doi.org/10.1242/jcs.060681] [PMID: 20103534]

[22]   Bliss TV, Collingridge GL. A synaptic model of memory: long-term potentiation in the hippocampus. Nature 1993; 361(6407): 31-9.
[http://dx.doi.org/10.1038/361031a0] [PMID: 8421494]

[23]   Malinow R, Malenka RC. AMPA receptor trafficking and synaptic plasticity. Annu Rev Neurosci 2002; 25: 103-26.
[http://dx.doi.org/10.1146/annurev.neuro.25.112701.142758] [PMID: 12052905]

[24]   Renner M, Choquet D, Triller A. Control of the postsynaptic membrane viscosity. J Neurosci 2009; 29(9): 2926-37.
[http://dx.doi.org/10.1523/JNEUROSCI.4445-08.2009] [PMID: 19261888]

[25]   Björkhem I. Crossing the barrier: oxysterols as cholesterol transporters and metabolic modulators in the brain. J Intern Med 2006; 260(6): 493-508.
[http://dx.doi.org/10.1111/j.1365-2796.2006.01725.x] [PMID: 17116000]

[26]   Radhakrishnan A, Ikeda Y, Kwon HJ, Brown MS, Goldstein JL. Sterol-regulated transport of SREBPs from endoplasmic reticulum to Golgi: oxysterols block transport by binding to Insig. Proc Natl Acad Sci USA 2007; 104(16): 6511-8.
[http://dx.doi.org/10.1073/pnas.0700899104] [PMID: 17428920]

[27]   Foster TC. Biological markers of age-related memory deficits: treatment of senescent physiology. CNS Drugs 2006; 20(2): 153-66.
[http://dx.doi.org/10.2165/00023210-200620020-00006] [PMID: 16478290]

[28]   Kivipelto M, Helkala E-L, Hänninen T, *et al.* Midlife vascular risk factors and late-life mild cognitive impairment: A population-based study. Neurology 2001; 56(12): 1683-9.
[http://dx.doi.org/10.1212/WNL.56.12.1683] [PMID: 11425934]

[29]   Näslund J, Haroutunian V, Mohs R, *et al.* Correlation between elevated levels of amyloid β-peptide in the brain and cognitive decline. JAMA 2000; 283(12): 1571-7.
[http://dx.doi.org/10.1001/jama.283.12.1571] [PMID: 10735393]

[30]   Solomon A, Kåreholt I, Ngandu T, *et al.* Serum cholesterol changes after midlife and late-life cognition: twenty-one-year follow-up study. Neurology 2007; 68(10): 751-6.
[http://dx.doi.org/10.1212/01.wnl.0000256368.57375.b7] [PMID: 17339582]

[31]   Yaffe K, Barrett-Connor E, Lin F, Grady D. Serum lipoprotein levels, statin use, and cognitive function in older women. Arch Neurol 2002; 59(3): 378-84.
[http://dx.doi.org/10.1001/archneur.59.3.378] [PMID: 11890840]

[32]   Solomon A, Kivipelto M, Wolozin B, Zhou J, Whitmer RA. Midlife serum cholesterol and increased risk of Alzheimer's and vascular dementia three decades later. Dement Geriatr Cogn Disord 2009; 28(1): 75-80. b
[http://dx.doi.org/10.1159/000231980] [PMID: 19648749]

[33]   Whitmer RA, Sidney S, Selby J, Johnston SC, Yaffe K. Midlife cardiovascular risk factors and risk of dementia in late life. Neurology 2005; 64(2): 277-81.
[http://dx.doi.org/10.1212/01.WNL.0000149519.47454.F2] [PMID: 15668425]

[34]   Teunissen CE, De Vente J, von Bergmann K, *et al.* Serum cholesterol, precursors and metabolites and cognitive performance in an aging population. Neurobiol Aging 2003; 24(1): 147-55.
[http://dx.doi.org/10.1016/S0197-4580(02)00061-1] [PMID: 12493560]

[35]   Atzmon G, Gabriely I, Greiner W, Davidson D, Schechter C, Barzilai N. Plasma HDL levels highly correlate with cognitive function in exceptional longevity. J Gerontol A Biol Sci Med Sci 2002; 57(11): M712-5.
[http://dx.doi.org/10.1093/gerona/57.11.M712] [PMID: 12403798]

[36]   van Exel E, de Craen AJM, Gussekloo J, *et al.* Association between high-density lipoprotein and

cognitive impairment in the oldest old. Ann Neurol 2002; 51(6): 716-21.
[http://dx.doi.org/10.1002/ana.10220] [PMID: 12112077]

[37]   Solomon A, Kåreholt I, Ngandu T, *et al.* Serum total cholesterol, statins and cognition in non-demented elderly. Neurobiol Aging 2009; 30(6): 1006-9. a
[http://dx.doi.org/10.1016/j.neurobiolaging.2007.09.012] [PMID: 18022292]

[38]   West R, Beeri MS, Schmeidler J, *et al.* Better memory functioning associated with higher total and low-density lipoprotein cholesterol levels in very elderly subjects without the apolipoprotein e4 allele. Am J Geriatr Psychiatry 2008; 16(9): 781-5.
[http://dx.doi.org/10.1097/JGP.0b013e3181812790] [PMID: 18757771]

[39]   Singh-Manoux A, Gimeno D, Kivimaki M, Brunner E, Marmot MG. Low HDL cholesterol is a risk factor for deficit and decline in memory in midlife: the Whitehall II study. Arterioscler Thromb Vasc Biol 2008; 28(8): 1556-62.
[http://dx.doi.org/10.1161/ATVBAHA.108.163998] [PMID: 18591462]

[40]   Mielke MM, Zandi PP, Sjögren M, *et al.* High total cholesterol levels in late life associated with a reduced risk of dementia. Neurology 2005; 64(10): 1689-95.
[http://dx.doi.org/10.1212/01.WNL.0000161870.78572.A5] [PMID: 15911792]

[41]   Panza F, D'Introno A, Colacicco AM, *et al.* Lipid metabolism in cognitive decline and dementia. Brain Res Brain Res Rev 2006; 51(2): 275-92.
[http://dx.doi.org/10.1016/j.brainresrev.2005.11.007] [PMID: 16410024]

[42]   Elias PK, Elias MF, D'Agostino RB, Sullivan LM, Wolf PA. Serum cholesterol and cognitive performance in the Framingham Heart Study. Psychosom Med 2005; 67(1): 24-30.
[http://dx.doi.org/10.1097/01.psy.0000151745.67285.c2] [PMID: 15673620]

[43]   Miller S, Wehner JM. Cholesterol treatment facilitates spatial learning performance in DBA/2Ibg mice. Pharmacol Biochem Behav 1994; 49(1): 257-61.
[http://dx.doi.org/10.1016/0091-3057(94)90487-1] [PMID: 7816886]

[44]   Upchurch M, Wehner JM. DBA/2Ibg mice are incapable of cholinergically-based learning in the Morris water task. Pharmacol Biochem Behav 1988; 29(2): 325-9.
[http://dx.doi.org/10.1016/0091-3057(88)90164-5] [PMID: 3362927]

[45]   Dufour F, Liu Q-Y, Gusev P, Alkon D, Atzori M. Cholesterol-enriched diet affects spatial learning and synaptic function in hippocampal synapses. Brain Res 2006; 1103(1): 88-98.
[http://dx.doi.org/10.1016/j.brainres.2006.05.086] [PMID: 16814755]

[46]   Endo Y, Nishimura J-I, Kimura F. Impairment of maze learning in rats following long-term glucocorticoid treatments. Neurosci Lett 1996; 203(3): 199-202.
[http://dx.doi.org/10.1016/0304-3940(95)12296-6] [PMID: 8742027]

[47]   O'Brien WT, Xu G, Batta A, *et al.* Developmental sensitivity of associative learning to cholesterol synthesis inhibitors. Behav Brain Res 2002; 129(1-2): 141-52.
[http://dx.doi.org/10.1016/S0166-4328(01)00342-4] [PMID: 11809505]

[48]   Võikar V, Rauvala H, Ikonen E. Cognitive deficit and development of motor impairment in a mouse model of Niemann-Pick type C disease. Behav Brain Res 2002; 132(1): 1-10.
[http://dx.doi.org/10.1016/S0166-4328(01)00380-1] [PMID: 11853852]

[49]   Schreurs BG, Smith-Bell CA, Lochhead J, Sparks DL. Cholesterol modifies classical conditioning of the rabbit (Oryctolagus cuniculus) nictitating membrane response. Behav Neurosci 2003; 117(6): 1220-32.
[http://dx.doi.org/10.1037/0735-7044.117.6.1220] [PMID: 14674842]

[50]   Schreurs BG, Smith-Bell CA, Darwish DS, Stankovic G, Sparks DL. High dietary cholesterol facilitates classical conditioning of the rabbit's nictitating membrane response. Nutr Neurosci 2007; 10(1-2): 31-43. b
[http://dx.doi.org/10.1080/10284150701232034] [PMID: 17539481]

[51] Schreurs BG, Smith-Bell CA, Darwish DS, Stankovic G, Sparks DL. Classical conditioning of the rabbit's nictitating membrane response is a function of the duration of dietary cholesterol. Nutr Neurosci 2007; 10(3-4): 159-68. a
[http://dx.doi.org/10.1080/10284150701565540] [PMID: 18019398]

[52] Kotti TJ, Ramirez DM, Pfeiffer BE, Huber KM, Russell DW. Brain cholesterol turnover required for geranylgeraniol production and learning in mice. Proc Natl Acad Sci USA 2006; 103(10): 3869-74.
[http://dx.doi.org/10.1073/pnas.0600316103] [PMID: 16505352]

[53] Blalock EM, Chen KC, Sharrow K, *et al.* Gene microarrays in hippocampal aging: statistical profiling identifies novel processes correlated with cognitive impairment. J Neurosci 2003; 23(9): 3807-19.
[http://dx.doi.org/10.1523/JNEUROSCI.23-09-03807.2003] [PMID: 12736351]

[54] Brankatschk M, Eaton S. Lipoprotein particles cross the blood-brain barrier in Drosophila. J Neurosci 2010; 30(31): 10441-7.
[http://dx.doi.org/10.1523/JNEUROSCI.5943-09.2010] [PMID: 20685986]

[55] Strittmatter WJ, Weisgraber KH, Huang DY, *et al.* Binding of human apolipoprotein E to synthetic amyloid beta peptide: isoform-specific effects and implications for late-onset Alzheimer disease. Proc Natl Acad Sci USA 1993; 90(17): 8098-102.
[http://dx.doi.org/10.1073/pnas.90.17.8098] [PMID: 8367470]

[56] Tokuda T, Tamaoka A, Matsuno S, *et al.* Plasma levels of amyloid beta proteins did not differ between subjects taking statins and those not taking statins. Ann Neurol 2001; 49(4): 546-7.
[http://dx.doi.org/10.1002/ana.112] [PMID: 11310640]

[57] Utermann G, Langenbeck U, Beisiegel U, Weber W. Genetics of the apolipoprotein E system in man. Am J Hum Genet 1980; 32(3): 339-47.
[PMID: 7386461]

[58] Waelsch H, Sperry WM, Stoyanoff VA. A study of the synthesis and deposition of lipids in brain and other tissues with deuterium as an indicator. J Biol Chem 1940; 135: 291-6.

[59] Poirier J. M. Panisset M. Apolipoprotein E: a novel therapeutic target for the treatment of Alzheimer's disease. Adv Exp Med Biol 2002; •••: 36-42.

[60] Smiljanic K, Vanmierlo T, Djordjevic AM, *et al.* Aging induces tissue-specific changes in cholesterol metabolism in rat brain and liver. Lipids 2013; 48(11): 1069-77.
[http://dx.doi.org/10.1007/s11745-013-3836-9] [PMID: 24057446]

[61] Poirier J, Baccichet A, Dea D, Gauthier S. Cholesterol synthesis and lipoprotein reuptake during synaptic remodelling in hippocampus in adult rats. Neuroscience 1993; 55(1): 81-90.
[http://dx.doi.org/10.1016/0306-4522(93)90456-P] [PMID: 8350994]

[62] Nieweg K, Schaller H, Pfrieger FW. Marked differences in cholesterol synthesis between neurons and glial cells from postnatal rats. J Neurochem 2009; 109(1): 125-34.
[http://dx.doi.org/10.1111/j.1471-4159.2009.05917.x] [PMID: 19166509]

[63] Björkhem I, Lütjohann D, Diczfalusy U, Ståhle L, Ahlborg G, Wahren J. Cholesterol homeostasis in human brain: turnover of 24S-hydroxycholesterol and evidence for a cerebral origin of most of this oxysterol in the circulation. J Lipid Res 1998; 39(8): 1594-600.
[PMID: 9717719]

[64] Korade Z, Mi Z, Portugal C, Schor NF. Expression and p75 neurotrophin receptor dependence of cholesterol synthetic enzymes in adult mouse brain. Neurobiol Aging 2007; 28(10): 1522-31.
[http://dx.doi.org/10.1016/j.neurobiolaging.2006.06.026] [PMID: 16887237]

[65] Herz J, Beffert U. Apolipoprotein E receptors: linking brain development and Alzheimer's disease. Nat Rev Neurosci 2000; 1(1): 51-8.
[http://dx.doi.org/10.1038/35036221] [PMID: 11252768]

[66] Ong WY, Kim JH, He X, Chen P, Farooqui AA, Jenner AM. Changes in brain cholesterol metabolome

after excitotoxicity. Mol Neurobiol 2010; 41(2-3): 299-313.
[http://dx.doi.org/10.1007/s12035-010-8099-3] [PMID: 20140539]

[67]   Zhang Y, Appelkvist EL, Kristensson K, Dallner G. The lipid compositions of different regions of rat brain during development and aging. Neurobiol Aging 1996; 17(6): 869-75.
[http://dx.doi.org/10.1016/S0197-4580(96)00076-0] [PMID: 9363798]

[68]   Saito M, Benson EP, Saito M, Rosenberg A. Metabolism of cholesterol and triacylglycerol in cultured chick neuronal cells, glial cells, and fibroblasts: accumulation of esterified cholesterol in serum-free culture. J Neurosci Res 1987; 18(2): 319-25.
[http://dx.doi.org/10.1002/jnr.490180208] [PMID: 3694714]

[69]   Suzuki S, Kiyosue K, Hazama S, *et al.* Brain-derived neurotrophic factor regulates cholesterol metabolism for synapse development. J Neurosci 2007; 27(24): 6417-27.
[http://dx.doi.org/10.1523/JNEUROSCI.0690-07.2007] [PMID: 17567802]

[70]   Pfrieger FW. Outsourcing in the brain: do neurons depend on cholesterol delivery by astrocytes? BioEssays 2003; 25(1): 72-8.
[http://dx.doi.org/10.1002/bies.10195] [PMID: 12508285]

[71]   Valdez CM, Smith MA, Perry G, Phelix CF, Santamaria F. Cholesterol homeostasis markers are localized to mouse hippocampal pyramidal and granule layers. Hippocampus 2010; 20(8): 902-5.
[PMID: 20054815]

[72]   Segatto M, Leboffe L, Trapani L, Pallottini V. Cholesterol homeostasis failure in the brain: implications for synaptic dysfunction and cognitive decline. Curr Med Chem 2014; 21(24): 2788-802.
[http://dx.doi.org/10.2174/0929867321666140303142902] [PMID: 24606521]

[73]   Goldstein JL, Brown MS. Regulation of the mevalonate pathway. Nature 1990; 343(6257): 425-30.
[http://dx.doi.org/10.1038/343425a0] [PMID: 1967820]

[74]   Lütjohann D, Brzezinka A, Barth E, *et al.* Profile of cholesterol-related sterols in aged amyloid precursor protein transgenic mouse brain. J Lipid Res 2002; 43(7): 1078-85.
[http://dx.doi.org/10.1194/jlr.M200071-JLR200] [PMID: 12091492]

[75]   Pfrieger FW, Ungerer N. Cholesterol metabolism in neurons and astrocytes. Prog Lipid Res 2011; 50(4): 357-71.
[http://dx.doi.org/10.1016/j.plipres.2011.06.002] [PMID: 21741992]

[76]   Boyles JK, Zoellner CD, Anderson LJ, *et al.* A role for apolipoprotein E, apolipoprotein A-I, and low density lipoprotein receptors in cholesterol transport during regeneration and remyelination of the rat sciatic nerve. J Clin Invest 1989; 83(3): 1015-31.
[http://dx.doi.org/10.1172/JCI113943] [PMID: 2493483]

[77]   Posse De Chaves EI, Vance DE, Campenot RB, Kiss RS, Vance JE. Uptake of lipoproteins for axonal growth of sympathetic neurons. J Biol Chem 2000; 275(26): 19883-90.
[http://dx.doi.org/10.1074/jbc.275.26.19883] [PMID: 10867025]

[78]   Bu G. Apolipoprotein E and its receptors in Alzheimer's disease: pathways, pathogenesis and therapy. Nat Rev Neurosci 2009; 10(5): 333-44.
[http://dx.doi.org/10.1038/nrn2620] [PMID: 19339974]

[79]   Holtzman DM, Herz J, Bu G. Apolipoprotein E and apolipoprotein E receptors: normal biology and roles in Alzheimer disease. Cold Spring Harb Perspect Med 2012; 2(3): a006312.
[http://dx.doi.org/10.1101/cshperspect.a006312] [PMID: 22393530]

[80]   Mauch DH, Nägler K, Schumacher S, *et al.* CNS synaptogenesis promoted by glia-derived cholesterol. Science 2001; 294(5545): 1354-7.
[http://dx.doi.org/10.1126/science.294.5545.1354] [PMID: 11701931]

[81]   Vance JE, Hayashi H. Formation and function of apolipoprotein E-containing lipoproteins in the nervous system. Biochim Biophys Acta 2010; 1801(8): 806-18.
[http://dx.doi.org/10.1016/j.bbalip.2010.02.007] [PMID: 20170744]

[82] Hayashi H, Campenot RB, Vance DE, Vance JE. Glial lipoproteins stimulate axon growth of central nervous system neurons in compartmented cultures. J Biol Chem 2004; 279(14): 14009-15.
[http://dx.doi.org/10.1074/jbc.M313828200] [PMID: 14709547]

[83] Hayashi H, Campenot RB, Vance DE, Vance JE. Apolipoprotein E-containing lipoproteins protect neurons from apoptosis *via* a signaling pathway involving low-density lipoprotein receptor-related protein-1. J Neurosci 2007; 27(8): 1933-41.
[http://dx.doi.org/10.1523/JNEUROSCI.5471-06.2007] [PMID: 17314289]

[84] Ignatius MJ, Gebicke-Härter PJ, Skene JH, *et al.* Expression of apolipoprotein E during nerve degeneration and regeneration. Proc Natl Acad Sci USA 1986; 83(4): 1125-9.
[http://dx.doi.org/10.1073/pnas.83.4.1125] [PMID: 2419900]

[85] Espenshade PJ, Hughes AL. Regulation of sterol synthesis in eukaryotes. Annu Rev Genet 2007; 41: 401-27.
[http://dx.doi.org/10.1146/annurev.genet.41.110306.130315] [PMID: 17666007]

[86] McPherson R, Gauthier A. Molecular regulation of SREBP function: the Insig-SCAP connection and isoform-specific modulation of lipid synthesis. Biochem Cell Biol 2004; 82(1): 201-11.
[http://dx.doi.org/10.1139/o03-090] [PMID: 15052338]

[87] Goldstein JL, DeBose-Boyd RA, Brown MS. Protein sensors for membrane sterols. Cell 2006; 124(1): 35-46.
[http://dx.doi.org/10.1016/j.cell.2005.12.022] [PMID: 16413480]

[88] Gong Y, Lee JN, Lee PC, Goldstein JL, Brown MS, Ye J. Sterol-regulated ubiquitination and degradation of Insig-1 creates a convergent mechanism for feedback control of cholesterol synthesis and uptake. Cell Metab 2006; 3(1): 15-24.
[http://dx.doi.org/10.1016/j.cmet.2005.11.014] [PMID: 16399501]

[89] Hong C, Tontonoz P. Liver X receptors in lipid metabolism: opportunities for drug discovery. Nat Rev Drug Discov 2014; 13(6): 433-44.
[http://dx.doi.org/10.1038/nrd4280] [PMID: 24833295]

[90] Lund EG, Guileyardo JM, Russell DW. cDNA cloning of cholesterol 24-hydroxylase, a mediator of cholesterol homeostasis in the brain. Proc Natl Acad Sci USA 1999; 96(13): 7238-43.
[http://dx.doi.org/10.1073/pnas.96.13.7238] [PMID: 10377398]

[91] Li-Hawkins J, Lund EG, Bronson AD, Russell DW. Expression cloning of an oxysterol 7alpha-hydroxylase selective for 24-hydroxycholesterol. J Biol Chem 2000; 275(22): 16543-9.
[http://dx.doi.org/10.1074/jbc.M001810200] [PMID: 10748047]

[92] Ramirez DM, Andersson S, Russell DW. Neuronal expression and subcellular localization of cholesterol 24-hydroxylase in the mouse brain. J Comp Neurol 2008; 507(5): 1676-93.
[http://dx.doi.org/10.1002/cne.21605] [PMID: 18241055]

[93] Russell DW, Halford RW, Ramirez DMO, Shah R, Kotti T. Cholesterol 24-hydroxylase: an enzyme of cholesterol turnover in the brain. Annu Rev Biochem 2009; 78: 1017-40.
[http://dx.doi.org/10.1146/annurev.biochem.78.072407.103859] [PMID: 19489738]

[94] Heverin M, Meaney S, Lütjohann D, Diczfalusy U, Wahren J, Björkhem I. Crossing the barrier: net flux of 27-hydroxycholesterol into the human brain. J Lipid Res 2005; 46(5): 1047-52.
[http://dx.doi.org/10.1194/jlr.M500024-JLR200] [PMID: 15741649]

[95] Lund EG, Xie C, Kotti T, Turley SD, Dietschy JM, Russell DW. Knockout of the cholesterol 24-hydroxylase gene in mice reveals a brain-specific mechanism of cholesterol turnover. J Biol Chem 2003; 278(25): 22980-8.
[http://dx.doi.org/10.1074/jbc.M303415200] [PMID: 12686551]

[96] Janowski BA, Willy PJ, Devi TR, Falck JR, Mangelsdorf DJ. An oxysterol signalling pathway mediated by the nuclear receptor LXR α. Nature 1996; 383(6602): 728-31.
[http://dx.doi.org/10.1038/383728a0] [PMID: 8878485]

[97]   Lehmann JM, Kliewer SA, Moore LB, *et al.* Activation of the nuclear receptor LXR by oxysterols defines a new hormone response pathway. J Biol Chem 1997; 272(6): 3137-40.
[http://dx.doi.org/10.1074/jbc.272.6.3137] [PMID: 9013544]

[98]   Brown J III, Theisler C, Silberman S, *et al.* Differential expression of cholesterol hydroxylases in Alzheimer's disease. J Biol Chem 2004; 279(33): 34674-81.
[http://dx.doi.org/10.1074/jbc.M402324200] [PMID: 15148325]

[99]   Whitney KD, Watson MA, Collins JL, *et al.* Regulation of cholesterol homeostasis by the liver X receptors in the central nervous system. Mol Endocrinol 2002; 16(6): 1378-85.
[http://dx.doi.org/10.1210/mend.16.6.0835] [PMID: 12040022]

[100] Fryer JD, Demattos RB, McCormick LM, *et al.* The low density lipoprotein receptor regulates the level of central nervous system human and murine apolipoprotein E but does not modify amyloid plaque pathology in PDAPP mice. J Biol Chem 2005; 280(27): 25754-9.
[http://dx.doi.org/10.1074/jbc.M502143200] [PMID: 15888448]

[101] Hughes TM, Rosano C, Evans RW, Kuller LH. Brain cholesterol metabolism, oxysterols, and dementia. J Alzheimers Dis 2013; 33(4): 891-911.
[http://dx.doi.org/10.3233/JAD-2012-121585] [PMID: 23076077]

[102] Papassotiropoulos A, Lütjohann D, Bagli M, *et al.* 24S-hydroxycholesterol in cerebrospinal fluid is elevated in early stages of dementia. J Psychiatr Res 2002; 36(1): 27-32.
[http://dx.doi.org/10.1016/S0022-3956(01)00050-4] [PMID: 11755458]

[103] Heverin M, Bogdanovic N, Lütjohann D, *et al.* Changes in the levels of cerebral and extracerebral sterols in the brain of patients with Alzheimer's disease. J Lipid Res 2004; 45(1): 186-93.
[http://dx.doi.org/10.1194/jlr.M300320-JLR200] [PMID: 14523054]

[104] Shafaati M, Solomon A, Kivipelto M, Björkhem I, Leoni V. Levels of ApoE in cerebrospinal fluid are correlated with Tau and 24S-hydroxycholesterol in patients with cognitive disorders. Neurosci Lett 2007; 425(2): 78-82.
[http://dx.doi.org/10.1016/j.neulet.2007.08.014] [PMID: 17822846]

[105] Schönknecht P, Lütjohann D, Pantel J, *et al.* Cerebrospinal fluid 24S-hydroxycholesterol is increased in patients with Alzheimer's disease compared to healthy controls. Neurosci Lett 2002; 324(1): 83-5.
[http://dx.doi.org/10.1016/S0304-3940(02)00164-7] [PMID: 11983301]

[106] Besga A, Cedazo-Minguez A, Kåreholt I, *et al.* Differences in brain cholesterol metabolism and insulin in two subgroups of patients with different CSF biomarkers but similar white matter lesions suggest different pathogenic mechanisms. Neurosci Lett 2012; 510(2): 121-6.
[http://dx.doi.org/10.1016/j.neulet.2012.01.017] [PMID: 22281444]

[107] Bretillon L, Lütjohann D, Ståhle L, *et al.* Plasma levels of 24S-hydroxycholesterol reflect the balance between cerebral production and hepatic metabolism and are inversely related to body surface. J Lipid Res 2000; 41(5): 840-5.
[PMID: 10787445]

[108] Leoni V, Caccia C. 24S-hydroxycholesterol in plasma: a marker of cholesterol turnover in neurodegenerative diseases. Biochimie 2013; 95(3): 595-612.
[http://dx.doi.org/10.1016/j.biochi.2012.09.025] [PMID: 23041502]

[109] Leoni V, Caccia C. Potential diagnostic applications of side chain oxysterols analysis in plasma and cerebrospinal fluid. Biochem Pharmacol 2013; 86(1): 26-36.
[http://dx.doi.org/10.1016/j.bcp.2013.03.015] [PMID: 23541982]

[110] Lütjohann D, Papassotiropoulos A, Björkhem I, *et al.* Plasma 24S-hydroxycholesterol (cerebrosterol) is increased in Alzheimer and vascular demented patients. J Lipid Res 2000; 41(2): 195-8.
[PMID: 10681402]

[111] Martin MG, Perga S, Trovò L, *et al.* Cholesterol loss enhances TrkB signaling in hippocampal neurons aging *in vitro.* Mol Biol Cell 2008; 19(5): 2101-12.

[http://dx.doi.org/10.1091/mbc.e07-09-0897] [PMID: 18287532]

[112] Sodero AO, Weissmann C, Ledesma MD, Dotti CG. Cellular stress from excitatory neurotransmission contributes to cholesterol loss in hippocampal neurons aging *in vitro*. Neurobiol Aging 2011; 32(6): 1043-53.
[http://dx.doi.org/10.1016/j.neurobiolaging.2010.06.001] [PMID: 20663588]

[113] Söderberg M, Edlund C, Kristensson K, Dallner G. Lipid compositions of different regions of the human brain during aging. J Neurochem 1990; 54(2): 415-23.
[http://dx.doi.org/10.1111/j.1471-4159.1990.tb01889.x] [PMID: 2299344]

[114] Svennerholm L, Gottfries CG. Membrane lipids, selectively diminished in Alzheimer brains, suggest synapse loss as a primary event in early-onset form (type I) and demyelination in late-onset form (type II). J Neurochem 1994; 62(3): 1039-47.
[http://dx.doi.org/10.1046/j.1471-4159.1994.62031039.x] [PMID: 8113790]

[115] Igbavboa U, Avdulov NA, Schroeder F, Wood WG. Increasing age alters transbilayer fluidity and cholesterol asymmetry in synaptic plasma membranes of mice. J Neurochem 1996; 66(4): 1717-25.
[http://dx.doi.org/10.1046/j.1471-4159.1996.66041717.x] [PMID: 8627330]

[116] Thelen KM, Falkai P, Bayer TA, Lütjohann D. Cholesterol synthesis rate in human hippocampus declines with aging. Neurosci Lett 2006; 403(1-2): 15-9.
[http://dx.doi.org/10.1016/j.neulet.2006.04.034] [PMID: 16701946]

[117] Jira PE, Waterham HR, Wanders RJA, Smeitink JAM, Sengers RCA, Wevers RA. Smith-Lemli-Opitz syndrome and the DHCR7 gene. Ann Hum Genet 2003; 67(Pt 3): 269-80.
[http://dx.doi.org/10.1046/j.1469-1809.2003.00034.x] [PMID: 12914579]

[118] Fitzky BU, Moebius FF, Asaoka H, *et al.* 7-Dehydrocholesterol-dependent proteolysis of HMG-CoA reductase suppresses sterol biosynthesis in a mouse model of Smith-Lemli-Opitz/RSH syndrome. J Clin Invest 2001; 108(6): 905-15.
[http://dx.doi.org/10.1172/JCI200112103] [PMID: 11560960]

[119] Xu X, Bittman R, Duportail G, Heissler D, Vilcheze C, London E. Effect of the structure of natural sterols and sphingolipids on the formation of ordered sphingolipid/sterol domains (rafts). Comparison of cholesterol to plant, fungal, and disease-associated sterols and comparison of sphingomyelin, cerebrosides, and ceramide. J Biol Chem 2001; 276(36): 33540-6.
[http://dx.doi.org/10.1074/jbc.M104776200] [PMID: 11432870]

[120] Leoni V, Mariotti C, Nanetti L, *et al.* Whole body cholesterol metabolism is impaired in Huntington's disease. Neurosci Lett 2011; 494(3): 245-9.
[http://dx.doi.org/10.1016/j.neulet.2011.03.025] [PMID: 21406216]

[121] Sipione S, Rigamonti D, Valenza M, *et al.* Early transcriptional profiles in huntingtin-inducible striatal cells by microarray analyses. Hum Mol Genet 2002; 11(17): 1953-65.
[http://dx.doi.org/10.1093/hmg/11.17.1953] [PMID: 12165557]

[122] Block RC, Dorsey ER, Beck CA, Brenna JT, Shoulson I. Altered cholesterol and fatty acid metabolism in Huntington disease. J Clin Lipidol 2010; 4(1): 17-23.
[http://dx.doi.org/10.1016/j.jacl.2009.11.003] [PMID: 20802793]

[123] Valenza M, Carroll JB, Leoni V, *et al.* Cholesterol biosynthesis pathway is disturbed in YAC128 mice and is modulated by huntingtin mutation. Hum Mol Genet 2007; 16(18): 2187-98.
[http://dx.doi.org/10.1093/hmg/ddm170] [PMID: 17613541]

[124] Rohanizadegan M, Sacharow S. Desmosterolosis presenting with multiple congenital anomalies. Eur J Med Genet 2018; 61(3): 152-6.
[http://dx.doi.org/10.1016/j.ejmg.2017.11.009] [PMID: 29175559]

[125] Arenas F, Garcia-Ruiz C, Fernandez-Checa JC. Intracellular Cholesterol Trafficking and Impact in Neurodegeneration. Front Mol Neurosci 2017; 10: 382.
[http://dx.doi.org/10.3389/fnmol.2017.00382] [PMID: 29204109]

[126] Hu G, Antikainen R, Jousilahti P, Kivipelto M, Tuomilehto J. Total cholesterol and the risk of Parkinson disease. Neurology 2008; 70: 1972-e1979.
[http://dx.doi.org/10.1212/01.wnl.0000312511.62699.a8]

[127] Miyake Y, Sasaki S, Tanaka K, *et al.* Fukuoka Kinki Parkinson's Disease Study Group, Dietary fat intake and risk of Parkinson's disease: a case-control study in Japan. J Neurol Sci 2010; 288: 117-e122.

[128] Huang X, Auinger P, Eberly S, *et al.* Serum cholesterol and the progression of Parkinson's disease: results from DATATOP. PLoS One 2011; 6(8): e22854.
[http://dx.doi.org/10.1371/journal.pone.0022854] [PMID: 21853051]

[129] Gudala K, Bansal D, Muthyala H. Role of serum cholesterol in Parkinson's disease: a meta-analysis of evidence. J Park Dis 2013; 3: 363-e370.

[130] Tan LC, Methawasin K, Tan EK, *et al.* Dietary cholesterol, fats and risk of Parkinson's disease in the Singapore Chinese Health Study. J Neurol Neurosurg Psychiatry 2015; pii: jnnp-2014e310065.

[131] Paul R, Choudhury A, Borah A. Cholesterol - A putative endogenous contributor towards Parkinson's disease. Neurochem Int 2015; 90: 125-33.
[http://dx.doi.org/10.1016/j.neuint.2015.07.025] [PMID: 26232622]

[132] Choi JY, Jang EH, Park CS, Kang JH. Enhanced susceptibility to 1-methyl-4-phenyl-1,2-3,6-tetrahydropyridine neurotoxicity in high-fat diet-induced obesity. Free Radic Biol Med 2005; 38(6): 806-16.
[http://dx.doi.org/10.1016/j.freeradbiomed.2004.12.008] [PMID: 15721991]

[133] Morris JK, Bomhoff GL, Stanford JA, Geiger PC. Neurodegeneration in an animal model of Parkinson's disease is exacerbated by a high-fat diet. Am J Physiol Regul Integr Comp Physiol 2010; 299: R1082-eR1090.

[134] Bousquet M, St-Amour I, Vandal M, Julien P, Cicchetti F, Calon F. High-fat diet exacerbates MPTP-induced dopaminergic degeneration in mice. Neurobiol Dis 2012; 45: 529-e538.

[135] Lee CY, Seet RC, Huang SH, Long LH, Halliwell B. Different patterns of oxidized lipid products in plasma and urine of dengue fever, stroke, and Parkinson's disease patients: cautions in the use of biomarkers of oxidative stress. Antioxid Redox 2009; Signal 11: 407-e420.
[http://dx.doi.org/10.1089/ars.2008.2179]

[136] Marwarha G, Ghribi O. Does the oxysterol 27-hydroxycholesterol underlie Alzheimer's disease-Parkinson's disease overlap?. Exp Gerontol 2014; 68: 13-e18.

[137] Bosco DA, Fowler DM, Zhang Q, *et al.* Elevated levels of oxidized cholesterol metabolites in Lewy body disease brains accelerate alpha-synuclein fibrilization. Nat Chem Biol 2006; 2: 249-e253.

[138] Paul R, Choudhury A, Kumar S, Giri A, Sandhir R, Borah A. Cholesterol contributes to dopamine-neuronal loss in MPTP mouse model of Parkinson's disease: Involvement of mitochondrial dysfunctions and oxidative stress. PLoS One 2017; 12(2): e0171285.
[http://dx.doi.org/10.1371/journal.pone.0171285] [PMID: 28170429]

[139] Paul R, Choudhury A, Chandra Boruah D, *et al.* Hypercholesterolemia causes psychomotor abnormalities in mice and alterations in cortico-striatal biogenic amine neurotransmitters: Relevance to Parkinson's disease. Neurochem Int 2017; 108: 15-26.
[http://dx.doi.org/10.1016/j.neuint.2017.01.021] [PMID: 28167224]

[140] Plassman BL, Havlik RJ, Steffens DC, *et al.* Documented head injury in early adulthood and risk of Alzheimer's disease and other dementias. Neurology 2000; 55(8): 1158-66.
[http://dx.doi.org/10.1212/WNL.55.8.1158] [PMID: 11071494]

[141] Goldman SM, Tanner CM, Oakes D, Bhudhikanok GS, Gupta A, Langston JW. Head injury and Parkinson's disease risk in twins. Ann Neurol 2006; 60(1): 65-72.
[http://dx.doi.org/10.1002/ana.20882] [PMID: 16718702]

[142] Rigg JL, Zafonte RD. Corticosteroids in TBI: is the story closed? J Head Trauma Rehabil 2006; 21(3): 285-8.
[http://dx.doi.org/10.1097/00001199-200605000-00010] [PMID: 16717507]

[143] Khatri N, Thakur M, Pareek V, Kumar S, Sharma S, Datusalia AK. Oxidative stress: Major threat in traumatic brain injury. CNS Neurol Disord Drug Targets 2018; 17(9): 689-95.
[http://dx.doi.org/10.2174/1871527317666180627120501] [PMID: 29952272]

[144] Hoane MR, Pierce JL, Holland MA, *et al.* The novel apolipoprotein E-based peptide COG1410 improves sensorimotor performance and reduces injury magnitude following cortical contusion injury. J Neurotrauma 2007; 24(7): 1108-18.
[http://dx.doi.org/10.1089/neu.2006.0254] [PMID: 17610351]

[145] Laskowitz DT, Grocott H, Hsia A, Copeland KR. Serum markers of cerebral ischemia. J Stroke Cerebrovasc Dis 1998; 7(4): 234-41.
[http://dx.doi.org/10.1016/S1052-3057(98)80032-3] [PMID: 17895090]

[146] Pan DS, Liu WG, Yang XF, Cao F. Inhibitory effect of progesterone on inflammatory factors after experimental traumatic brain injury. Biomed Environ Sci 2007; 20(5): 432-8.
[PMID: 18188998]

[147] Loncarevic-Vasiljkovic N, Pesic V, Todorovic S, *et al.* Caloric restriction suppresses microglial activation and prevents neuroapoptosis following cortical injury in rats. PLoS One 2012; 7(5): e37215.
[http://dx.doi.org/10.1371/journal.pone.0037215] [PMID: 22615943]

[148] Akiyama H, Barger S, Barnum S, *et al.* Inflammation and Alzheimer's disease. Neurobiol Aging 2000; 21(3): 383-421.
[http://dx.doi.org/10.1016/S0197-4580(00)00124-X] [PMID: 10858586]

[149] Bogdanovic N, Bretillon L, Lund EG, *et al.* On the turnover of brain cholesterol in patients with Alzheimer's disease. Abnormal induction of the cholesterol-catabolic enzyme CYP46 in glial cells. Neurosci Lett 2001; 314(1-2): 45-8.
[http://dx.doi.org/10.1016/S0304-3940(01)02277-7] [PMID: 11698143]

[150] Fassbender K, Simons M, Bergmann C, *et al.* Simvastatin strongly reduces levels of Alzheimer's disease beta -amyloid peptides Abeta 42 and Abeta 40 *in vitro* and *in vivo*. Proc Natl Acad Sci USA 2001; 98(10): 5856-61.
[http://dx.doi.org/10.1073/pnas.081620098] [PMID: 11296263]

[151] Bonotis K, Krikki E, Holeva V, Aggouridaki C, Costa V, Baloyannis S. Systemic immune aberrations in Alzheimer's disease patients. J Neuroimmunol 2008; 193(1-2): 183-7.
[http://dx.doi.org/10.1016/j.jneuroim.2007.10.020] [PMID: 18037502]

[152] Rojo LE, Fernández JA, Maccioni AA, Jimenez JM, Maccioni RB. Neuroinflammation: implications for the pathogenesis and molecular diagnosis of Alzheimer's disease. Arch Med Res 2008; 39(1): 1-16.
[http://dx.doi.org/10.1016/j.arcmed.2007.10.001] [PMID: 18067990]

[153] McGeer PL, McGeer EG. Glial reactions in Parkinson's disease. Mov Disord 2008; 23(4): 474-83.
[http://dx.doi.org/10.1002/mds.21751] [PMID: 18044695]

[154] Reynolds AD, Glanzer JG, Kadiu I, *et al.* Nitrated alpha-synuclein-activated microglial profiling for Parkinson's disease. J Neurochem 2008; 104(6): 1504-25.
[http://dx.doi.org/10.1111/j.1471-4159.2007.05087.x] [PMID: 18036154]

[155] Bar-On P, Crews L, Koob AO, *et al.* Statins reduce neuronal alpha-synuclein aggregation in *in vitro* models of Parkinson's disease. J Neurochem 2008; 105(5): 1656-67.
[http://dx.doi.org/10.1111/j.1471-4159.2008.05254.x] [PMID: 18248604]

[156] Kim HJ, Fan X, Gabbi C, *et al.* Liver X receptor beta (LXRbeta): a link between beta-sitosterol and amyotrophic lateral sclerosis-Parkinson's dementia. Proc Natl Acad Sci USA 2008; 105(6): 2094-9.
[http://dx.doi.org/10.1073/pnas.0711599105] [PMID: 18238900]

[157] Cernak I, Vink R, Zapple DN, *et al.* The pathobiology of moderate diffuse traumatic brain injury as identified using a new experimental model of injury in rats. Neurobiol Dis 2004; 17(1): 29-43.
[http://dx.doi.org/10.1016/j.nbd.2004.05.011] [PMID: 15350963]

[158] Di Giovanni S, Movsesyan V, Ahmed F, *et al.* Cell cycle inhibition provides neuroprotection and reduces glial proliferation and scar formation after traumatic brain injury. Proc Natl Acad Sci USA 2005; 102(23): 8333-8.
[http://dx.doi.org/10.1073/pnas.0500989102] [PMID: 15923260]

[159] Kay AD, Day SP, Kerr M, Nicoll JA, Packard CJ, Caslake MJ. Remodeling of cerebrospinal fluid lipoprotein particles after human traumatic brain injury. J Neurotrauma 2003; 20(8): 717-23.
[http://dx.doi.org/10.1089/089771503767869953] [PMID: 12965051]

[160] Faden AI. Pharmacologic treatment of acute traumatic brain injury. JAMA 1996; 276(7): 569-70.
[http://dx.doi.org/10.1001/jama.1996.03540070065034] [PMID: 8709409]

[161] Gasparovic C, Rosenberg GA, Wallace JA, *et al.* Magnetic resonance lipid signals in rat brain after experimental stroke correlate with neutral lipid accumulation. Neurosci Lett 2001; 301(2): 87-90.
[http://dx.doi.org/10.1016/S0304-3940(01)01616-0] [PMID: 11248429]

[162] Kamada H, Sato K, Iwai M, *et al.* Temporal and spatial changes of free cholesterol and neutral lipids in rat brain after transient middle cerebral artery occlusion. Neurosci Res 2003; 45(1): 91-100.
[http://dx.doi.org/10.1016/S0168-0102(02)00203-1] [PMID: 12507728]

[163] Pascotini ET, Flores AE, Kegler A, *et al.* Apoptotic markers and DNA damage are related to late phase of stroke: Involvement of dyslipidemia and inflammation. Physiol Behav 2015; 151: 369-78.
[http://dx.doi.org/10.1016/j.physbeh.2015.08.005] [PMID: 26253215]

[164] Hansson GK, Robertson AK, Söderberg-Nauclér C. Inflammation and atherosclerosis. Annu Rev Pathol 2006; 1: 297-329.
[http://dx.doi.org/10.1146/annurev.pathol.1.110304.100100] [PMID: 18039117]

[165] Ansell BJ, Fonarow GC, Navab M, Fogelman AM. Modifying the anti-inflammatory effects of high-density lipoprotein. Curr Atheroscler Rep 2007; 9(1): 57-63.
[http://dx.doi.org/10.1007/BF02693941] [PMID: 17169248]

[166] Navab M, Yu R, Gharavi N, *et al.* High-density lipoprotein: antioxidant and anti-inflammatory properties. Curr Atheroscler Rep 2007; 9(3): 244-8.
[http://dx.doi.org/10.1007/s11883-007-0026-3] [PMID: 18241620]

[167] Sanossian N, Saver JL, Navab M, Ovbiagele B. High-density lipoprotein cholesterol: an emerging target for stroke treatment. Stroke 2007; 38(3): 1104-9.
[http://dx.doi.org/10.1161/01.STR.0000258347.19449.0f] [PMID: 17255541]

[168] Cartagena CM, Ahmed F, Burns MP, *et al.* Cortical injury increases cholesterol 24S hydroxylase (Cyp46) levels in the rat brain. J Neurotrauma 2008; 25(9): 1087-98.
[http://dx.doi.org/10.1089/neu.2007.0444] [PMID: 18729719]

[169] Smiljanic K, Lavrnja I, Mladenovic Djordjevic A, *et al.* Brain injury induces cholesterol 24-hydroxylase (Cyp46) expression in glial cells in a time-dependent manner. Histochem Cell Biol 2010; 134(2): 159-69.
[http://dx.doi.org/10.1007/s00418-010-0718-6] [PMID: 20559650]

[170] Streit WJ. Microglia as neuroprotective, immunocompetent cells of the CNS. Glia 2002; 40(2): 133-9.
[http://dx.doi.org/10.1002/glia.10154] [PMID: 12379901]

[171] Moore KJ, Tabas I. The Cellular Biology of Macrophages in Atherosclerosis. Cell 2011; 145(3): 341-55.
[http://dx.doi.org/10.1016/j.cell.2011.04.005] [PMID: 21529710]

[172] Castranio EL, Wolfe CM, Nam KN, *et al.* ABCA1 haplodeficiency affects the brain transcriptome following traumatic brain injury in mice expressing human APOE isoforms. Acta Neuropathol

Commun 2018; 6(1): 69.
[http://dx.doi.org/10.1186/s40478-018-0569-2] [PMID: 30049279]

[173] Mahley RW, Huang Y. Apolipoprotein e sets the stage: response to injury triggers neuropathology. Neuron 2012; 76(5): 871-85.
[http://dx.doi.org/10.1016/j.neuron.2012.11.020] [PMID: 23217737]

[174] Fullerton SM, Shirman GA, Strittmatter WJ, Matthew WD. Impairment of the blood-nerve and blood-brain barriers in apolipoprotein e knockout mice. Exp Neurol 2001; 169(1): 13-22.
[http://dx.doi.org/10.1006/exnr.2001.7631] [PMID: 11312553]

[175] Jofre-Monseny L, Minihane AM, Rimbach G. Impact of apoE genotype on oxidative stress, inflammation and disease risk. Mol Nutr Food Res 2008; 52(1): 131-45.
[http://dx.doi.org/10.1002/mnfr.200700322] [PMID: 18203129]

[176] Shie FS, Jin LW, Cook DG, Leverenz JB, LeBoeuf RC. Diet-induced hypercholesterolemia enhances brain A beta accumulation in transgenic mice. Neuroreport 2002; 13(4): 455-9.
[http://dx.doi.org/10.1097/00001756-200203250-00019] [PMID: 11930160]

[177] Sparks DL. The early and ongoing experience with the cholesterol-fed rabbit as a model of Alzheimer's disease: the old, the new and the pilot. J Alzheimers Dis 2008; 15(4): 641-56.
[http://dx.doi.org/10.3233/JAD-2008-15410] [PMID: 19096162]

[178] Poirier J. Cholesterol transport and synthesis are compromised in the brain in sporadic Alzheimer's disease: From risk factors to therapeutic targets.Alzheimer's Disease and Related Disorders Annual. 2002; pp. 1-23.
[http://dx.doi.org/10.1201/b14342-2]

[179] Ledesma MD, Dotti CG. Peripheral cholesterol, metabolic disorders and Alzheimer's disease. Front Biosci (Elite Ed) 2012; 4: 181-94.
[http://dx.doi.org/10.2741/e368] [PMID: 22201863]

[180] Gamba P, Testa G, Sottero B, Gargiulo S, Poli G, Leonarduzzi G. The link between altered cholesterol metabolism and Alzheimer's disease. Ann N Y Acad Sci 2012; 1259: 54-64.
[http://dx.doi.org/10.1111/j.1749-6632.2012.06513.x] [PMID: 22758637]

[181] Ong WY, Tanaka K, Dawe GS, Ittner LM, Farooqui AA. Slow excitotoxicity in Alzheimer's disease. J Alzheimers Dis 2013; 35(4): 643-68.
[http://dx.doi.org/10.3233/JAD-121990] [PMID: 23481689]

[182] Vignini A, Giulietti A, Nanetti L, et al. Alzheimer's disease and diabetes: new insights and unifying therapies. Curr Diabetes Rev 2013; 9(3): 218-27.
[http://dx.doi.org/10.2174/1573399811309030003] [PMID: 23363296]

[183] Wood WG, Li L, Müller WE, Eckert GP. Cholesterol as a causative factor in Alzheimer's disease: a debatable hypothesis. J Neurochem 2014; 129(4): 559-72.
[http://dx.doi.org/10.1111/jnc.12637] [PMID: 24329875]

[184] Umeda T, Tomiyama T, Kitajima E, et al. Hypercholesterolemia accelerates intraneuronal accumulation of Aβ oligomers resulting in memory impairment in Alzheimer's disease model mice. Life Sci 2012; 91(23-24): 1169-76.
[http://dx.doi.org/10.1016/j.lfs.2011.12.022] [PMID: 22273754]

[185] Anstey KJ, Lipnicki DM, Low LF. Cholesterol as a risk factor for dementia and cognitive decline: a systematic review of prospective studies with meta-analysis. Am J Geriatr Psychiatry 2008; 16(5): 343-54.
[http://dx.doi.org/10.1097/01.JGP.0000310778.20870.ae] [PMID: 18448847]

[186] Mielke MM, Zandi PP, Sjögren M, et al. High total cholesterol levels in late life associated with a reduced risk of dementia. Neurology 2005; 64(10): 1689-95.
[http://dx.doi.org/10.1212/01.WNL.0000161870.78572.A5] [PMID: 15911792]

[187] Reitz C, Tang MX, Luchsinger J, Mayeux R. Relation of plasma lipids to Alzheimer disease and

vascular dementia. Arch Neurol 2004; 61(5): 705-14.
[http://dx.doi.org/10.1001/archneur.61.5.705] [PMID: 15148148]

[188] van Vliet P. Cholesterol and late-life cognitive decline. J Alzheimers Dis 2012; 30 (Suppl. 2): S147-62.
[http://dx.doi.org/10.3233/JAD-2011-111028] [PMID: 22269162]

[189] Ma C, Yin Z, Zhu P, Luo J, Shi X, Gao X. Blood cholesterol in late-life and cognitive decline: a longitudinal study of the Chinese elderly. Mol Neurodegener 2017; 12(1): 24.
[http://dx.doi.org/10.1186/s13024-017-0167-y] [PMID: 28270179]

[190] Cheng Y, Jin Y, Unverzagt FW, *et al.* The relationship between cholesterol and cognitive function is homocysteine-dependent. Clin Interv Aging 2014; 9: 1823-9.
[PMID: 25364240]

[191] Paul R, Borah A. Global loss of acetylcholinesterase activity with mitochondrial complexes inhibition and inflammation in brain of hypercholesterolemic mice. Sci Rep 2017; 7(1): 17922.
[http://dx.doi.org/10.1038/s41598-017-17911-z] [PMID: 29263397]

[192] de Oliveira J, Hort MA, Moreira EL, *et al.* Positive correlation between elevated plasma cholesterol levels and cognitive impairments in LDL receptor knockout mice: relevance of cortico-cerebral mitochondrial dysfunction and oxidative stress. Neuroscience 2011; 197(197): 99-106.
[http://dx.doi.org/10.1016/j.neuroscience.2011.09.009] [PMID: 21945034]

[193] Mulder M, Terwel D. Possible link between lipid metabolism and cerebral amyloid angiopathy in Alzheimer's disease: A role for high-density lipoproteins? Haemostasis 1998; 28(3-4): 174-94.
[PMID: 10420065]

[194] Catafau AM, Bullich S, Seibyl JP, *et al.* Cerebellar Amyloid-β Plaques: How Frequent Are They, and Do They Influence 18F-Florbetaben SUV Ratios? J Nucl Med 2016; 57(11): 1740-5.
[http://dx.doi.org/10.2967/jnumed.115.171652] [PMID: 27363836]

[195] Maślińska D, Laure-Kamionowska M, Szukiewicz D, Maśliński S, Księżopolska-Orłowska K. Commitment of protein p53 and amyloid-beta peptide (Aβ) in aging of human cerebellum. Folia Neuropathol 2017; 55(2): 161-7.
[http://dx.doi.org/10.5114/fn.2017.68583] [PMID: 28677373]

[196] Calderon-Garcidueñas AL, Duyckaerts C. Alzheimer disease. Handb Clin Neurol 2017; 145: 325-37.
[http://dx.doi.org/10.1016/B978-0-12-802395-2.00023-7] [PMID: 28987180]

[197] Perovic M, Mladenovic Djordjevic A, Smiljanic K, *et al.* Expression of cholesterol homeostasis genes in the brain of the male rat is affected by age and dietary restriction. Biogerontology 2009; 10(6): 735-45.
[http://dx.doi.org/10.1007/s10522-009-9220-8] [PMID: 19267214]

[198] Liu CC, Liu CC, Kanekiyo T, Xu H, Bu G. Apolipoprotein E and Alzheimer disease: risk, mechanisms and therapy. Nat Rev Neurol 2013; 9(2): 106-18.
[http://dx.doi.org/10.1038/nrneurol.2012.263] [PMID: 23296339]

[199] Yu JT, Tan L, Hardy J. Apolipoprotein E in Alzheimer's disease: an update. Annu Rev Neurosci 2014; 37: 79-100.
[http://dx.doi.org/10.1146/annurev-neuro-071013-014300] [PMID: 24821312]

[200] Salameh TS, Rhea EM, Banks WA, Hanson AJ. Insulin resistance, dyslipidemia, and apolipoprotein E interactions as mechanisms in cognitive impairment and Alzheimer's disease. Exp Biol Med (Maywood) 2016; 241(15): 1676-83.
[http://dx.doi.org/10.1177/1535370216660770] [PMID: 27470930]

[201] Shen L, Jia J. An Overview of Genome-Wide Association Studies in Alzheimer's Disease. Neurosci Bull 2016; 32(2): 183-90.
[http://dx.doi.org/10.1007/s12264-016-0011-3] [PMID: 26810783]

[202] Rosenberg ME, Girton R, Finkel D, *et al.* Apolipoprotein J/clusterin prevents a progressive

glomerulopathy of aging. Mol Cell Biol 2002; 22(6): 1893-902.
[http://dx.doi.org/10.1128/MCB.22.6.1893-1902.2002] [PMID: 11865066]

[203] Lee JY, Parks JS. ATP-binding cassette transporter AI and its role in HDL formation. Curr Opin Lipidol 2005; 16(1): 19-25.
[http://dx.doi.org/10.1097/00041433-200502000-00005] [PMID: 15650559]

[204] Koldamova R, Fitz NF, Lefterov I. The role of ATP-binding cassette transporter A1 in Alzheimer's disease and neurodegeneration. Biochim Biophys Acta 2010; 1801(8): 824-30.
[http://dx.doi.org/10.1016/j.bbalip.2010.02.010] [PMID: 20188211]

[205] Kim WS, Hill AF, Fitzgerald ML, Freeman MW, Evin G, Garner B. Wild type and Tangier disease ABCA1 mutants modulate cellular amyloid-β production independent of cholesterol efflux activity. J Alzheimers Dis 2011; 27(2): 441-52.
[http://dx.doi.org/10.3233/JAD-2011-110521] [PMID: 21860089]

[206] Wahrle SE, Jiang H, Parsadanian M, *et al.* ABCA1 is required for normal central nervous system ApoE levels and for lipidation of astrocyte-secreted apoE. J Biol Chem 2004; 279(39): 40987-93.
[http://dx.doi.org/10.1074/jbc.M407963200] [PMID: 15269217]

[207] Wahrle SE, Jiang H, Parsadanian M, *et al.* Overexpression of ABCA1 reduces amyloid deposition in the PDAPP mouse model of Alzheimer disease. J Clin Invest 2008; 118(2): 671-82.
[PMID: 18202749]

[208] Canepa E, Borghi R, Viña J, *et al.* Cholesterol and amyloid-β: evidence for a cross-talk between astrocytes and neuronal cells. J Alzheimers Dis 2011; 25(4): 645-53.
[http://dx.doi.org/10.3233/JAD-2011-110053] [PMID: 21483097]

[209] Farfel JM, Yu L, De Jager PL, Schneider JA, Bennett DA. Association of APOE with tau-tangle pathology with and without β-amyloid. Neurobiol Aging 2016; 37: 19-25.
[http://dx.doi.org/10.1016/j.neurobiolaging.2015.09.011] [PMID: 26481403]

[210] Coon KD, Myers AJ, Craig DW, *et al.* A high-density whole-genome association study reveals that APOE is the major susceptibility gene for sporadic late-onset Alzheimer's disease. J Clin Psychiatry 2007; 68(4): 613-8.
[http://dx.doi.org/10.4088/JCP.v68n0419] [PMID: 17474819]

[211] Parasuraman R, Greenwood PM, Sunderland T. The apolipoprotein E gene, attention, and brain function. Neuropsychology 2002; 16(2): 254-74.
[http://dx.doi.org/10.1037/0894-4105.16.2.254] [PMID: 11949718]

[212] Poirier J. Apolipoprotein E, cholesterol transport and synthesis in sporadic Alzheimer's disease. Neurobiol Aging 2005; 26(3): 355-61.
[http://dx.doi.org/10.1016/j.neurobiolaging.2004.09.003] [PMID: 15639314]

[213] Polvikoski T, Sulkava R, Haltia M, *et al.* Apolipoprotein E, dementia, and cortical deposition of beta-amyloid protein. N Engl J Med 1995; 333(19): 1242-7.
[http://dx.doi.org/10.1056/NEJM199511093331902] [PMID: 7566000]

[214] Holtzman DM, Fagan AM, Mackey B, *et al.* Apolipoprotein E facilitates neuritic and cerebrovascular plaque formation in an Alzheimer's disease model. Ann Neurol 2000; 47(6): 739-47.
[http://dx.doi.org/10.1002/1531-8249(200006)47:6<739::AID-ANA6>3.0.CO;2-8] [PMID: 10852539]

[215] Carter DB, Dunn E, McKinley DD, *et al.* Human apolipoprotein E4 accelerates beta-amyloid deposition in APPsw transgenic mouse brain. Ann Neurol 2001; 50(4): 468-75.
[http://dx.doi.org/10.1002/ana.1134] [PMID: 11601499]

[216] Ye S, Huang Y, Müllendorff K, *et al.* Apolipoprotein (apo) E4 enhances amyloid beta peptide production in cultured neuronal cells: apoE structure as a potential therapeutic target. Proc Natl Acad Sci USA 2005; 102(51): 18700-5.
[http://dx.doi.org/10.1073/pnas.0508693102] [PMID: 16344478]

[217] Ji ZS, Miranda RD, Newhouse YM, Weisgraber KH, Huang Y, Mahley RW. Apolipoprotein E4

potentiates amyloid beta peptide-induced lysosomal leakage and apoptosis in neuronal cells. J Biol Chem 2002; 277(24): 21821-8.
[http://dx.doi.org/10.1074/jbc.M112109200] [PMID: 11912196]

[218] Deane R, Sagare A, Hamm K, *et al.* apoE isoform-specific disruption of amyloid beta peptide clearance from mouse brain. J Clin Invest 2008; 118(12): 4002-13.
[http://dx.doi.org/10.1172/JCI36663] [PMID: 19033669]

[219] Jarvik GP, Wijsman EM, Kukull WA, Schellenberg GD, Yu C, Larson EB. Interactions of apolipoprotein E genotype, total cholesterol level, age, and sex in prediction of Alzheimer's disease: a case-control study. Neurology 1995; 45(6): 1092-6.
[http://dx.doi.org/10.1212/WNL.45.6.1092] [PMID: 7783869]

[220] Wisniewski T, Frangione B. Apolipoprotein E: a pathological chaperone protein in patients with cerebral and systemic amyloid. Neurosci Lett 1992; 135(2): 235-8.
[http://dx.doi.org/10.1016/0304-3940(92)90444-C] [PMID: 1625800]

[221] Harris FM, Brecht WJ, Xu Q, *et al.* Carboxyl-terminal-truncated apolipoprotein E4 causes Alzheimer's disease-like neurodegeneration and behavioral deficits in transgenic mice. Proc Natl Acad Sci USA 2003; 100(19): 10966-71.
[http://dx.doi.org/10.1073/pnas.1434398100] [PMID: 12939405]

[222] Brecht WJ, Harris FM, Chang S, *et al.* Neuron-specific apolipoprotein e4 proteolysis is associated with increased tau phosphorylation in brains of transgenic mice. J Neurosci 2004; 24(10): 2527-34.
[http://dx.doi.org/10.1523/JNEUROSCI.4315-03.2004] [PMID: 15014128]

[223] Tai LM, Ghura S, Koster KP, *et al.* APOE-modulated Aβ-induced neuroinflammation in Alzheimer's disease: current landscape, novel data, and future perspective. J Neurochem 2015; 133(4): 465-88.
[http://dx.doi.org/10.1111/jnc.13072] [PMID: 25689586]

[224] Arendt T, Schindler C, Brückner MK, *et al.* Plastic neuronal remodeling is impaired in patients with Alzheimer's disease carrying apolipoprotein epsilon 4 allele. J Neurosci 1997; 17(2): 516-29.
[http://dx.doi.org/10.1523/JNEUROSCI.17-02-00516.1997] [PMID: 8987775]

[225] Beffert U, Danik M, Krzywkowski P, Ramassamy C, Berrada F, Poirier J. The neurobiology of apolipoproteins and their receptors in the CNS and Alzheimer's disease. Brain Res Brain Res Rev 1998; 27(2): 119-42.
[http://dx.doi.org/10.1016/S0165-0173(98)00008-3] [PMID: 9622609]

[226] Friedman G, Froom P, Sazbon L, *et al.* Apolipoprotein E-epsilon4 genotype predicts a poor outcome in survivors of traumatic brain injury. Neurology 1999; 52(2): 244-8.
[http://dx.doi.org/10.1212/WNL.52.2.244] [PMID: 9932938]

[227] Lichtman SW, Seliger G, Tycko B, Marder K. Apolipoprotein E and functional recovery from brain injury following postacute rehabilitation. Neurology 2000; 55(10): 1536-9.
[http://dx.doi.org/10.1212/WNL.55.10.1536] [PMID: 11094110]

[228] Lin YT, Seo J, Gao F, *et al.* APOE4 Causes Widespread Molecular and Cellular Alterations Associated with Alzheimer's Disease Phenotypes in Human iPSC-Derived Brain Cell Types. Neuron 2018; 98(6): 1294.
[http://dx.doi.org/10.1016/j.neuron.2018.06.011] [PMID: 29953873]

[229] Barbero-Camps E, Fernández A, Baulies A, Martinez L, Fernández-Checa JC, Colell A. Endoplasmic reticulum stress mediates amyloid β neurotoxicity *via* mitochondrial cholesterol trafficking. Am J Pathol 2014; 184(7): 2066-81.
[http://dx.doi.org/10.1016/j.ajpath.2014.03.014] [PMID: 24815354]

[230] Spell C, Kölsch H, Lütjohann D, *et al.* SREBP-1a polymorphism influences the risk of Alzheimer's disease in carriers of the ApoE4 allele. Dement Geriatr Cogn Disord 2004; 18(3-4): 245-9.
[http://dx.doi.org/10.1159/000080023] [PMID: 15286454]

[231] Pierrot N, Tyteca D, D'auria L, *et al.* Amyloid precursor protein controls cholesterol turnover needed

for neuronal activity. EMBO Mol Med 2013; 5(4): 608-25.
[http://dx.doi.org/10.1002/emmm.201202215] [PMID: 23554170]

[232] Avila-Muñoz E, Arias C. Cholesterol-induced astrocyte activation is associated with increased amyloid precursor protein expression and processing. Glia 2015; 63(11): 2010-22.
[http://dx.doi.org/10.1002/glia.22874] [PMID: 26096015]

[233] Mohamed A, Viveiros A, Williams K, Posse de Chaves E. Aβ inhibits SREBP-2 activation through Akt inhibition. J Lipid Res 2018; 59(1): 1-13.
[http://dx.doi.org/10.1194/jlr.M076703] [PMID: 29122977]

[234] Mohamed A, Saavedra L, Di Pardo A, Sipione S, Posse de Chaves E. β-amyloid inhibits protein prenylation and induces cholesterol sequestration by impairing SREBP-2 cleavage. J Neurosci 2012; 32(19): 6490-500.
[http://dx.doi.org/10.1523/JNEUROSCI.0630-12.2012] [PMID: 22573671]

[235] Djelti F, Braudeau J, Hudry E, et al. CYP46A1 inhibition, brain cholesterol accumulation and neurodegeneration pave the way for Alzheimer's disease. Brain 2015; 138(Pt 8): 2383-98.
[http://dx.doi.org/10.1093/brain/awv166] [PMID: 26141492]

[236] Hudry E, Van Dam D, Kulik W, et al. Adeno-associated virus gene therapy with cholesterol 24-hydroxylase reduces the amyloid pathology before or after the onset of amyloid plaques in mouse models of Alzheimer's disease. Mol Ther 2010; 18(1): 44-53.
[http://dx.doi.org/10.1038/mt.2009.175] [PMID: 19654569]

[237] Bryleva EY, Rogers MA, Chang CC, et al. ACAT1 gene ablation increases 24(S)-hydroxycholesterol content in the brain and ameliorates amyloid pathology in mice with AD. Proc Natl Acad Sci USA 2010; 107(7): 3081-6.
[http://dx.doi.org/10.1073/pnas.0913828107] [PMID: 20133765]

[238] Urano Y, Ochiai S, Noguchi N. Suppression of amyloid-β production by 24S-hydroxycholesterol via inhibition of intracellular amyloid precursor protein trafficking. FASEB J 2013; 27(10): 4305-15.
[http://dx.doi.org/10.1096/fj.13-231456] [PMID: 23839932]

[239] Fan QW, Yu W, Senda T, Yanagisawa K, Michikawa M. Cholesterol-dependent modulation of tau phosphorylation in cultured neurons. J Neurochem 2001; 76(2): 391-400.
[http://dx.doi.org/10.1046/j.1471-4159.2001.00063.x] [PMID: 11208902]

[240] Burlot MA, Braudeau J, Michaelsen-Preusse K, et al. Cholesterol 24-hydroxylase defect is implicated in memory impairments associated with Alzheimer-like Tau pathology. Hum Mol Genet 2015; 24(21): 5965-76.
[http://dx.doi.org/10.1093/hmg/ddv268] [PMID: 26358780]

[241] Gratuze M, Julien J, Morin F, et al. High-fat, high-sugar, and high-cholesterol consumption does not impact tau pathogenesis in a mouse model of Alzheimer's disease-like tau pathology. Neurobiol Aging 2016; 47: 71-3.
[http://dx.doi.org/10.1016/j.neurobiolaging.2016.07.016] [PMID: 27565300]

[242] Shobab LA, Hsiung GY, Feldman HH. Cholesterol in Alzheimer's disease. Lancet Neurol 2005; 4(12): 841-52.
[http://dx.doi.org/10.1016/S1474-4422(05)70248-9] [PMID: 16297842]

[243] Ehehalt R, Keller P, Haass C, Thiele C, Simons K. Amyloidogenic processing of the Alzheimer beta-amyloid precursor protein depends on lipid rafts. J Cell Biol 2003; 160(1): 113-23.
[http://dx.doi.org/10.1083/jcb.200207113] [PMID: 12515826]

[244] Xiong H, Callaghan D, Jones A, et al. Cholesterol retention in Alzheimer's brain is responsible for high beta- and gamma-secretase activities and Abeta production. Neurobiol Dis 2008; 29(3): 422-37.
[http://dx.doi.org/10.1016/j.nbd.2007.10.005] [PMID: 18086530]

[245] Harris JR. Cholesterol binding to amyloid-beta fibrils: a TEM study. Micron 2008; 39(8): 1192-6.
[http://dx.doi.org/10.1016/j.micron.2008.05.001] [PMID: 18586500]

[246] Beel AJ, Sakakura M, Barrett PJ, Sanders CR. Direct binding of cholesterol to the amyloid precursor protein: An important interaction in lipid-Alzheimer's disease relationships? Biochim Biophys Acta 2010; 1801(8): 975-82.
[http://dx.doi.org/10.1016/j.bbalip.2010.03.008] [PMID: 20304095]

[247] Schneider A, Rajendran L, Honsho M, *et al.* Flotillin-dependent clustering of the amyloid precursor protein regulates its endocytosis and amyloidogenic processing in neurons. J Neurosci 2008; 28(11): 2874-82.
[http://dx.doi.org/10.1523/JNEUROSCI.5345-07.2008] [PMID: 18337418]

[248] Grimm MO, Grimm HS, Tomic I, Beyreuther K, Hartmann T, Bergmann C. Independent inhibition of Alzheimer disease beta- and gamma-secretase cleavage by lowered cholesterol levels. J Biol Chem 2008; 283(17): 11302-11.
[http://dx.doi.org/10.1074/jbc.M801520200] [PMID: 18308724]

[249] Reid PC, Urano Y, Kodama T, Hamakubo T. Alzheimer's disease: cholesterol, membrane rafts, isoprenoids and statins. J Cell Mol Med 2007; 11(3): 383-92.
[http://dx.doi.org/10.1111/j.1582-4934.2007.00054.x] [PMID: 17635634]

[250] Taghibiglou C, Khalaj S. Drug Discovery Approaches for the Treatment of Neurodegenerative Disorders Alzheimer's Disease. Cholesterol and Fat Metabolism in Alzheimer's Disease 2017; pp. 161-93.

[251] Saavedra L, Mohamed A, Ma V, Kar S, de Chaves EP. Internalization of beta-amyloid peptide by primary neurons in the absence of apolipoprotein E. J Biol Chem 2007; 282(49): 35722-32.
[http://dx.doi.org/10.1074/jbc.M701823200] [PMID: 17911110]

[252] Kawarabayashi T, Shoji M, Younkin LH, *et al.* Dimeric amyloid beta protein rapidly accumulates in lipid rafts followed by apolipoprotein E and phosphorylated tau accumulation in the Tg2576 mouse model of Alzheimer's disease. J Neurosci 2004; 24(15): 3801-9.
[http://dx.doi.org/10.1523/JNEUROSCI.5543-03.2004] [PMID: 15084661]

[253] Puglielli L, Konopka G, Pack-Chung E, *et al.* Acyl-coenzyme A: cholesterol acyltransferase modulates the generation of the amyloid beta-peptide. Nat Cell Biol 2001; 3(10): 905-12.
[http://dx.doi.org/10.1038/ncb1001-905] [PMID: 11584272]

[254] Shibuya Y, Chang CC, Chang TY. ACAT1/SOAT1 as a therapeutic target for Alzheimer's disease. Future Med Chem 2015; 7(18): 2451-67.
[http://dx.doi.org/10.4155/fmc.15.161] [PMID: 26669800]

[255] Bhattacharyya R, Kovacs DM. ACAT inhibition and amyloid beta reduction. Biochim Biophys Acta 2010; 1801(8): 960-5.
[http://dx.doi.org/10.1016/j.bbalip.2010.04.003] [PMID: 20398792]

[256] Avigan J, Steinberg D, Thompson MJ, Mosettig E. The mechanism of action of MER-29. Prog Cardiovasc Dis 1960; 2: 525-30.
[http://dx.doi.org/10.1016/S0033-0620(60)80024-2] [PMID: 13795275]

[257] Steinberg D, Avigan J, Feigelson EB. Identification of 24-dehydrocholesterol in the serum of patients treated with MER-29. Prog Cardiovasc Dis 1960; 2: 586-92.
[http://dx.doi.org/10.1016/S0033-0620(60)80038-2] [PMID: 13834161]

[258] Endo A, Kuroda M. Citrinin, an inhibitor of cholesterol synthesis. J Antibiot (Tokyo) 1976; 29(8): 841-3.
[http://dx.doi.org/10.7164/antibiotics.29.841] [PMID: 791911]

[259] Tanzawa K, Kuroda M, Endo A. Time-dependent, irreversible inhibition of 3-hydroxy-3-methylglutaryl-coenzyme A reductase by the antibiotic citrinin. Biochim Biophys Acta 1977; 488(1): 97-101.
[http://dx.doi.org/10.1016/0005-2760(77)90126-6] [PMID: 889862]

[260] Hoyer S, Riederer P. Alzheimer disease--no target for statin treatment. A mini review. Neurochem Res

2007; 32(4-5): 695-706.
[http://dx.doi.org/10.1007/s11064-006-9168-x] [PMID: 17063393]

[261] Fonseca AC, Resende R, Oliveira CR, Pereira CM. Cholesterol and statins in Alzheimer's disease: current controversies. Exp Neurol 2010; 223(2): 282-93.
[http://dx.doi.org/10.1016/j.expneurol.2009.09.013] [PMID: 19782682]

[262] McGuinness B, Passmore P. Can statins prevent or help treat Alzheimer's disease? J Alzheimers Dis 2010; 20(3): 925-33.
[http://dx.doi.org/10.3233/JAD-2010-091570] [PMID: 20182019]

[263] Shepardson NE, Shankar GM, Selkoe DJ. Cholesterol level and statin use in Alzheimer disease: I. Review of epidemiological and preclinical studies. Arch Neurol 2011; 68(10): 1239-44.
[http://dx.doi.org/10.1001/archneurol.2011.203] [PMID: 21987540]

[264] Zacco A, Togo J, Spence K, *et al.* 3-hydroxy-3-methylglutaryl coenzyme A reductase inhibitors protect cortical neurons from excitotoxicity. J Neurosci 2003; 23(35): 11104-11.
[http://dx.doi.org/10.1523/JNEUROSCI.23-35-11104.2003] [PMID: 14657168]

[265] Kojro E, Füger P, Prinzen C, *et al.* Statins and the squalene synthase inhibitor zaragozic acid stimulate the non-amyloidogenic pathway of amyloid-beta protein precursor processing by suppression of cholesterol synthesis. J Alzheimers Dis 2010; 20(4): 1215-31.
[http://dx.doi.org/10.3233/JAD-2010-091621] [PMID: 20413873]

[266] Serrano-Pozo A, Vega GL, Lütjohann D, *et al.* Effects of simvastatin on cholesterol metabolism and Alzheimer disease biomarkers. Alzheimer Dis Assoc Disord 2010; 24(3): 220-6.
[PMID: 20473136]

[267] Feldman HH, Doody RS, Kivipelto M, *et al.* Randomized controlled trial of atorvastatin in mild to moderate Alzheimer disease: LEADe. Neurology 2010; 74(12): 956-64.
[http://dx.doi.org/10.1212/WNL.0b013e3181d6476a] [PMID: 20200346]

[268] McGuinness B, Craig D, Bullock R, Malouf R, Passmore P. Statins for the treatment of dementia. Cochrane Database Syst Rev 2014; (7): CD007514.
[PMID: 25004278]

[269] Cibičková L. Statins and their influence on brain cholesterol. J Clin Lipidol 2011; 5(5): 373-9.
[http://dx.doi.org/10.1016/j.jacl.2011.06.007] [PMID: 21981838]

[270] Smiljanic K, Vanmierlo T, Mladenovic Djordjevic A, *et al.* Cholesterol metabolism changes under long-term dietary restrictions while the cholesterol homeostasis remains unaffected in the cortex and hippocampus of aging rats. Age (Dordr) 2014; 36(3): 9654.
[http://dx.doi.org/10.1007/s11357-014-9654-z] [PMID: 24756765]

[271] Kirsch C, Eckert GP, Mueller WE. Statin effects on cholesterol micro-domains in brain plasma membranes. Biochem Pharmacol 2003; 65(5): 843-56.
[http://dx.doi.org/10.1016/S0006-2952(02)01654-4] [PMID: 12628479]

[272] Liao JK, Laufs U. Pleiotropic effects of statins. Annu Rev Pharmacol Toxicol 2005; 45: 89-118.
[http://dx.doi.org/10.1146/annurev.pharmtox.45.120403.095748] [PMID: 15822172]

[273] He X, Jenner AM, Ong WY, Farooqui AA, Patel SC. Lovastatin modulates increased cholesterol and oxysterol levels and has a neuroprotective effect on rat hippocampal neurons after kainate injury. J Neuropathol Exp Neurol 2006; 65(7): 652-63.
[http://dx.doi.org/10.1097/01.jnen.0000225906.82428.69] [PMID: 16825952]

[274] Lizard G, Miguet C, Bésséde G, *et al.* Impairment with various antioxidants of the loss of mitochondrial transmembrane potential and of the cytosolic release of cytochrome c occuring during 7-ketocholesterol-induced apoptosis. Free Radic Biol Med 2000; 28(5): 743-53.
[http://dx.doi.org/10.1016/S0891-5849(00)00163-5] [PMID: 10754270]

[275] Ong WY, Goh EW, Lu XR, Farooqui AA, Patel SC, Halliwell B. Increase in cholesterol and cholesterol oxidation products, and role of cholesterol oxidation products in kainate-induced neuronal

injury. Brain Pathol 2003; 13(3): 250-62.
[http://dx.doi.org/10.1111/j.1750-3639.2003.tb00026.x] [PMID: 12946016]

[276] Kölsch H, Ludwig M, Lütjohann D, Rao ML. Neurotoxicity of 24-hydroxycholesterol, an important cholesterol elimination product of the brain, may be prevented by vitamin E and estradiol-17beta. J Neural Transm (Vienna) 2001; 108(4): 475-88.
[http://dx.doi.org/10.1007/s007020170068] [PMID: 11475014]

[277] Wolozin B, Manger J, Bryant R, Cordy J, Green RC, McKee A. Re-assessing the relationship between cholesterol, statins and Alzheimer's disease. Acta Neurol Scand Suppl 2006; 185: 63-70.
[http://dx.doi.org/10.1111/j.1600-0404.2006.00687.x] [PMID: 16866913]

[278] Thirumangalakudi L, Prakasam A, Zhang R, *et al.* High cholesterol-induced neuroinflammation and amyloid precursor protein processing correlate with loss of working memory in mice. J Neurochem 2008; 106(1): 475-85.
[http://dx.doi.org/10.1111/j.1471-4159.2008.05415.x] [PMID: 18410513]

[279] Geifman N, Brinton RD, Kennedy RE, Schneider LS, Butte AJ. Evidence for benefit of statins to modify cognitive decline and risk in Alzheimer's disease. Alzheimers Res Ther 2017; 9(1): 10.
[http://dx.doi.org/10.1186/s13195-017-0237-y] [PMID: 28212683]

[280] Saheki A, Terasaki T, Tamai I, Tsuji A. *In vivo* and *in vitro* blood-brain barrier transport of 3-hydrox-3-methylglutaryl coenzyme A (HMG-CoA) reductase inhibitors. Pharm Res 1994; 11(2): 305-11.
[http://dx.doi.org/10.1023/A:1018975928974] [PMID: 8165193]

[281] Wagstaff LR, Mitton MW, Arvik BM, Doraiswamy PM. Statin-associated memory loss: analysis of 60 case reports and review of the literature. Pharmacotherapy 2003; 23(7): 871-80.
[http://dx.doi.org/10.1592/phco.23.7.871.32720] [PMID: 12885101]

[282] Johnson-Anuna LN, Eckert GP, Keller JH, *et al.* Chronic administration of statins alters multiple gene expression patterns in mouse cerebral cortex. J Pharmacol Exp Ther 2005; 312(2): 786-93.
[http://dx.doi.org/10.1124/jpet.104.075028] [PMID: 15358814]

[283] Martini C, Pallottini V, Cavallini G, Donati A, Bergamini E, Trentalance A. Caloric restrictions affect some factors involved in age-related hypercholesterolemia. J Cell Biochem 2007; 101(1): 235-43.
[http://dx.doi.org/10.1002/jcb.21158] [PMID: 17203467]

[284] Martini C, Pallottini V, De Marinis E, *et al.* Omega-3 as well as caloric restriction prevent the age-related modifications of cholesterol metabolism. Mech Ageing Dev 2008; 129(12): 722-7.
[http://dx.doi.org/10.1016/j.mad.2008.09.010] [PMID: 18930075]

[285] Pallottini V, Marino M, Cavallini G, Bergamini E, Trentalance A. Age-related changes of isoprenoid biosynthesis in rat liver and brain. Biogerontology 2003; 4(6): 371-8.
[http://dx.doi.org/10.1023/B:BGEN.0000006557.92558.60] [PMID: 14739708]

[286] Mulas MF, Demuro G, Mulas C, *et al.* Dietary restriction counteracts age-related changes in cholesterol metabolism in the rat. Mech Ageing Dev 2005; 126(6-7): 648-54.
[http://dx.doi.org/10.1016/j.mad.2004.11.010] [PMID: 15888318]

[287] Stranahan AM, Cutler RG, Button C, Telljohann R, Mattson MP. Diet-induced elevations in serum cholesterol are associated with alterations in hippocampal lipid metabolism and increased oxidative stress. J Neurochem 2011; 118(4): 611-5.
[http://dx.doi.org/10.1111/j.1471-4159.2011.07351.x] [PMID: 21682722]

[288] Camargo N, Brouwers JF, Loos M, Gutmann DH, Smit AB, Verheijen MH. High-fat diet ameliorates neurological deficits caused by defective astrocyte lipid metabolism. FASEB J 2012; 26(10): 4302-15.
[http://dx.doi.org/10.1096/fj.12-205807] [PMID: 22751013]

[289] Fon Tacer K, Pompon D, Rozman D. Adaptation of cholesterol synthesis to fasting and TNF-alpha: profiling cholesterol intermediates in the liver, brain, and testis. J Steroid Biochem Mol Biol 2010; 121(3-5): 619-25.
[http://dx.doi.org/10.1016/j.jsbmb.2010.02.026] [PMID: 20206258]

[290]   Hayakawa K, Mishima K, Nozako M, *et al.* High-cholesterol feeding aggravates cerebral infarction *via* decreasing the CB1 receptor. Neurosci Lett 2007; 414(2): 183-7.
[http://dx.doi.org/10.1016/j.neulet.2006.12.022] [PMID: 17208374]

[291]   Smiljanic K, Todorovic S, Mladenovic Djordjevic A, *et al.* Limited daily feeding and intermittent feeding have different effects on regional brain energy homeostasis during aging. Biogerontology 2018; 19(2): 121-32.
[http://dx.doi.org/10.1007/s10522-018-9743-y] [PMID: 29340834]

[292]   Notkola IL, Sulkava R, Pekkanen J, *et al.* Serum total cholesterol, apolipoprotein E epsilon 4 allele, and Alzheimer's disease. Neuroepidemiology 1998; 17(1): 14-20.
[http://dx.doi.org/10.1159/000026149] [PMID: 9549720]

[293]   Kivipelto M, Laakso MP, Tuomilehto J, Nissinen A, Soininen H. Hypertension and hypercholesterolaemia as risk factors for Alzheimer's disease: potential for pharmacological intervention. CNS Drugs 2002; 16(7): 435-44. [Review].
[http://dx.doi.org/10.2165/00023210-200216070-00001] [PMID: 12056919]

[294]   Kalmijn S. Fatty acid intake and the risk of dementia and cognitive decline: a review of clinical and epidemiological studies. J Nutr Health Aging 2000; 4(4): 202-7.
[PMID: 11115801]

[295]   Tan ZS, Seshadri S, Beiser A, *et al.* Plasma total cholesterol level as a risk factor for Alzheimer disease: the Framingham Study. Arch Intern Med 2003; 163(9): 1053-7.
[http://dx.doi.org/10.1001/archinte.163.9.1053] [PMID: 12742802]

[296]   Picard C, Julien C, Frappier J, Miron J, Théroux L, Dea D. United Kingdom Brain Expression Consortium and for the Alzheimer's Disease Neuroimaging Initiative, Breitner JCS, Poirier J. Alterations in cholesterol metabolism–related genes in sporadic Alzheimer's disease. Neurobiol Aging 2018; 180: e1-180.e9.

[297]   Ungar L, Altmann A, Greicius MD, Apolipoprotein E. Apolipoprotein E, gender, and Alzheimer's disease: an overlooked, but potent and promising interaction. Brain Imaging Behav 2014; 8(2): 262-73.
[http://dx.doi.org/10.1007/s11682-013-9272-x] [PMID: 24293121]

[298]   McGuinness B, O'Hare J, Craig D, Bullock R, Malouf R, Passmore P. Cochrane review on 'Statins for the treatment of dementia'. Int J Geriatr Psychiatry 2013; 28(2): 119-26.
[http://dx.doi.org/10.1002/gps.3797] [PMID: 22473869]

[299]   Hirsch-Reinshagen V, Burgess BL, Wellington CL. Why lipids are important for Alzheimer disease? Mol Cell Biochem 2009; 326(1-2): 121-9.
[http://dx.doi.org/10.1007/s11010-008-0012-2] [PMID: 19116777]

[300]   Di Paolo G, Kim TW. Linking lipids to Alzheimer's disease: cholesterol and beyond. Nat Rev Neurosci 2011; 12(5): 284-96.
[http://dx.doi.org/10.1038/nrn3012] [PMID: 21448224]

[301]   Barres BA, Smith SJ. Neurobiology. Cholesterol--making or breaking the synapse. Science 2001; 294(5545): 1296-7.
[http://dx.doi.org/10.1126/science.1066724] [PMID: 11701918]

[302]   Nicholson AM, Ferreira A. Cholesterol and neuronal susceptibility to beta-amyloid toxicity. Cogn Sci (Hauppauge) 2010; 5(1): 35-56.
[PMID: 25339981]

[303]   Cartocci V, Servadio M, Trezza V, Pallottini V. Can Cholesterol Metabolism Modulation Affect Brain Function and Behavior? J Cell Physiol 2017; 232(2): 281-6.
[http://dx.doi.org/10.1002/jcp.25488] [PMID: 27414240]

[304]   Vance JE. Dysregulation of cholesterol balance in the brain: contribution to neurodegenerative diseases. Dis Model Mech 2012; 5(6): 746-55.

[http://dx.doi.org/10.1242/dmm.010124] [PMID: 23065638]

[305] Sato N, Morishita R. The roles of lipid and glucose metabolism in modulation of β-amyloid, tau, and neurodegeneration in the pathogenesis of Alzheimer disease. Front Aging Neurosci 2015; 7: 199.
[http://dx.doi.org/10.3389/fnagi.2015.00199] [PMID: 26557086]

[306] Kang J, Rivest S. Lipid metabolism and neuroinflammation in Alzheimer's disease: a role for liver X receptors. Endocr Rev 2012; 33(5): 715-46.
[http://dx.doi.org/10.1210/er.2011-1049] [PMID: 22766509]

[307] Wolozin B, Kellman W, Ruosseau P, Celesia GG, Siegel G. Decreased prevalence of Alzheimer disease associated with 3-hydroxy-3-methyglutaryl coenzyme A reductase inhibitors. Arch Neurol 2000; 57(10): 1439-43.
[http://dx.doi.org/10.1001/archneur.57.10.1439] [PMID: 11030795]

[308] Wolozin B. Cholesterol and the biology of Alzheimer's disease. Neuron 2004; 41(1): 7-10.
[http://dx.doi.org/10.1016/S0896-6273(03)00840-7] [PMID: 14715130]

[309] Fernandez ML. Rethinking dietary cholesterol. Curr Opin Clin Nutr Metab Care 2012; 15(2): 117-21.
[http://dx.doi.org/10.1097/MCO.0b013e32834d2259] [PMID: 22037012]

CHAPTER 3

Advances in Treatment of Mild Cognitive Impairment (MCI) and Dementia: A Review of Promising Non-pharmaceutical Modalities

Zahra Ayati[1,2], Dennis Chang[2] and James Lake[2,3,*]

[1] *Department of Traditional Pharmacy, School of Pharmacy, Mashhad University of Medical Sciences, Mashhad, Iran*

[2] *NICM Health Research Institute, Western Sydney University, Westmead, New South Wales, Australia*

[3] *Department of Psychiatry, University of Arizona College of Medicine, Tucson, Arizona, USA*

Abstract: Currently, available mainstream approaches used to treat mild cognitive impairment (MCI) and dementia are limited. In view of the high prevalence rate of Alzheimer's disease (AD) and other forms of dementia and the enormous social and financial burden associated with dementia on a global scale, developing more effective and more cost-effective ways to treat MCI and dementia is an urgent priority. This chapter reviews research findings on promising non-pharmaceutical approaches being investigated for their potential clinical applications in treating symptoms of cognitive impairment and behavioral dysregulation associated with dementia, reducing the risk of developing dementia, and slowing rate of progression of cognitive decline in individuals with Mild Cognitive Impairment (MCI) or dementia. Non-pharmaceutical treatment approaches covered include diet, exercise, single and compound herbal formulas used in Asian medicine, herbals used in Western countries, select other natural products including dehydroepiandrosterone, idebenone, acetyl-L-carnitine, alpha lipoic acid, phosphatidylserine and phosphatidic acid, select vitamins, nPUFAs and probiotics. Other non-pharmaceutical modalities reviewed include chelating agents; non-invasive brain stimulation techniques employing weak electrical current, sound and light; music therapy; cognitive training; electroencephalography (EEG) biofeedback; multi-modal interventions; Wander gardens; sensory stimulation interventions; massage, mindfulness, and energetic therapies (Healing Touch, Therapeutic Touch, taichi and qigong). Although select natural products are supported by compelling research evidence, most modalities reviewed in this chapter are supported by limited findings. Large prospective placebo-controlled studies are needed to further elucidate mechanisms of action, verify the efficacy of the various non-pharmaceutical modalities, and identify safe and appropriate treatment protocols for MCI and dementia.

*** Corresponding author James Lake:** Department of Psychiatry, University of Arizona College of Medicine, Tucson, Arizona, USA; E-mail: jameslakemd@gmail.com

Atta-ur-Rahman (Ed.)
All rights reserved-© 2020 Bentham Science Publishers

Keywords: Dementia, Healing Touch and Therapeutic Touch, Lifestyle Changes, Mild Cognitive Impairment (MCI), Multi-Modal Interventions, Natural Products, Non-Pharmaceutical Treatment, Taichi and Qigong.

PROMISING NON-PHARMACEUTICAL TREATMENTS OF MCI AND DEMENTIA

- Limitations of conventional mainstream treatments
- Reviews of non-pharmaceutical treatments
  - Lifestyle changes: diet, exercise and sleep
  - Natural Products
    - Herbals
      - Herbals used in Asian systems of medicine (single herbs, complex herbal formulas)
      - Other herbals (*Rhodiola rosea, Melissa officinalis, Salvia officinalis*)
  - Non-herbal natural products: acetyl-l-carnitine, folate, B-12, B-6, thiamin, niacin, vitamin C, vitamin E, DHEA and testosterone, estrogen, n-PUFA (omega-3s), phosphatidylserine and phosphatidic acid, CDP-choline, essential oils, probiotics)
    - Other non-pharmaceutical treatment modalities
  - non-invasive brain stimulation (TMS, tDCS, gamma-band stimulation, photobiomodulation)
  - Music therapy
  - Cognitive training
  - Bright light
  - EEG biofeedback
  - Sensory stimulation interventions
  - Massage
  - Wander gardens
  - Mindfulness
  - Multi-modal interventions
  - Energetic therapies
- Healing Touch and Therapeutic Touch
- Taichi and Qigong

LIMITATIONS OF CONVENTIONAL MAINSTREAM TREATMENTS

A systematic review found inconclusive evidence that non-steroidal anti-inflammatory agents (NSAID), antihypertensives, cholinesterase inhibitors, hormones and other medications are effective for preventing or reducing the risk of cognitive decline during healthy aging, developing mild cognitive impairment (MCI) or reducing the risk of developing Alzheimer's disease (AD) [1]. Current mainstream treatments of AD work by inhibiting the enzyme that breaks down acetylcholine, thus increasing available levels of the neurotransmitter that is critical for learning and memory. Although available drugs sometimes lessen the severity of cognitive decline and behavioral disturbances that accompany AD, they do not address its root causes. Only five drugs have been approved by the U.S. Food and Drug Administration (FDA) to treat Alzheimer's disease (AD) [2]. Among these the cholinesterase inhibitors include tacrine, donepezil, galantamine and rivastigmine (Alzheimer's Association. FDA-Approved Treatment for Alzheimer's) [2]. Another drug, memantine, works by antagonizing glutamate receptors and is in a class by itself. Second-generation acetylcholinesterase inhibitors (donepezil, rivastigmine, and galantamine) are no more effective than tacrine but require less frequent dosing.

All currently available pharmaceuticals have associated side effects that can be very distressing to individuals struggling with dementia, including vomiting, diarrhea and appetite loss. Tacrine, the first cholinesterase inhibitor, was removed from the market in 2013 because of concerns over hepatotoxicity.

Other drug classes that have been investigated for possible cognitive-enhancing benefits in AD include the monoamine oxidase inhibitors (MAOIs), estrogen replacement therapy (*i.e.*, in cognitively impaired postmenopausal women), naloxone, and different neuropeptides including vasopressin and somatostatin [3]. Promising novel treatments of Alzheimer's disease currently being investigated in clinical trials include a vaccine that may immunize individuals against formation of amyloid-beta, secretase inhibitors, anti-inflammatory agents, statins and a variety of nano-medicinal approaches [4 - 6]. Findings of studies on statins in dementia are inconsistent. However, a 2017 meta-analysis of 31 studies that met inclusion criteria for size and rigor found that regular statin use is associated with significant reduction of risk of developing dementia [7].

In addition to cognitive impairment, individuals struggling with MCI and dementia frequently experience depressed mood, anxiety, and psychosis. Conventional allopathic management of complex clinical presentations uses combinations of drugs that increase risk of unsafe interactions. Behavioral disturbances, including agitation and aggressive behavior toward caregivers, are commonly encountered in demented individuals. Although cholinesterase inhibitors result in only transient improvement in the early stages of dementia, they have become the standard conventional treatment of AD and other forms of dementia in industrialized countries because of findings of reduced agitation. In addition to pharmacological management, behavioral interventions, environmental enrichment, and social support may lessen the cognitive and behavioral symptoms of dementia.

The limitations of available pharmaceuticals comprise an urgent mandate to develop more effective treatments of AD and other forms of dementia.

NON-PHARMACEUTICAL TREATMENTS OF MILD COGNITIVE IMPAIRMENT [MCI] AND DEMENTIA

Lifestyle Changes

As this chapter was going to press, the findings of a landmark study were announced confirming that lifestyle significantly affects risk of developing dementia [8]. The 10-year retrospective cohort study followed almost 200,000 healthy adults (aged 60 and older) of European ancestry monitoring them for smoking, physical activity, diet, and alcohol consumption. Individuals at high genetic risk for dementia who adhered to a favorable lifestyle were significantly less likely to develop dementia compared to individuals with less favorable lifestyle habits.

Diet

Diet may be the most important preventable risk factor in AD. Foods that increase the risk of Alzheimer's include red meat, foods with high sugar content, and high-fat dairy products. Individuals who consume a high-fat, high-calorie diet are at significantly greater risk of developing AD compared with individuals who have moderate fat intake and restrict total calories. High red meat consumption also impacts health in general by increasing the risk of several types of cancer, diabetes, obesity, kidney disease and stroke. Foods known to reduce the risk of developing AD include vegetables, grains, fatty fish and fruits. Individuals who consume a traditional Mediterranean diet are at roughly one half the risk of developing Alzheimer's disease compared to individuals who consumer a high-fat high-calorie diet, and individuals in countries with very low meat consumption such as Japan, are at even lower risk of developing Alzheimer's disease [9].

Relationships between dietary preferences and alcohol consumption and dementia risk are complex and probably involve multiple mechanisms of action including oxidative stress caused by red meat, atherosclerosis caused by high cholesterol, and formation of damaging molecules caused by dysregulation of insulin secretion. High caloric intake and high fat intake promote formation of damaging free radicals that cause diffuse neuropathological changes in the brain, eventually manifesting as AD. A meta-analysis of findings from 18 community-wide studies concluded that the risk of AD increased linearly at a rate of 0.3% with every 100-calorie increase in daily intake [10]. The same meta-analysis showed that fish consumption was the only specific dietary factor associated with a measurable reduction in the risk of developing AD. While heavy chronic drinking significantly increases risk of vascular dementia (VaD) moderate alcohol consumption (2–4 glasses of wine per day) is associated with reduced risk of AD [11, 12].

Specific dietary interventions have been investigated for their possible beneficial effects on reducing AD risk. The most promising results have come from studies on the Mediterranean diet which emphasizes fish, fruits and vegetables and olive oil [13]. Preliminary findings suggest that the so-called MIND diet (Mediterranean-DASH Intervention for Neurodegenerative Delay) may reduce the risk of AD by up to 50% [14]. Recent research findings suggest that changes in the brain associated with increased risk of developing AD start many years before the onset of cognitive decline. Hence proactive dietary changes constitute an important strategy for delaying or preventing AD [15]. Findings of epidemiological studies suggest that moderate alcohol consumption and regular adherence to a Mediterranean diet high in polyunsaturated fats may reduce risk of both mild cognitive impairment (MCI) and AD [16]. However, a systematic review of 64 studies covering 141 dietary patterns [total N=132,491] did not find consistent relationships between AD risk, food measures and sample size [17]. Further studies are needed to determine whether and to what extent specific dietary patterns are significant risk or protective factors for AD.

Findings on studies on the relationship between high fish consumption in the diet and AD risk are inconsistent. It has been suggested that regular consumption of fatty fish [and other foods rich in omega-3s] reduces oxidative stress and associated atherosclerotic changes in the brain, indirectly lowering the risk of cognitive decline due to cerebrovascular disease. In contrast, high dietary intake of omega-6 polyunsaturated fatty acids, including linoleic acid, may contribute to increased oxidative stress in the brain, indirectly promoting atherosclerosis and thrombosis, eventually manifesting as declines in global cognitive functioning. A large epidemiologic study concluded that fish consumption 2 to 3 times weekly significantly reduces the risk of cognitive decline in elderly populations [18]. Cognitive impairment scores were analyzed for two groups of elderly men (ages 69–89) with different dietary preferences. In that study high fish consumption (containing large amounts of omega-3 fatty acids) was inversely correlated with overall cognitive impairment. In contrast, a preference for foods rich in linoleic acid (an omega-6 fatty acid) was associated with significantly higher rates of cognitive decline. Findings of a prospective cohort study [19] suggested that individuals who consume fish at least weekly have a 60% lower risk of developing AD compared with individuals who seldom eat fish. However, a similar study failed to show a correlation between fish consumption and the risk of developing AD [20]. Another cohort study found that enhanced cognitive performance in nonimpaired middle-aged individuals is correlated with high intake of fatty fish and other foods rich in omega-3 fatty acids [21]. Although the above findings suggest that fish consumption has beneficial protective effects, reduced risk of AD among individuals who regularly consume fish cannot be ascribed only to fish consumption because disparate lifestyle factors and educational level are correlated with healthy dietary preferences [22].

Exercise

There have been important advances in understanding causal mechanisms linking regular physical activity to reduced risk of MCI and dementia. In contrast, clinical trials have reported disparate findings which probably reflect differences in study populations and research designs (see below).

Physical exercise increases levels of brain-derived neurotrophic factor, probably enhancing neural plasticity and new synapse formation. Regular exercise is associated with increases in the relative size of the frontotemporal and parietal lobes, important centers for learning, memory, and executive functioning [23]. In older individuals the hippocampus loses volume at the rate of 1 to 2% annually reflecting neuronal cell loss resulting in age-related decline in memory and increased risk of dementia. Animal studies have established that exercise is associated with growth and development of neurons in the hippocampus and reduction in the rate of age-related loss of hippocampal neurons [24, 25]. Findings of human trials show a correlation between the level of physical activity and the volume of the hippocampus [26, 27]. A randomized controlled trial in healthy older adults (N=120) found that regular aerobic exercise over a 1-year period reversed age-related loss in the anterior hippocampus but *not* the posterior hippocampus. The anterior hippocampus is known to play a central role in spatial memory acquisition, and is subject to greater deterioration with normal aging than the posterior hippocampus. The finding of increased anterior hippocampal volume was associated with increased levels of brain derived neurotrophic factor (BDNF) an important mediator of neurogenesis and new synapse formation [28].

A recent systematic review and meta-analysis of human studies found a consistent relationship between aerobic exercise and age-related decreases in the volume of the hippocampus over time [29]. These findings constitute strong evidence that regular aerobic exercise helps prevent age-related neuronal loss in the hippocampus and slows the rate of cognitive decline that takes place with normal aging. Long-term prospective studies are needed to determine whether different types or durations of exercise have greater preventive effects on memory loss than others.

After decades of research on the relationship between exercise and dementia risk there is still no consensus on the optimal type, frequency or duration of exercise

[30]. While many studies have reported significant beneficial effects of regular exercise for both prevention and treatment of AD and other forms of dementia, other studies have reported equivocal findings. Combining a Mediterranean diet with regular exercise may reduce AD risk more than diet or exercise alone [31]. In one study, over 2000 physically nonimpaired men aged 71 to 93 years were followed with routine neurological assessments at 2-year intervals [34]. At the end of the study period, men who walked less than 0.25 miles daily had an almost twofold greater probability of being diagnosed with any category of dementia [including Alzheimer's disease and vascular dementia] compared with men who walked at least 2 miles each day. The authors accounted for potentially confounding factors other than the level of physical activity including the possibility that limited activity could be a result of early undiagnosed dementia. Findings of the Nurses' Health Study based on biannual mailed surveys over 10 years showed that elderly women ages 70 to 81 years who engaged in regular vigorous physical activity were significantly less likely to have been diagnosed with dementia compared with women with more sedentary lifestyles [35]. A randomized controlled trial showed that regular daily exercise in moderately demented individuals receiving in-home care reduces depressed mood, but it does not improve cognitive functioning [36].

Because studies on the effects of exercise and dementia risk use different exercise protocols and outcome measures in individuals with differing levels of severity and types of cognitive impairment, meta-analyses report disparate findings depending on kinds of studies examined. A 2019 systematic review found inconclusive evidence for preventive effects of regular exercise in non-demented individuals and in individuals already diagnosed with dementia [32]. In contrast, a separate systematic review of randomized controlled trials on exercise in dementia found evidence for beneficial effects on activities of daily living (ADL) but inconclusive evidence for cognitive enhancing effects of regular exercise [33]. Prospective long-term studies examining homogeneous populations using standard protocols are needed to clarify the role of exercise in reducing risk of AD and other forms of dementia and slowing down the rate of cognitive decline in individuals with MCI and early AD.

Sleep

Complex relationships exist between sleep disturbances and risk of developing both MCI and AD. There is not a direct relationship between total sleep duration and cognition [37], however fragmented sleep and prolonged periods of middle waking are associated with cognitive decline. Elderly individuals who report long wake intervals or a fragmented sleep pattern are at relatively greater risk of decline in global cognitive function including executive function [37, 38]. In an observational study, non-demented elderly adults (75 years and older) who reported chronic sleep problems were at 70 to 100% increased risk of developing all-cause dementia one decade later compared with elderly individuals with normal sleep patterns [39]. The authors identified a strong association between cognitive decline and a high rate of depressed mood and respiratory problems in subjects reporting sleep problems. In a large cohort study (N>17,000) older adults who reported frequent sleep problems, fatigue, or regular use of a hypnotic agent, were at significantly greater risk of dementia within 4 years compared to matched subjects who reported normal sleep [40].

A recent study using positron emission tomography (PET) imaging found that healthy elderly individuals who reported shorter sleep duration and poorer sleep quality had relatively greater uptake of amyloid-beta in the precuneus, a brain region where amyloid beta accumulates early in the course of AD [41, 42]. This finding suggests that fragmented sleep is causally related to a principle underlying mechanism in the pathogenesis of AD. There is emerging evidence for a relationship between genetic risk of developing AD and disordered sleep. Non-demented older adults with the APOE-4 genotype (a genotype known to increase risk of developing AD) who reported fragmented sleep were found to have a greater density of neurofibrillary tangles post-mortem [43].

Sleep apnea and other disorders that affect breathing during sleep [*i.e.* 'sleep disordered breathing'] are associated with increased risk of developing MCI or all-cause dementia 5 years later [44, 45]. For unclear reasons, the relationship between 'sleep disordered breathing' and risk of cognitive decline was statistically significant only in women, and stronger in individuals aged 50 to 59 than in younger or older individuals. Research findings on the relationship between disordered sleep and risk of cognitive decline are difficult to interpret because of methodological differences between studies, the fact that most studies are observational and do not allow inferences about causation. Also, most studies do not take into account confounding factors such as demographic variables, depressed mood [or other mental health problems] that may be related to both disordered sleep and cognitive function in complex ways.

Because most studies on sleep and dementia risk are cross-sectional or evaluate symptoms of cognitive decline at a single point in time it is impossible to make inferences about discrete causal relationships between sleep quality and risk of developing MCI or AD. Sleep and cognitive function are influenced by multifactorial relationships between genetic factors, demographic variables, medical and mental health problems, and lifestyle variables. Long-term prospective studies are needed to clarify these complex relationships and provide insights into practical strategies for preventing or reducing risk of MCI and AD.

Natural Products

Introduction

This section contains concise evidence reviews of natural products currently used to treat MCI and dementia. Despite decades of research and, in some cases, extensive research supporting efficacy and safety, the use of herbals and other natural products to treat cognitive disorders remains controversial. A systematic review of placebo-controlled studies on over-the-counter (OTC) supplements and herbals found insufficient evidence to recommend any supplement for adults with normal cognition or MCI [46]. However, accumulating research evidence supports that select nutraceuticals have beneficial effects on cognitive functioning in individuals with MCI and dementia. Research highlights are reviewed below.

Herbals

Herbal Medicines used in Asian Traditional Medicine Systems

Many herbals used in Chinese medicine, Ayurveda and traditional Persian medicine have been investigated alone and in different combinations for their potential beneficial effects on symptoms of AD and vascular dementia (VaD), including *Ginkgo biloba, Huperzia serrata, Curcuma longa, Panax ginseng, Panax notoginseng, Bacopa monnieri, Salvia miltiorrhiza, Crocus sativus,* and *Camellia sinensis.* Below is a concise review of studies on single herbs and complex herbal formulas being investigated for their potential cognitive enhancing benefits in healthy adults and individuals with dementia.

Single Herbs

Recent findings suggest that *G. biloba* extracts improve learning and memory in animal models of VaD [47 - 52]. Large placebo-controlled studies and meta-analyses of studies that meet strict inclusion criteria for rigor support that *G. biloba* extract slows the rate of decline in cognition, executive functioning and behavior in individuals diagnosed with AD and VaD [53 - 56]. Proposed mechanisms by which *G. biloba* enhances brain function resulting in improved memory and cognitive functioning include decreased activity of pro-inflammatory macrophages, improved blood flow, reduced activity of platelet activating factor (which reduces stroke risk), reduced corticosteroid production and increased glucose uptake, enhancing neural stem cell proliferation, accelerating synaptic plasticity following brain injury, reducing circulating free cholesterol, and reducing brain β-amyloid precursor protein production [57 - 60]. All practitioners should become familiar with important safety issues before recommending *G. biloba* to patients. Informed consent should be documented in the patient's chart at the time of the consultation when *G. biloba* is discussed. Because of its strong anti-platelet activating factor (PAF) profile, *G. biloba* extract increases the risk of bleeding, and concurrent use should be avoided in patients taking aspirin, warfarin, heparin, or other medications that interfere with platelet activity and increase bleeding time. Because of the risk of increased bleeding, *G. biloba* preparations should be discontinued at least 2 weeks prior to surgery. G. biloba preparations have been reported to result in the elevation of hepatic enzymes, and there are case reports of possible serotonin syndrome when ginkgo is taken with selective serotonin reuptake inhibitors (SSRIs). Mild transient adverse effects of *G. biloba* include upset stomach, dizziness, and headaches.

Curcuma longa [turmeric] has been used for centuries in Chinese, Hindu, and Ayurvedic medicine for centuries to treat medical disorders including pancreatitis, arthritis, cancer, and inflammatory, neurodegenerative, and digestive disorders. Animal and *in vitro* studies suggest that cognitive enhancing benefits of curcumin are based on multiple mechanisms of action including inhibition of lipid peroxidation, scavenging reactive oxygen species [ROS], and reactive nitrogen species, inhibition of NF-kB activation, and its anti-inflammatory actions [61, 62]. Curcumin may also directly bind small beta-amyloid species to block aggregation and the formation fibrillary tangles [63]. In a 24-month RCT 36 patients with mild to moderate AD randomized to curcumin (2 and 4gm/day) *vs* placebo experienced equivalent non-significant changes in cognition and memory [64]. These findings may be due in part to the low bioavailability of the curcumin preparation used in the study.

More recently a form of curcumin with greater bioavailability was investigated in a double-blind, placebo-controlled 18 month trial. Forty non-demented adults were randomized to receive curcumin (90 mg, twice per day) *versus* placebo. Individuals receiving curcumin reported significantly greater improvements in verbal memory, visual memory and attention compared to the placebo goup. In the curcumin group amyloid beta and tau accumulation were decreased in brain regions involved in mood regulation and memory [65].

Findings of animal studies suggest that bioactive constituents of *Panax ginseng* may improve cognition and memory in patients with dementia. Ginsenoside Rg5 reduces amyloid-β and cholinesterase activity [66], and ginsenoside Rg3 promotes β-amyloid peptide degradation *via* enhancing gene expression [67 - 70]. *Panax ginseng* may also decrease blood pressure and improves blood circulation by enhancing vasodilation [71]. Two open 12-week trials suggest that ginseng may improve cognition in individuals diagnosed with AD [72, 73]. In two recent small open trials individuals diagnosed with AD who received *P. ginseng* in doses of 4.5 and 9gm/day, experienced significant improvement in cognition and memory

[73, 74]. Findings of two small placebo-controlled trials suggest that *Panax notoginseng* improves cerebral blood flow and enhances memory in individuals diagnosed with VaD [75, 76]. Huperzine A is an alkaloid derivative of the herb *Huperzia serrata*, and is a central ingredient of many compound herbal formulas used in Chinese medicine to treat cognitive impairment related to normal aging. Huperzine A reversibly inhibits acetylcholinesterase, the enzyme that breaks down the neurotransmitter acetylcholine which plays a critical role in memory. Huperzine may also slow production of nitric oxide in the brain, reducing age-related neurotoxicity [77]. Findings from animal studies suggest that huperzine A may be a more potent and more specific inhibitor of acetylcholinesterase than some conventional cholinesterase inhibitors. Early controlled trials reported consistent beneficial effects in both age-related memory loss (*i.e.*, benign senescent forgetfulness) and Alzheimer's disease at doses between 0.2 and 0.8 milligrams per day [78, 79]. Huperzine A is more effective than piracetam for age-related memory loss [78]. In frequent adverse effects include transient dizziness, nausea, and diarrhea.

Two systematic reviews of placebo-controlled randomized trials on Huperzine A in patients diagnosed with AD found inconclusive evidence that Huperzine A ameliorates symptoms of memory loss and other cognitive problems. Methodological problems and small study size were cited as issues in both reviews [80, 81]. Although one review of RCTs on huperzine A in AD found evidence for improved cognitive function the authors reported inconclusive findings because of methodological flaws and heterogeneity in study designs [82]. Because most studies on Huperzine A are of short duration, 8 to 12 weeks, it is possible that observed beneficial effects might reflect temporary symptomatic improvement [83]. Huperzine A is generally well tolerated [84]. However, as most studies do not report adverse effects thus the reviewers could not comment on its safety profile.

Bacopa monnieri [Brahmi] has neuroprotective and antioxidant effects, works as a free radical scavenger and may increase cerebral blood flow [85 - 87]. The herbal is widely used in Ayurvedic medicine for memory problems [88]. Studies on cognitive enhancing benefits of the herbal in healthy adults and individuals diagnosed with AD are currently ongoing [89].

Saffron [*Crocus sativus*] is widely used in both traditional Persian medicine and Chinese medicine as antidepressant, antispasmodic, an anti-catarrhal and for enhancing and restoring memory. Extracts of saffron containing crocin that have antioxidant and antiplatelet properties [90, 91] have been shown to improve learning and memory in animal models of dementia [92, 93]. Findings of animal and human trials support that saffron is a promising therapy for both MDI and AD

[94 - 97]. A 1-year single-blind placebo-controlled clinical trial on saffron in MCI patients found general beneficial effects of daily saffron on cognitive functioning [98]. MRI images showed a small volume increase in the left inferior temporal gyrus in the saffron group and no changes in the placebo group. In a 16-week double-blind trial, AD patients who received saffron responded significantly better than the placebo group [99]. In a 22-week double-blind RCT AD patients randomized to saffron 30mg/day (15mg twice daily) and the cholinesterase inhibitor donepezil 10mg/day (5mg twice daily) showed improvements in cognition while saffron was better tolerated [100]. Significantly greater improvements in cognitive functioning had occurred in the saffron group at study end point. In a 1-year double-blind trial (N=68) individuals with moderate to severe AD randomized to a standardized saffron extract (30 mg/day) *versus* the NMDA antagonist memantine (20 mg/day) showed similar non-significant improvements in cognitive functioning by study end point [101].

Tea (*Camellia sinensis)* is widely consumed for health, contains epigallocatechin-3-gallate [EGCG], which has neuroprotective benefits mediated by anti-inflammatory effects, its role as a free-radical scavenger, and others [61, 102, 103]. Epigallocatechin gallate (EGCG), a naturally occurring molecule in tea, has attracted recent attention because it prevents neuronal cell death caused by amyloid beta neurotoxicity in cell cultures and transgenic mice [104, 105]. Individuals who frequently drink tea may have a reduced risk of developing AD [103]. Two prospective studies found that regular consumption of green tea in the elderly is associated with relatively lower risk of cognitive impairment and dementia [106, 107].

Findings of studies on single herbals in dementia are limited by small sample sizes of individual clinical trials, poor methodological quality, and short study duration. Further, plasma concentrations of bioactive constituents of many single herbs may be too low to have beneficial effects suggesting that observed improvements in cognition may be related to synergistic interactions between two or more bioactive constituents. Chinese medicine, Ayurveda and other Asian systems of medicine often employ combinations of herbals, possibly resulting in synergistic interactions between discrete bioactive constituents that may more effectively target diseases with complex etiologies such as AD and VaD [108, 109]. A novel research method called system-to-system analysis has recently been applied to the study of complex synergistic interactions in herbal formulas [109].

Complex Herbal Formulas

Only a few studies have been done on complex herbal formulas in VaD. While some have reported positive findings, the significance of findings is limited by small study size and methodological flaws [110, 111]. A 2012 systematic review of studies on complex herbal formulas in VaD reported that the majority of formulas examined resulted in significantly greater improvements in cognitive functioning compared to conventionally used medicines or placebo [112]. Four studies in which herbal medicines were combined with conventional medicines reported better cognitive functioning compared to conventional medicines alone, however the significance of these findings is limited by serious methodological flaws. A more recent meta-analysis [113] included 24 randomized clinical trials [all conducted in China] on individuals diagnosed with VaD. In a subgroup analyses, complex herbal interventions significantly enhanced cognitive function when compared to piracetam (in 10 studies) or placebos [in 3 studies]. Individuals receiving herbal medicine experienced greater improvements in activities of daily living (ADLs) compared to those treated with piracetam. However, as in the above studies, the significance of findings was limited by methodological flaws.

In response to the above challenges for over a decade a collaborative effort has been ongoing between the Academy of Chinese Medical Sciences and Western Sydney University to develop a standardized complex herbal formulation for the treatment of VaD. The formula, called SLT, contains standardized preparations of *Ginkgo biloba* (ginkgo), *Panax ginseng* (ginseng), and *Crocus sativus* (saffron) extracts.

The optimal ratio of bioactive constituents and the optimal dosage of SLT were determined through a series of animal studies. Preclinical trials demonstrated significant improvement in learning and memory, markers of neuropathology and antioxidant activity in animal models of dementia [67, 114 - 116]. At the time of writing two large phase III studies are under way to establish efficacy in individuals diagnosed with VaD. Cumulative findings from preclinical trials have demonstrated numerous cerebrovascular benefits of SLT including decreased areas of focal cerebral ischemia/reperfusion injury, decreased platelet aggregation, and increased free radical scavenging activity [117, 118].

Individuals treated with SLT or placebo have the same risk of adverse effects [119]. In a small 1-week RCT 16 healthy adults randomized to SLT experienced improvements in working memory [120]. In a small phase II study individuals diagnosed with probable VaD randomized to SLT showed significantly greater improvement in cognitive functioning, and a sub-set showed increased blood flow in brain regions associated with memory, auditory and speech processing [121, 122]. A second 12-month phase II study on 325 individuals with probable VaD found similar cognitive improvements [123]. This 59-week, phase II, randomized placebo-controlled double blind, parallel-arm study was performed at 16 academic centers throughout China. Participants were randomly assigned to four groups: group A, SLT 360 mg, and group B, 240 mg SLT, for 52 weeks; group C (C1 and C2), placebo for the first 26 weeks and switched to SLT 360 mg (C1) and 240 mg (C2), respectively, for the next 26 weeks. The findings suggest that SLT significantly improved cognition and daily functioning in patients with mild to moderate VaD following 6 months of treatment. Neither phase II study reported serious adverse effects related to SLT. At the time of writing two multi-center phase III trials are ongoing. Pending confirmation by phase III findings, SLT may emerge as an evidence-based herbal treatment of VaD, a neurodegenerative disorder for which there is presently no effective treatment.

Kami-untan-to

Kami-untan-to is a compound herbal formula consisting of 13 different herbs that is used in Japanese traditional healing (Kampo) to treat cognitive impairment and frank dementia, as well as other psychiatric symptoms. Animal studies suggest that KUT increases brain levels of both nerve growth factor (NGF) and choline acetyltransferase, the enzyme that makes acetylcholine [124, 125]. The putative mechanism of action of KUT is thus the converse of conventional cholinesterase inhibitors, which inhibit the enzyme that degrades acetylcholine in the synaptic cleft. KUT also protects against cognitive impairment due to thiamine deficiency in mice, suggesting possible beneficial effects in delirium tremens and other syndromes of cognitive impairment related to thiamine deficiency [126]. In a 12-month open trial, 20 moderately demented Alzheimer's patients treated with KUT alone and 7 treated with a combined regimen of vitamin E, estrogen, and a nonsteroidal anti-inflammatory drug deteriorated at a significantly slower rate compared with 32 moderately demented control patients who received no treatment. The beneficial effects of KUT were most notable 3 months into the study [127]. In a single-blind 12-week trial (N=38) AD patients were randomized to receive donepezil *vs* donepezil plus KUT [128]. Significant improvements in MMSE and ADAS-cog scores occurred in the combination group only suggesting that combined therapy may be more effective than a cholinesterase inhibitor alone.

Withania somnifera (Ashwagandha)

Ashwagandha (*Withania somnifera*) is widely used in traditional Ayurvedic medicine as a restorative and is being investigated in Western countries because of strong anecdotal evidence for its reported memory-enhancing, antiseizure, anticancer, and anti-inflammatory benefits. Mechanisms of action underlying beneficial cognitive effects include increased acetylcholine activity in the cerebral cortex and basal forebrain, inhibition of amyloid beta formation, clearance of amyloid beta in the brain, and attenuation of pro-inflammatory cytokines such as tumor necrosis factor (TNF) [129]. *In vitro* and animal studies point to another possible mechanism of action involving stimulation of dendritic and axonal sprouting in human nerve cell [130]. Case reports and human trials suggest that Ashwagandha extract at 50 mg/kg improves short- and long-term memory and executive functioning in cognitively impaired individuals [131]. In a double-blind placebo-controlled pilot study, individuals with MCI treated with a standardized Ashwagandha root extract 300mg twice daily had significantly greater improvement in executive function, sustained attention, and information-processing speed compared to the placebo group [132].

Select Korean, Chinese and Ayurvedic Herbals

Finding of animal studies suggest that dehydroevodiamine (DHED) a biologically active constituent of *Evodia rutaecarpa*, used in traditional Korean medicine, may be an effective treatment of age-related cognitive decline and AD [133, 134]. Qian Jin Yi Fang is a compound herbal formula used in Chinese medicine that reportedly reduces symptoms of dementia [135]. Trasina and Mentat are compound herbal formulas used in Ayurvedic medicine to treat symptoms that resemble Alzheimer's disease and age-related cognitive impairment [131, 136]. Animal studies suggest that Trasina has beneficial effects on experimental models of Alzheimer's disease [136]. Preparations of the herb lobeline have been studied as a memory-enhancing agent in rats with excellent results [137]. The above findings should be regarded as preliminary pending confirmation by large placebo-controlled human clinical trials are needed to determine whether the above herbals are safe and effective treatments of memory loss, AD or VaD.

Herbals used in other Systems of Medicine

Rhodiola rosea (Golden root)

Golden root was the object of intense research interest in the former Soviet Union because of its use as an adaptogen and performance enhancer in athletes, soldiers, and cosmonauts. In traditional Russian society, the herb is prepared as a tea, it is widely consumed and is believed to contribute to improved general health and longevity. Information on the diverse medical benefits of golden root has only recently been available in Western countries, but the herbal is already in widespread use in EU countries and North America. Different constituents of *R. rosea* including salidroside, rosavins and p-tyrosol probably contribute to its antioxidant, anti-inflammatory, anticancer, cardioprotective, and neuroprotective effects [138]. Mental health benefits are believed to be related to increased dopamine, serotonin, and norepinephrine [139], and include improved memory, increased mental stamina, and a general calming effect. Findings of open studies suggest that golden root at 500 mg improves overall mental performance and stamina in normal individuals [140], and may accelerate return to normal cognitive functioning following traumatic brain injury. Findings of animal studies support that *R. rosea* has cognitive enhancing effects however to date human clinical trials on *R. rosea* in dementia have not been conducted [141]. There are no reports of toxicities or serious drug–drug interactions; however, practitioners should advise patients diagnosed with bipolar disorder to avoid use of this herb because of reports of possible induction of mania [142].

Melissa officinalis (Lemon Balm) and Salvia officinalis (Common Sage)

Teas made from lemon balm and common sage have been used traditionally for centuries to enhance cognition. Animal studies have confirmed that components of both plant extracts have high binding affinities for brain acetylcholine receptors. In a 4-month randomized, double-blind study, mildly demented individuals received a standardized lemon balm extract (60 drops/day) *versus* placebo. By the end of the study, the group taking lemon balm extract showed significantly less decline in global cognitive functioning and less agitation compared with the control group [143]. In a separate 4-month study, 30 mildly demented patients were randomized to receive an extract of common sage (60 drops/day) *versus* placebo [144]. Findings were similar to those seen with lemon balm. No adverse effects were reported in either study. Large prospective studies are needed to confirm these preliminary findings and evaluate the potential therapeutic benefits of both herbal preparations in Alzheimer's disease and other neurodegenerative disorders.

Non-Herbal Natural Products

Acetyl-l-Carnitine

Acetyl-L-carnitine (ALC) occurs naturally in the brain and liver. Findings of studies suggest that ALC may stabilize nerve cell membranes, stimulate synthesis of acetylcholine, and increase the efficiency of mitochondrial energy production. A small 18-month open study on individuals with early mild AD found that the perfusion of the precuneus is increased in individuals who received ALC in doses of 1.5m/day and their cognitive and neuropsychiatric symptoms are not aggravated. The precuneus is involved in visuo-spatial imagery, retrieval of episodic memory and other higher cognitive processes [145]. The neuroprotective effects of ALC may be enhanced when combined with other natural products, including alpha-lipoic acid, CoQ10, or omega-3 fatty acids [146].

Findings of placebo-controlled studies suggest that ALC 1.5 to 3.0 g/day improves age-related symptoms of cognitive impairment in healthy nondemented elderly individuals and depressed elderly patients [147]. ALC is widely used to treat cognitive impairments in dementia and other neurodegenerative disorders, however findings of human clinical trials are inconsistent [148]. Double-blind, placebo-controlled studies support that ALC at 1500 to 3000 mg/day may improve overall performance on tests of reaction time, memory, and cognitive performance in demented patients and may slow down the overall rate of progression of cognitive impairment [149 - 151]. For unclear reasons, younger demented individuals (*i.e.*, individuals with presenile dementia) may benefit more than older individuals and 62 years may be the optimal age beyond which ALC loses much of its efficacy [148]. A systematic review of 11 double-blind, placebo-controlled studies of ALC in dementia found significant positive effects at weeks 12 and 24 that were not sustained [as measured by the Clinical Global Impression scale] with continued treatment at 1 year [152]. At high doses (2 g/day), ALC has been shown to improve memory, word recall, and visuospatial deficits in cognitively impaired abstinent alcoholics [153]. At doses commonly used to treat cognitive impairment ALC is well tolerated, and there are few reports of adverse effects.

Alpha Lipoic Acid (ALA)

Alpha lipoic acid (ALA) is a naturally occurring molecule in humans and many animals that plays essential roles in metabolism. ALA is found in red meat, broccoli, tomatoes, spinach and Brussels sprouts. It is widely used for weight loss and is prescribed in some European countries to treat diabetes, including a painful complication called diabetic neuropathy.

Findings from animal studies suggest that lipoic acid may slow down the rate of cognitive deterioration in the early stages of Alzheimer's disease. Beneficial effects are mediated by several mechanisms including increasing synthesis of acetylcholine, an important neurotransmitter required for the formation of new memories; increasing the activity of enzymes needed to synthesize glutathione, an important antioxidant; and savaging free radicals thus reducing inflammation in the brain that may increase the risk of developing Alzheimer's disease [154].

Few human clinical trials have been done on ALA in individuals diagnosed with Alzheimer's disease but most published findings are positive. In a small open pilot study 9 patients with probable AD were treated with lipoic acid 600mg daily for one year while taking cholinesterase inhibitors. All patients experienced stabilization in cognitive functioning which had previously been declining at a steady rate [155]. That open study was extended to 48 months and enlarged to include 43 patients diagnosed with mild Alzheimer's disease, all of whom experienced significant slowing in the rate of cognitive deterioration during the study period [156]. Although promising, the significance of the above findings is limited by the absence of a placebo control arm and small study size, hence they should be regarded as preliminary pending confirmation by a large double-blind placebo-controlled trial. ALA is generally safe even when taken at high doses (2 to 2 grams per day) and may cause mild nausea, itching or rash. Absorption is best when taken on an empty stomach, 30 to 60 minutes before eating or a few hours after eating.

Combinations of ALA and Other Nutraceuticals

In addition to studies on lipoic acid only, researchers have investigated nutraceutical formulas that include ALA in combination with other natural products known to have beneficial antioxidant or neuroprotective properties. As discussed in this chapter, epidemiologic findings support that individuals who regularly consume fruits and vegetables rich in polyphenols have a significantly reduced risk of developing AD [157]. Polyphenols are often recommended by health care providers because of their established protective effects against cancer and cardiovascular disease. Elderly individuals who frequently consume curry [which contains the antioxidant curcumin] or foods rich in the omega-3 fatty acid DHA, may have a significantly decreased risk of AD [19, 158, 159]. Curcumin is a more potent free radical scavenger than vitamin E and may inhibit build-up of amyloid beta plaque, one of the main causes of AD [63, 160]. Mice that were genetically modified (*i.e., transgenic mice*) to develop AD were found to have significantly reduced amyloid levels after 6 months on a high curcumin diet [63].

As a variety of nutraceuticals may prevent or delay progression of neuropathological changes that cause AD *via* different mechanisms, combining different substances into a single formula might be more effective than any single nutraceutical. Findings of a recent animal study bear this out. Transgenic mice predisposed to develop AD treated with a combination of ALA, curcumin and the omega-3 fatty acid DHA were found to have greater anti-inflammatory and neuroprotective effects compared to mice receiving only one nutraceutical [161]. Placebo-controlled trials in humans diagnosed with AD are needed to determine whether combining multiple nutraceuticals is more effective than single nutraceuticals, and whether particular combinations and dosages of nutraceuticals result in optimal effects on cognitive functioning. This is the domain of multi-modal interventions (see below).

Folic Acid, B-12, B-6 and Thiamine

Many B vitamins are essential enzyme cofactors in the synthesis of neurotransmitters. Chronic dietary deficiencies in folic acid, B-12 and thiamine result in abnormal low serum levels and are associated with increased risk of cognitive impairment [162]. A diet low in folic acid and vitamin B_{12} leads to elevated blood levels of homocysteine and decreased synthesis of *S*-adenosylmethionine (SAMe), resulting in reduced synthesis of several neurotransmitters essential for normal cognitive functioning. Although folate, B-12 and B-6 have been shown to reduce serum levels of homocysteine this does not result in improved cognitive functioning [163].

Folate plays a central role in DNA methylation possibly mediating epigenetic changes that result in AD and other neurodegenerative disorders [164] however the relationships between cognitive functioning and folate, B-12, B-6 and thiamin have not been clearly elucidated. Vitamin B_{12} is frequently recommended to elderly patients who complain of impaired cognition however research findings are inconclusive. In a large cohort study, there was no association between intake of vitamin B-12 or B-6 and AD risk [165]. A systematic review of four studies concluded that there is insufficient evidence to support the use of folic acid with or without B_{12} as a treatment of dementia or other forms of severe cognitive impairment [166].

Thiamin-dependent enzymes play a central role in the brain's energy metabolism and their functioning is often impaired in individuals with AD [167] however findings of studies on thiamin supplementation in AD are inconsistent [168]. In a small 12-week open trial daily supplementation with thiamine 100 mg/day resulted in mild improvement in cognitive impairment in individuals with mild AD [169]. In placebo-controlled study individuals treated with high doses of oral thiamin between 3 to 8 g/day showed mild improvement in symptoms of AD [170].

A systematic review of randomized controlled trials on B vitamin supplementation in individuals with MCI or AD found inconclusive evidence for folate alone or in combination with other B vitamins [171]. Although individuals with AD are frequently advised to take B vitamins together with their prescribed medications, a placebo-controlled double-blind study found no evidence that daily augmentation with folate, B-12 and B-6 enhances the efficacy of cholinesterase inhibitors in mild to moderate AD [172].

Niacin

Niacin (especially in the form of nicotinamide) mediates different biological processes including energy metabolism, calcium homeostasis and mitochondrial function that may be involved in the pathogenesis of AD [173]. Maintaining adequate dietary niacin or taking niacin as a supplement would be expected to have protective benefits however, research findings are limited. In a 6-year prospective study, elderly individuals who consumed the highest amount of niacin in the form of supplements or food had a 70% reduced risk of developing

Alzheimer's disease (*i.e.*, during the study period) compared with individuals who used the least amount of niacin [174]. Cognitively intact individuals who consumed the highest amount of niacin experienced less than half of the average cognitive decline observed in the group with the lowest niacin intake. There was no difference in protective benefits of niacin obtained through diet or supplements. Large prospective placebo-controlled studies are needed to confirm a causal relationship between high niacin intake and a reduced risk of developing AD or age-related cognitive decline.

Vitamins C and E

Vitamins C (ascorbic acid) and E are important antioxidants and function as free-radical scavengers throughout the body and brain, possibly slowing progression of Alzheimer's disease and other neurodegenerative diseases. Ascorbic acid plays important protective roles that may reduce AD risk including suppression of neuroinflammation, suppression of pathological changes that lead to formation of amyloid-beta, and its role as a chelating agent of iron, copper and zinc. However, findings of human clinical trials on vitamin C supplementation in MCI and AD are inconsistent [175]. The findings of a large epidemiologic study showed a correlation between intake of vitamins C and E in the form of supplements and reduced risk of Alzheimer's disease [176]. This effect was greatest for vitamin E. Anecdotal reports suggest that supplementation with vitamins C and E improves cognitive functioning in AD patients.

A systematic review of placebo-controlled studies on vitamin E in dementia found no evidence that vitamin E (*i.e.*, in the form of alpha-tocopherol) prevents progression from MCI to dementia or improves cognitive function in people with MCI or AD [177]. Combining a cholinesterase inhibitor with vitamin E may delay the rate of cognitive decline by as much as 20% per year [178].

Combining vitamin E and vitamin C may reduce the risk of developing AD. At the end of a 5-year study on healthy elderly adults (N=4740) a strong inverse correlation was found between the incidence and prevalence of AD and combined use of vitamin C [at least 500 mg/day] and vitamin E (at least 400 IU/day) [179]. There was no association between the use of vitamin C alone, vitamin E alone, or a multivitamin alone and AD risk. The significance of these findings is limited by the fact that almost 1500 individuals in the study were lost to 5-year follow-up, suggesting that the incidence of new cases of Alzheimer's disease could have been higher than reported.

Large doses of vitamin E are associated with an increased risk of bleeding. Individuals who are at increased risk of stroke should consult their physician before starting a high-dose vitamin E regimen.

DHEA and Testosterone

Dehydroepiandrosterone (DHEA) is a precursor of testosterone and other androgenic hormones and has many important roles in physiology including modulation of cognition, appetite and sexual behavior. Serum levels of DHEA and its sulfate form DHEA-S gradually decline with aging and reach their lowest levels in the 7th decade. Both DHEA and DHEA-S have beneficial effects on the immune and the cardiovascular systems. Both prohormones have neuroprotective, antioxidant and anti-inflammatory effects in the brain, promote neurogenesis and increase the synthesis and release of the catecholamine neurotransmitters norepinephrine and dopamine [180]. DHEA is widely used in Europe and North America to self-treat decline in cognitive functioning associated with normal aging.

DHEA binds to both γ-aminobutyric acid (GABA) receptors and *N*-methyl-D-asparate (NMDA) receptors, but it is not clear whether these receptor affinities are related to its cognitive-enhancing effects [181]. Research findings on the effects of DHEA on memory in healthy and cognitively impaired adults are inconsistent [182]. DHEA may improve memory in elderly patients who have low DHEA serum levels more than in younger adults with normal DHEA levels. A systematic review and meta-analysis found no support for DHEA as a cognitive enhancer in healthy older individuals [183]. The significance of findings was limited by the fact that only three studies met inclusion criteria, all studies reviewed were 3 months in duration or shorter, and doses tested ranged between 25 and 50 mg only. A systematic review and meta-analysis found that individuals with AD had consistently low serum DHEA-S levels but normal DHEA levels [184]. Further studies are needed to determine whether DHEA-S levels provide a reliable diagnostic tool of AD risk.

In a 6-month randomized placebo-controlled study unmedicated male patients with AD were randomized to DHEA alone (50 mg twice daily) *vs* placebo. At 3 months the DHEA group showed a trend toward superior cognitive performance compared to the placebo group which persisted until the end of the study [185]. In a small 4-week open study, 7 individuals with multi-infarct dementia who received daily intravenous administration of 200 mg DHEA-S exhibited significant increases in serum and CSF levels of DHEA-S, improvements in activities of daily living and less frequent emotional disturbances [186]. Further studies are needed to investigate possible cognitive enhancing effects of DHEA in AD and multi-infarct dementia and determine safe, optimal dosing strategies. Mild insomnia is an infrequent side effect of DHEA, which should be dosed in the morning. It is prudent to avoid DHEA when there is a history of benign prostatic hypertrophy or prostate cancer.

Lower circulating testosterone levels in older men are associated with increased risk of developing AD [187, 188]. In a 6-month double-blind, placebo-controlled study, 16 males with AD and 22 healthy adult males were randomized to receive testosterone 75 mg (in the form of a dermal gel) *versus* placebo together with their usual medications [189]. Global quality of life improved in both the demented group and healthy controls. No significant group differences were found at the end of the study however mildly demented patients who received testosterone experienced less decline in visuospatial abilities.

Estrogen

The use of hormones to treat mental health problems remains controversial because of associated significant medical risks. There is some evidence that estrogen replacement therapy reduces the rate of progression of Alzheimer's disease in postmenopausal women by interfering with synthesis of a protein called amyloid-precursor protein (APP) [190]. However, a meta-analysis of all published RCTs (total N=351) failed to show that hormonal replacement slows the rate of progression of dementia in postmenopausal women diagnosed with AD or other forms of dementia [191]. A separate meta-analysis of 9 trials concluded that there is little evidence for beneficial effects of hormone replacement therapy (HRT) with estrogens or combined estrogens and progestagens in the prevention of cognitive decline in postmenopausal women [192]. HRT significantly increases the risk of breast cancer, heart disease, and stroke [193].

Polyunsaturated Fatty Acids [PUFA]

Omega-3 polyunsaturated fatty acids [*i.e.,* eicosapentaenoic acid (EPA) and docosahexaenoic acid (DHA)] from certain fishes and plants are a widely used non-pharmacological treatment of cognitive impairment and dementia. Findings of animal studies suggest that DHA may help prevent AD by reducing amyloid beta accumulation thus decreasing the long-term neurotoxic effects of amyloid beta [194, 195]. However, findings of studies on omega-3s for prevention or treatment of dementia are inconsistent. In one systematic review 7 observational studies reported positive findings but no randomized placebo-controlled trials reported effects greater than placebo [196]. A meta-analysis of 21 cohort studies on fish intake and AD risk found an association between higher consumption of marine-derived DHA and reduced risk of cognitive impairment and AD, but did not find a linear dose-response relationship between fish consumption and AD risk [197]. A systematic review of placebo-controlled studies on n-3 PUFAs in AD identified only 3 studies that met inclusion criteria (total N=632) and found no convincing evidence for the efficacy of omega-3 PUFA supplements in the treatment of mild to moderate AD [198].

Phosphatidylserine and Phosphatidic Acid

The phospholipids phosphatidylserine (PS) and phosphatidic acid (PA) occur naturally in the brain and are essential to healthy neuronal functioning. Both molecules keep neuronal cell membranes flexible, decrease oxidative stress in the brain, stimulate neurotransmitter release and increase brain glucose metabolism [199]. Phosphatidylserine is one of the most important phospholipids in the brain and is an essential component of nerve cell membranes. Commercial phosphatidylserine products are derived from soy lecithin or bovine brains and are usually dosed at 300 mg/day. Findings of open-label and placebo-controlled trials on phosphatidylserine 300 mg/day (100 mg three times daily) in patients with mild cognitive impairment or very early mild dementia report significant improvements in learning, memory, visual learning, attention and socialization [200]. Only non-bovine brain-derived PS and PA have been approved as dietary supplements Because of safety concerns regarding bovine spongiform encephalopathy.

CDP-choline

Phosphatidylcholine citicoline (cytidine 5'-diphosphocholine or CDP-choline) is a naturally occurring molecule in all mammals and is a necessary precursor for synthesis of phosphatidylcholine and other phospholipids that comprise neuronal membranes. Like acetyl-L-carnitine, CDP-choline increases mitochondrial energy production and is widely used to treat cognitive impairments resulting from acute brain injury or neurodegenerative diseases [201]. Systematic reviews of placebo-controlled trials support that CDP-choline has positive effects on the rate of recovery in post-stroke patients, as well as in elderly individuals who are cognitively impaired due to cerebrovascular disease [202 - 205]. An interesting finding was the absence of beneficial effects on attention in spite of significant improvements in global functioning and memory. There is preliminary but promising evidence of a beneficial effect following traumatic brain injury [206]. Uncommon adverse effects of CDP-choline include headache, vertigo, and changes in blood pressure, nausea, vomiting and rash.

Essential Oils

Accumulating findings from *in vitro*, animal and human trials supports that beneficial effects of essential oils on AD may be mediated by anti-amyloid, antioxidant, anticholinesterase, and memory-enhancement activity. In addition to their effects on cognition, essential oils when used as aromatherapy or applied directly to the skin may have beneficial calming effects on agitation in individuals with AD or other forms of dementia. The findings of a small 4-week study of aromatherapy in individuals with AD suggest that certain essential oils experience improvements in cognitive function [207].

The essential oils of lemon balm and lavender reduced agitated behavior in demented individuals when topically applied directly to the face and arms [208, 209]. A systematic review identified only one study that met inclusion criteria for quality and size [210]. Although the outcome of that study was positive the reviewers concluded that methodological problems limited the significance of findings. A recent narrative review of all *in vitro*, animal and human studies on essential oils in AD published between 1998 and 2018 identified 55 essential oils used as interventions in AD studies and provided detailed reviews of findings on 28 essential oils [211]. It is likely that essential oil therapies in agitated demented patients do not act primarily through the sense of smell, as olfaction is frequently impaired in severely demented individuals. In fact, a recent controlled study determined that aromatherapy alone in the absence of massage using essential oils was no more effective than placebo [unscented grape seed oil] in agitated demented patients [212]. This finding supports the hypothesis that essential oils are effective in agitated demented patients through systemic absorption or the combined effects of massage and olfaction, but not through olfaction alone.

Uncommon adverse effects of essential oils include skin allergies, phototoxic reactions, and potentiation of sedative-hypnotics when used with lavender or other oils known to have sedating effects. Pregnant women should exercise caution when considering aromatherapy because of possible effects on the fetus and uterus caused by systemic absorption of certain essential oils.

Chelating Agents

Abnormal high levels of zinc, copper and iron are frequently present in amyloid plaques in postmortem AD brains. According to the 'metal-based neurodegeneration hypothesis' abnormal high levels of one or more metallic ions can result in metal ion dyshomeostasis in the brain leading to the formation of reactive ion species (*i.e.* 'free radicals') which in turn cause errors in the normal molecular folding process involved in synthesis of proteins that play key roles in brain function [213]. Errors in protein folding result in formation of amyloid beta (Aβ) and other pathological products that manifest as neuropsychiatric disorders. This pathological process has been implicated in the pathogenesis of different neurodegenerative diseases including AD, Parkinson's disease, Huntington's disease, multiple sclerosis and others.

There is evidence that chelating agents that selectively bind to iron, copper and zinc reducing their CNS levels may result in a reversal of amyloid-beta plaque deposition in the brain [214]. The rationale for this approach is consistent with the hypothesis that historically recent increases in the prevalence of AD—at least in industrialized countries—is related to the ingestion of large amounts of inorganic copper in drinking water (*i.e.* from copper plumbing) and copper-containing supplements [215, 216]. Paradoxically, although abnormal high CNS levels of zinc may contribute to the pathogenesis of AD (above), supplementation with oral zinc may reduce the body's copper load, mitigating the impact of toxic copper ions on the brain. In a 6-month double blind placebo-controlled study, individuals diagnosed with AD who received a daily zinc supplement showed less rapid cognitive decline compared to a matched placebo group. A significant finding of the study was lowering of serum free copper levels in the group treated with zinc [214]. It has been suggested that beneficial effects of zinc on cognition in AD may be mediated both indirectly by lowering copper toxicity, directly through neuroprotective effects, or *via* both mechanisms [217].

An early placebo-controlled study [N=48] found that AD patients treated with iron chelation therapy (desferrioxamine 125mg IM twice daily five times weekly) for two years experienced significantly less decline in activities of daily living (ADL) compared to a control group [218]. At the time of writing only one metal binding agent, PBT which binds to excess copper, zinc and iron in the brain, has been investigated as a possible treatment of AD and other neurodegenerative disorders. A systematic review identified only two studies on PBT that met inclusion criteria and found equivocal evidence for efficacy against AD [219].

Probiotics

The gut-brain 'axis' is a complex system that includes complex neural, immune, endocrine and metabolic pathways. It has been hypothesized that dysregulation of the gut-brain axis or alternations in gut microbiota may contribute to pathogenesis of disorders of the central nervous system such as AD and other degenerative neurologic disorders [220]. Dysbiosis of the gut microbiota may result in increased permeability of both the intestinal lining and the blood-brain barrier leading to production of proinflammatory cytokines implicated in the pathogenesis of AD. Findings of early studies on probiotics aimed at treating cognitive impairment in AD by correcting gut dysbiosis are promising.

Findings of animal studies suggest that supplementation with probiotics has antidepressant and cognitive-enhancing effects [221, 222]. In a 12-week placebo-controlled trial (N=60) individuals with AD were randomized to receive either milk or milk containing probiotics (200 ml/day) containing *Lactobacillus acidophilus, Lactobacillus casei, Bifidobacterium bifidum, and Lactobacillus fermentum* [223]. At study endpoint individuals receiving probiotics experienced significantly greater improvement in global cognitive functioning (based on MMSE scores) compared to no improvement or cognitive decline in the control group. Individuals receiving probiotics showed improvements in some inflammatory biomarkers including C reactive protein. Further studies using more sensitive neuropsychological tests for dementia, are needed to confirm these findings, investigate therapeutic benefits of different probiotic formulas and determine optimal dosing regimens.

OTHER NON-PHARMACEUTICAL TREATMENT MODALITIES

Noninvasive Brain Stimulation

Non-invasive brain stimulation techniques such as transcranial magnetic stimulation (TMS), transcranial direct current stimulation (tDCS), gamma-band neurostimulation and near infrared transcranial stimulation are being investigated as alternatives to pharmacological interventions for neurologic and psychiatric disorders. Research on potential benefits of transcranial ultrasound stimulation, transcranial alternating current and gamma band stimulation in AD is at an early stage [224]. A meta-analysis of three studies on tDCS in dementia found evidence of significant but transient improvements in word recall, face recognition, and motivation immediately following treatment [225]. A review of studies on noninvasive brain stimulation found promising but preliminary evidence for therapeutic effects of both TMS and tDCS based largely on small 'proof of concept' studies [226]. Large prospective sham-controlled studies are needed to confirm efficacy, clarify which domains of cognitive functioning benefit most from noninvasive brain stimulation, and determine optimal protocols. Emerging findings from animal and early human trials suggest that brain stimulation using auditory or visual stimuli to modulate the gamma frequency band so-called 'gamma-band stimulation' (*i.e.,* brain electrical activity between 30 and 110Hz)—may be associated with improved memory and cognitive functioning [227].

Photobiomodulation is a noninvasive technique that uses near infrared light (1072 nm) to stimulate the brain on the hypothesis that photonic stimulation may increase ATP production in mitochondria and proteasomes within neurons that have been damaged by amyloid beta, resulting in improvements in mood, behavior and cognitive function in individuals with AD. Findings of a small double blind pilot study (N=11) individuals diagnosed with dementia who received 28 consecutive six-minute transcranial infrared light (1060 to 1080 nm wavelength) treatments delivered transcranially *via* light emitting diodes showed a non-significant trend toward improved measures of immediate recall, executive functioning, visual attention and task switching, and improved EEG amplitude and measures of functional brain connectivity [228].

Findings of enhanced cognitive functioning achieved through global or regional brain stimulation is consistent with the hypothesis that at least some forms of dementia are caused by dysregulation of mitochondria within neurons or disruption of communication between brain networks, and that regular exposure to particular frequencies and patterns of electricity, sound or light may have beneficial effects on ATP production or dynamic brain organization globally.

Music Therapy

Music is used in many healing traditions to quiet the mind and reduce agitated behavior. Calming background music has been shown to significantly reduce irritable behavior, anxiety, and depressed mood in demented nursing home patients [229]. Regular music therapy reduces irritability and improves expressive language in demented individuals [230]. Improvements in global behavior and sleep following music therapy may be mediated by increased melatonin levels [231]. An early meta-analysis of studies on music therapy in dementia found that different approaches—singing, dance, listening to music, and musical games—are associated with improvements in cognitive and behavioral functioning in severely demented individuals, including reduced agitation, reduced wandering, enhanced social interaction, improved mood, reduced irritability and anxiety, increased cooperative behavior, and increased performance on standardized scales such as the Mini-Mental State Examination [232]. In this study the significance of findings is limited by the absence of quantitative outcomes measures in many studies, failure to specify end points before trials began, small study sizes, and the absence of randomization or blinding in most studies. Systematic reviews of studies on music in elderly patients with dementia report consistent evidence that music reduces disruptive behavior, decreases anxiety, and may have beneficial effects on cognitive function, mood and quality of life [163, 233, 234].

Cognitive Training

Cognitive training is probably efficacious for preserving cognitive functioning in healthy older individuals, and may reduce the rate of progression of age-related cognitive decline leading to AD, however findings are inconsistent. Differences in outcomes following long-term memory training in the elderly may reflect differences in protocols used, differences in motivation of subjects enrolled [235, 236], and incipient age-related cognitive decline or early AD in some study populations [239].

The findings of a large-scale multi-site 10-year study called The Advanced Cognitive Training for Independent and Vital Elderly (ACTIVE) study demonstrated that cognitive training improves cognitive function in community-dwelling non-demented older adults up to 5 years and that enhanced cognitive function transfers to improvement in activities of daily living (ADL) [239]. The study sample consisted of 2,832 individuals [average age 73.6 years, average education 13 years] randomly assigned to one of three intervention groups (memory, reasoning, or speed-of-processing training) or a no-contact control group. Outcome assessments were conducted at study onset and 1, 2, 3, 5, and 10 years after the intervention. Training was conducted in small groups in ten 60-70 minute sessions over 5-6 weeks. Memory training was aimed at improving verbal episodic memory. Reasoning training focused on improving problem solving ability, and speed-of-processing training focused on training subjects how to process increasingly more complex information in successively shorter times. Four 75-minute booster sessions were provided at 11 and 35 months after training to a random subset of participants in each training group who completed at least 8 of 10 training sessions. At 10 years subjects in all three groups reported significant improvements in all three areas of cognitive function and had fewer difficulties performing ADLs compared to the control group. Improvements in cognitive function and ADLs gradually declined starting 2 years after training, but persisted to at least 5 years for memory training and to 10 years for reasoning and speed-of-processing training. At the 10-year mark 60 to 70% of participants in all training groups were functioning at or above their cognitive baseline when training started. The authors suggest that long-term stability of ADLs over 10 years, in contrast to gradually declining cognitive function after only 2 years, may reflect stable changes in behavior and social interaction that may be related to patterns of sustained neural activation achieved through cognitive training. These findings suggest that if done early, cognitive training interventions targeting multiple cognitive abilities are more likely to preserve IADLs and may significantly delay onset of functional impairment leading to MCI and AD.

Accumulating findings support that cognitive training is also efficacious for improving memory and cognitive functioning in individuals with MCI and early AD. A systematic review of studies on computerized cognitive training (CCT) in individuals with MCI found evidence for small to moderate improvements in global cognition, attention, working memory, learning and memory but not in non-verbal memory or psychosocial functioning [237]. A meta-analysis of findings of placebo-controlled trials on 4 types of interventions (physical exercise, music therapy, computerized cognitive training and nutrition therapy) for MCI and AD found that exercise and CCT were efficacious for improving cognitive and neuropsychiatric symptoms associated with both conditions, and that such non-pharmaceutical interventions might be more efficacious than available pharmacologic agents [238].

Bright Light Exposure

A significant percentage of individuals with AD have sleep disorders and the prevalence and severity of insomnia increase with AD severity [240]. Early findings suggested that bright light exposure improves sleep and reduces sundowning in AD patients [241]. In a controlled study, 92 severely demented institutionalized patients exposed to daily morning bright light experienced a delay of 1.5 hours in the peak time of agitated behavior, but significant reductions in overall agitated behavior were not observed [242]. In a small 7-day double-blind, placebo-controlled trial, six demented residential patients were randomly given melatonin 2.5 mg *versus* placebo and exposed to bright light [10,000 lux] every night at 10:00 P.M. for 30 minutes. Each patient served as his or her own control on random nights [243]. Patients taking a placebo during bright light exposure displayed significantly reduced motor restlessness, but no improvements in inappropriate behaviors including aggression, confusion or repetitive behavior. No significant changes in behavior or restlessness were observed in patients who took melatonin while exposed to bright light. Results of a meta-analysis of studies on bright light therapy in the management of sleep, behavior, and mood in demented individuals were inconclusive [244]. A more recent systematic review that included 32 studies on bright light therapy used as an intervention for insomnia, disruptive behavior or cognition in AD found mixed findings for therapeutic effects of bright light in this population [245]. In sum, bright light exposure may be beneficial for symptoms of restlessness and may delay the peak time of agitated behavior, however it probably does not significantly reduce the severity of agitation or other inappropriate behaviors or improve cognitive function in individuals with AD.

EEG-Biofeedback

EEG-biofeedback training—sometimes called neurotherapy or neurofeedback—employs light or sound as feedback to entrain desirable changes in brain electrical activity associated with improved cognitive functioning. Recent advances in brain-computer interface (BCI) and EEG biofeedback training support that both interventions have general beneficial effects on symptoms of inattention, memory loss, and other areas of cognitive functioning in healthy adults [246]. EEG-biofeedback training is sometimes done following initial assessment of brain electrical activity using a qEEG "map" [see discussion this chapter], but it is more often done without the benefit of an initial "Q." The findings of a case study suggest that EEG-biofeedback training improves the speed of cognitive rehabilitation following stroke, including enhanced word finding, attention, speech, and concentration [247], however a review of neurofeedback in stroke rehabilitation reported largely inconclusive findings [248]. EEG-biofeedback training has been established to improve memory following traumatic brain injury [249]. In a small open study 10 individuals diagnosed with AD who received regular neurofeedback training experienced stable or improved memory and other measures of cognitive functioning. Biofeedback training directed at improving coherence in heart rate variability (HRV) may improve performance in concentration tasks in healthy adults [250]. Target symptoms of both EEG and HRV biofeedback include inattention, global decline in cognitive functioning following stroke or head injury, anxiety states, and mood. Large long-term sham-controlled studies are needed to confirm these findings and identify optimal neurofeedback protocols for enhancing cognitive functioning in individuals with AD or other disorders associated with cognitive impairment.

Sensory Stimulation Interventions

Snoezelen is an integrative treatment of dementia and other forms of severe cognitive impairment in which music, light patterns, tactile surfaces, essential oils, and other forms of sensory stimulation are employed to stimulate sight, hearing, touch, taste, and smell. This multimodal therapy is based on the belief that stimulation of multiple senses improves global cognitive functioning resulting in adaptive behavioral changes. Snoezelen sessions are usually administered 2 or 3 times weekly to demented individuals in skilled nursing homes. A systematic review of controlled double-blind studies suggests that apathy, speech skills, and psychomotor agitation improve in demented patients after four to eight sessions of Snoezelen therapy [251]. The significance of these findings is limited by the fact that only two studies met inclusion criteria and those studies were difficult to compare because of differences in methodology and non-standardized control conditions. In a more recent review 3 out of 4 studies reported significant beneficial effects of Snoezelen including reductions in apathetic, rebellious, aggressive and depressive behaviors, and improvements in adapted behavior and well-being [252].

Massage Therapy

Skilled nursing homes frequently provide regular massage therapy to demented residents. Findings of open trials suggest that massage decreases agitated behavior in demented individuals [253, 254]. Tactile massage, which consisted of hand massage, showed a significant reduction in aggressiveness and stress level for the experimental group [254]. Findings of a randomized controlled trial comparing regular foot massage to 'quiet presence' of another person without massage found that both conditions promoted increased relaxation and associated improvements in blood pressure and heart rate [255].

Wander Gardens

Regularly spending time in nature is beneficial for physical and psychological well-being in healthy individuals and is probably therapeutic in AD and individuals recovering from a stroke. A wander garden is a protected outdoor area in which individuals can safely walk, socialize, or explore nature. Wander gardens are increasingly used to provide visual, auditory, and tactile stimulation as well as pleasure and autonomy for institutionalized demented patients. There is evidence that nature-related activities, including strolling in a park or gardening, can improve baseline functioning in the early stages of dementia and provide a source of pleasure and relaxation in the later stages of dementia [256, 257]. Findings of a case report suggest that time spent in a wander garden reduces anxiety in post-stroke patients [258]. A review of studies on sensory gardens and horticultural activities in individuals with AD found that both interventions improve well-being, reduce disruptive behavior, improve sleep, and may reduce use of psychotropic drugs, and decrease fall risk [259]. A review of studies on the effects of gardens on cognition and behavior in individuals with dementia in residential care facilities found that benefits of gardens are achieved through reminiscence and regular sensory stimulation [260]. The reviewers remarked that gardens provide a relaxing and calming environment, provide opportunities for maintaining life skills and healthy habits, were achieved through reminiscence and sensory stimulation.

Mindfulness

A review of studies on the effects of mediation on stress and AD found evidence for different styles of meditation including transcendental meditation, mindfulness-based stress reduction, and a meditation technique called Kirtan Kriya (KK) [261]. A brief daily practice of KK was found to improve sleep quality and enhance quality of life (QOL). Regular practice of mediation may reduce the rate of cognitive decline in healthy older adults [262]. A prospective controlled trial on mindfulness in AD found that an intervention adapted from mindfulness-based stress reduction (MBSR) slowed the rate of cognitive decline in individuals with mild AD [263]. Mindfulness probably affects cognitive functioning at many levels. For example, regular mindfulness practice may reduce different stressors that contribute to increased allostatic load and increased risk of developing dementia [264]. Long-term meditation may help slow the rate of degenerative neurologic changes associated with cognitive decline [265, 266].

Multi-Modal Interventions

Positive findings of studies on diet in dementia are confounded by the fact that individuals who follow healthy dietary preferences also engage in other health-promoting behaviors that reduce risk of dementia, for example regular exercise and moderate alcohol consumption [22]. These findings have led to recent studies on so-called 'multi-modal' interventions aimed at identifying specific combinations of lifestyle factors that play the most significant roles in preventing or delaying onset of Alzheimer's disease.

It is estimated that one third of Alzheimer's disease cases are caused by one or more lifestyle factors that can be modified suggesting that multi-modal interventions addressing many of these factors may have significant preventive benefits. Important modifiable lifestyle factors are low education, hypertension, diabetes, obesity, smoking, sedentary lifestyle and depressed mood. So far only one large multi-center study has investigated multi-modal interventions aimed at preventing Alzheimer's disease in elderly at-risk individuals [267]. That study found significant improvement in overall cognition, improved processing speed and executive functioning in the treatment group that were significantly greater than in the control group. Other multi-modal studies are ongoing at the time of writing.

Along similar lines, case reports have been published of dramatic improvement in individuals diagnosed with early Alzheimer's disease who adhere to multi-modal life style changes [268] aimed at enhancing cognitive performance and reducing metabolic risk factors. These findings show that, in at least some cases, symptoms of early Alzheimer's disease can be reversed within 6 months after initiating a comprehensive lifestyle regimen [268]. The goal of this approach is to normalize multiple metabolic parameters related to inflammation in the body, thus interrupting the pathological processes that eventually lead to Alzheimer's disease. Beneficial effects achieved through dietary modification, regular exercise, stress management or a mindfulness practice, and supplementation with natural supplements probably involve different mechanisms at multiple levels in the body and brain including enhanced immune function; reduced insulin resistance; reduced inflammation; reduced brain atrophy and stimulation of new synapse formation.

The protocol, called metabolic enhancement for neurodegeneration (MEND), entails comprehensive laboratory screening that may include serologic studies of inflammatory markers, functional brain scans, genetic analysis of risk, and cognitive testing. Personalized lifestyle changes and nutritional strategies are subsequently recommended to correct underlying causal factors of cognitive decline identified in screening. The MEND protocol includes the following specific recommendations:

• Strict adherence to a low-glycemic diet
• 12-hour fast each night [including 3 hours before bedtime]
• Consumption of probiotic-rich foods like plain Greek yogurt, kombucha, kefir,

and fermented foods like miso and sauerkraut, and antioxidant-rich foods like blueberries and blackberries

- 8 hours sleep every night, treating sleep apnea when it is present and use of melatonin if needed
- A regular mind-body or mindfulness practice
- Regular exercise 30 to 60 minutes 4 to six times a week
- Vitamin B-12 (goal is serum vitamin B-12 levels higher than 500)
- Curcumin 400 to 500mg 3 to 4 times daily (taken with meals for better absorption] [please see separate entry on Curcumin)
- Citicoline 1000 to 2000mg and the omega-3 fatty acid (DHA) (see separate entry on omega-3s)
- Daily supplementation with vitamin D3 in individuals with vitamin D deficiency vitamin E 400 mg (mixed tocopherols and tocotrienols); vitamin C 500-1,000 mg
- Alpha lipoic acid 200 mg

Many individuals with early Alzheimer's disease (including some individuals with the ApoE4 gene who are at very high risk of developing an early severe form of Alzheimer's disease) who adhere to the MEND protocol report sustained improvement in cognitive performance for several years and no longer meet criteria for a diagnosis of Alzheimer's disease. Large prospective controlled trials are needed to confirm findings of case reports and elucidate the roles of specific lifestyle changes, metabolic factors and natural supplements in reversing or slowing rate of progression of Alzheimer's disease.

Energetic Therapies

Healing Touch and Therapeutic Touch

Healing Touch (HT) and Therapeutic Touch (TT) are based on principles of energy medicine in which the healer is believed to influence the 'energy field' of the patient resulting in beneficial physiological and emotional changes. Whereas HT practitioners use light touch, TT practitioners typically have no direct physical contact with patients. Open studies and a limited number of double-blind trials suggest that both techniques have beneficial effects on agitation in demented patients. TT is widely accepted by nurses as an appropriate and effective treatment of agitation in individuals with dementia [269]. In a small pilot study, measures of agitation were significantly improved in 14 demented residential patients who received 3 HT treatments weekly over a 4-week period [270]. Diminished need for [presumably sedating] psychotropic medications was observed in three patients during the active treatment phase, and two residents required dose increases in the first 2 weeks after HT treatments were stopped. In another small sham-controlled study [271, 272] two weekly 10- to 20-minute HT treatments were administered to AD patients over a 5-week period. Patients who received regular HT treatments were found to have consistent reductions in disruptive behaviors, and globally improved emotional and cognitive functioning including enhanced socialization, a more regular sleep schedule, improved compliance with nursing home routines, greater emotional stability, and improved communication with staff. Of note, agitation as measured by the frequency and intensity of outbursts was significantly reduced in the HT group but did not change in the control group. Demented patients who received HT treatments complained of physical distress or discomfort significantly less often compared with patients in the control group. In a double-blind study ($N = 57$) that included mock TT in the control arm, agitated demented patients who received two brief TT treatments daily for 3 days exhibited significantly fewer behavioral symptoms of dementia, including reduced restlessness and fewer disruptive vocalizations, compared with patients who received mock TT [273]. Because this short study lasted only 3 days, findings cannot be generalized to possible long-term benefits of TT in agitated demented patients. Recent reviews support that regular TT reduces agitation in individuals with dementia [274, 275].

The significance of findings on TT and HT in dementia is limited by study design flaws including lack of blinding, likely beneficial confounding effects of frequent human contact, and small numbers of enrolled patients in most studies, the lack of standardized methods, and the absence of sham control groups receiving sham HT or TT treatments in most studies.

Taichi and Qigong

Taichi and qigong are ancient mind-body practices that combine relaxed body movements with meditation. In contrast to taichi which involves complex patterns of flowing movements, qigong involves repetition of simple movements, standing postures, and breathing practices. Findings of a 12-month controlled trial [276] suggest that practicing taichi three times weekly may reduce risk of developing AD [276]. However, a review of studies on taichi as an intervention for behavioral and psychological symptoms of dementia found inconclusive evidence for beneficial effects of this mind-body practice [277].

In a randomized controlled trial (N=93) individualized with vascular dementia randomized to receive qigong, cognitive therapy or qigong plus cognitive therapy showed equivalent and significant improvement in memory, executive functioning and daily problem solving following 12 weeks of regular qigong practice [278]. Findings of dramatic early case reports using functional brain-imaging methods suggest that the consistent practice of qigong may normalize EEG slowing and possibly reverse cerebral cortical atrophy associated with cognitive impairment in the elderly. Three-dimensional positron emission tomography (PET) and EEG were used to examine relationships between brain electrical activity and changes in regional cerebral blood flow in a qigong practitioner [279]. Findings included significant increases in high-frequency (alpha and beta) EEG domains and reduced slow-frequency (delta) activity following qigong. Enhanced beta activity in the frontal lobes following qigong practice was correlated with increased cerebral blood flow in the same region and relatively decreased regional blood flow in posterior brain regions. In another case report, researchers claimed that regular qigong practice resulted in the reversal of cerebral atrophy in a 79-yea--old male [280]. The patient had gradually lost his capacity to read and work, complained of dizziness and "inert thinking," and was observed to have "stupid facial expressions." A pretreatment CT scan reportedly confirmed the presence of significant generalized cerebral atrophy consistent with dementia. After failing to respond to Western medicines, Chinese acupuncture, and Chinese herbal treatments, the patient was advised to follow a twice-daily routine of a meditative qigong practice called Quan Zhen Gong. The patient also received an unspecified number of qigong treatments from a qigong master, reportedly with observable improvements in cognitive symptoms. After 6 months of daily qigong practice and regular qigong treatments, the patient had returned to his previous baseline of cognitive functioning and was no longer assessed as demented. The authors claimed that a repeat CT scan confirmed reversal of generalized cerebral atrophy identified in the initial scan following several more months of qigong practice but no further qigong treatments (*i.e.,* provided by a qigong master). No other case reports making similar claims of dramatic cognitive improvement in demented individuals in response to qigong have been published. Replication of these findings under controlled conditions in a population of cognitively impaired elderly patients using blind raters to assess pre- and post-treatment mental status and to establish clinical correlation between changes in cognitive functioning and EEG or CT findings would constitute a truly remarkable finding.

Research findings on taichi and qigong for dementia are difficult to interpret for many reasons. Controlled studies on taichi and qigong for mental health problems, including dementia, are limited by heterogeneity in study designs, small study sizes, and the absence of patients with confirmed psychiatric diagnoses in many studies.

CLOSING REMARKS

Currently available mainstream approaches used to assess and treat MCI and dementia are limited. There is accumulating evidence for a large variety of novel assessment and treatment approaches that are not widely used in the conventional model of care. Select natural products are supported by compelling evidence, however most non-pharmaceutical modalities reviewed in this chapter are supported by limited findings. Large placebo-controlled studies are needed to further elucidate mechanisms of action, verify the efficacy of non-pharmaceutical modalities, and identify safe and appropriate treatment protocols for MCI and dementia.

CONSENT FOR PUBLICATION

Not applicable.

CONFLICT OF INTEREST

The authors confirm that this chapter contents have no conflict of interest.

ACKNOWLEDGEMENTS

Declared none.

REFERENCES

[1]   Fink HA, Jutkowitz E, McCarten JR, *et al.* Pharmacologic interventions to prevent cognitive decline, mild cognitive impairment, and clinical Alzheimer-type dementia: a systematic review. Ann Intern Med 2018; 168(1): 39-51.
[http://dx.doi.org/10.7326/M17-1529] [PMID: 29255847]

[2]   http://www.alz.org

[3]   Zandi PP, Sparks DL, Khachaturian AS, *et al.* Do statins reduce risk of incident dementia and Alzheimer disease? The Cache County Study. Arch Gen Psychiatry 2005; 62(2): 217-24.
[http://dx.doi.org/10.1001/archpsyc.62.2.217] [PMID: 15699299]

[4]   Herline K, Drummond E, Wisniewski T. Recent advancements toward therapeutic vaccines against Alzheimer's disease. Expert Rev Vaccines 2018; 17(8): 707-21.
[http://dx.doi.org/10.1080/14760584.2018.1500905] [PMID: 30005578]

[5]   Cao J, Hou J, Ping J, Cai D. Advances in developing novel therapeutic strategies for Alzheimer's disease. Mol Neurodegener 2018; 13(1): 64.
[http://dx.doi.org/10.1186/s13024-018-0299-8] [PMID: 30541602]

[6]   Ovais M, Zia N, Ahmad I, *et al.* Phyto-Therapeutic and Nanomedicinal Approaches to Cure Alzheimer's Disease: Present Status and Future Opportunities. Front Aging Neurosci 2018; 10: 284.
[http://dx.doi.org/10.3389/fnagi.2018.00284] [PMID: 30405389]

[7]   Zhang X, Wen J, Zhang Z. Statins use and risk of dementia: A dose-response meta analysis. Medicine (Baltimore) 2018; 97(30)e11304
[http://dx.doi.org/10.1097/MD.0000000000011304] [PMID: 30045255]

[8]   Lourida I, Hannon E, Littlejohns TJ, *et al.* Association of Lifestyle and Genetic Risk With Incidence of Dementia. JAMA 2019.
[http://dx.doi.org/10.1001/jama.2019.9879] [PMID: 31302669]

[9]   Grant WB. Using multicountry ecological and observational studies to determine dietary risk factors for Alzheimer's disease. J Am Coll Nutr 2016; 35(5): 476-89.
[http://dx.doi.org/10.1080/07315724.2016.1161566] [PMID: 27454859]

[10]   Grant WB. Dietary links to Alzheimer's disease. Alz Dis Rev 1997; 2: 42-55.

[11]   Orgogozo J-M, Dartigues J-F, Lafont S, Letenneur L, Commenges D, Salamon R, *et al.* Wine consumption and dementia in the elderly: a prospective community study in the Bordeaux area 1997.

[12]   Letenneur L. Risk of dementia and alcohol and wine consumption: a review of recent results. Biol Res 2004; 37(2): 189-93.
[http://dx.doi.org/10.4067/S0716-97602004000200003] [PMID: 15455646]

[13]   Féart C, Samieri C, Rondeau V, *et al.* Adherence to a Mediterranean diet, cognitive decline, and risk of dementia. JAMA 2009; 302(6): 638-48.
[http://dx.doi.org/10.1001/jama.2009.1146] [PMID: 19671905]

[14]   Morris MC, Tangney CC, Wang Y, Sacks FM, Bennett DA, Aggarwal NT. MIND diet associated with reduced incidence of Alzheimer's disease. Alzheimers Dement 2015; 11(9): 1007-14.
[http://dx.doi.org/10.1016/j.jalz.2014.11.009] [PMID: 25681666]

[15]   Rodriguez-Vieitez E, Saint-Aubert L, Carter SF, *et al.* Diverging longitudinal changes in astrocytosis and amyloid PET in autosomal dominant Alzheimer's disease. Brain 2016; 139(Pt 3): 922-36.
[http://dx.doi.org/10.1093/brain/awv404] [PMID: 26813969]

[16]   Solfrizzi V, Panza F, Frisardi V, *et al.* Diet and Alzheimer's disease risk factors or prevention: the current evidence. Expert Rev Neurother 2011; 11(5): 677-708.
[http://dx.doi.org/10.1586/ern.11.56] [PMID: 21539488]

[17]   Yusufov M, Weyandt LL, Piryatinsky I. Alzheimer's disease and diet: a systematic review. Int J Neurosci 2017; 127(2): 161-75.
[http://dx.doi.org/10.3109/00207454.2016.1155572] [PMID: 26887612]

[18]   Kalmijn S, Feskens EJ, Launer LJ, Kromhout D. Polyunsaturated fatty acids, antioxidants, and cognitive function in very old men. Am J Epidemiol 1997; 145(1): 33-41.
[http://dx.doi.org/10.1093/oxfordjournals.aje.a009029] [PMID: 8982020]

[19]   Morris MC, Evans DA, Bienias JL, *et al.* Consumption of fish and n-3 fatty acids and risk of incident Alzheimer disease. Arch Neurol 2003; 60(7): 940-6.
[http://dx.doi.org/10.1001/archneur.60.7.940] [PMID: 12873849]

[20]   Engelhart MJ, Geerlings MI, Ruitenberg A, *et al.* Diet and risk of dementia: Does fat matter?: The Rotterdam Study. Neurology 2002; 59(12): 1915-21.
[http://dx.doi.org/10.1212/01.WNL.0000038345.77753.46] [PMID: 12499483]

[21]   Kalmijn S, van Boxtel MP, Ocké M, Verschuren WM, Kromhout D, Launer LJ. Dietary intake of fatty acids and fish in relation to cognitive performance at middle age. Neurology 2004; 62(2): 275-80.
[http://dx.doi.org/10.1212/01.WNL.0000103860.75218.A5] [PMID: 14745067]

[22]   Barberger-Gateau P, Letenneur L, Deschamps V, Pérès K, Dartigues J-F, Renaud S. Fish, meat, and risk of dementia: cohort study. BMJ 2002; 325(7370): 932-3.
[http://dx.doi.org/10.1136/bmj.325.7370.932] [PMID: 12399342]

[23]   Cass SP. 56. Cass SP. Alzheimer's disease and exercise: a literature review. Curr Sports Med Rep 2017; 16(1): 19-22.
[http://dx.doi.org/10.1249/JSR.0000000000000332] [PMID: 28067736]

[24]   van Praag H, Shubert T, Zhao C, Gage FH. Exercise enhances learning and hippocampal neurogenesis in aged mice. J Neurosci 2005; 25(38): 8680-5.
[http://dx.doi.org/10.1523/JNEUROSCI.1731-05.2005] [PMID: 16177036]

[25]   van Praag H. Neurogenesis and exercise: past and future directions. Neuromolecular Med 2008; 10(2): 128-40.
[http://dx.doi.org/10.1007/s12017-008-8028-z] [PMID: 18286389]

[26]   Erickson KI, Raji CA, Lopez OL, *et al.* Physical activity predicts gray matter volume in late adulthood: the Cardiovascular Health Study. Neurology 2010; 75(16): 1415-22.

[http://dx.doi.org/10.1212/WNL.0b013e3181f88359] [PMID: 20944075]

[27] Sexton CE, Betts JF, Demnitz N, Dawes H, Ebmeier KP, Johansen-Berg H. A systematic review of MRI studies examining the relationship between physical fitness and activity and the white matter of the ageing brain. Neuroimage 2016; 131: 81-90.
       [http://dx.doi.org/10.1016/j.neuroimage.2015.09.071] [PMID: 26477656]

[28] Erickson KI, Voss MW, Prakash RS, Basak C, Szabo A, *et al.* 2011.Exercise training increases size of hippocampus and improves memory
       [http://dx.doi.org/10.1073/pnas.1015950108]

[29] Firth J, Stubbs B, Vancampfort D, *et al.* Effect of aerobic exercise on hippocampal volume in humans: A systematic review and meta-analysis. Neuroimage 2018; 166(166): 230-8.
       [http://dx.doi.org/10.1016/j.neuroimage.2017.11.007] [PMID: 29113943]

[30] Stephen R, Hongisto K, Solomon A, Lönnroos E. Physical Activity and Alzheimer's Disease: A Systematic Review 2017.
       [http://dx.doi.org/10.1093/gerona/glw251]

[31] Scarmeas N, Luchsinger JA, Schupf N, *et al.* Physical activity, diet, and risk of Alzheimer disease. JAMA 2009; 302(6): 627-37.
       [http://dx.doi.org/10.1001/jama.2009.1144] [PMID: 19671904]

[32] Agüera Sánchez MÁ, Barbancho Ma MÁ, García-Casares N. [Effect of physical exercise on Alzheimer's disease. A systematic review]. Aten Primaria 2019.S0212-6567(18)30468-2
       [PMID: 31153668]

[33] Forbes D, Forbes SC, Blake CM, Thiessen EJ, Forbes S. Exercise programs for people with dementia 2015.
       [http://dx.doi.org/10.1002/14651858.CD006489.pub4]

[34] Abbott RD, White LR, Ross GW, Masaki KH, Curb JD, Petrovitch H. Walking and dementia in physically capable elderly men. JAMA 2004; 292(12): 1447-53.
       [http://dx.doi.org/10.1001/jama.292.12.1447] [PMID: 15383515]

[35] Weuve J, Kang JH, Manson JE, Breteler MM, Ware JH, Grodstein F. Physical activity, including walking, and cognitive function in older women. JAMA 2004; 292(12): 1454-61.
       [http://dx.doi.org/10.1001/jama.292.12.1454] [PMID: 15383516]

[36] Teri L, Gibbons LE, McCurry SM, *et al.* Exercise plus behavioral management in patients with Alzheimer disease: a randomized controlled trial. JAMA 2003; 290(15): 2015-22.
       [http://dx.doi.org/10.1001/jama.290.15.2015] [PMID: 14559955]

[37] Blackwell T, Yaffe K, Laffan A, *et al.* Associations of objectively and subjectively measured sleep quality with subsequent cognitive decline in older community-dwelling men: the MrOS sleep study. Sleep (Basel) 2014; 37(4): 655-63.
       [http://dx.doi.org/10.5665/sleep.3562] [PMID: 24899757]

[38] Lim AS, Kowgier M, Yu L, Buchman AS, Bennett DA. Sleep Fragmentation and the Risk of Incident Alzheimer's Disease and Cognitive Decline in Older Persons. Sleep (Basel) 2013; 36(7): 1027-32. a
       [http://dx.doi.org/10.5665/sleep.2802] [PMID: 23814339]

[39] Hahn EA, Wang HX, Andel R, Fratiglioni L. A Change in Sleep Pattern May Predict Alzheimer Disease. Am J Geriatr Psychiatry 2013.
       [PMID: 23954041]

[40] Sterniczuk R, Theou O, Rusak B, Rockwood K. Sleep disturbance is associated with incident dementia and mortality. Curr Alzheimer Res 2013; 10(7): 767-75.
       [http://dx.doi.org/10.2174/15672050113109990134] [PMID: 23905991]

[41] Spira AP, Gamaldo AA, An Y, *et al.* Self-reported sleep and β-amyloid deposition in community-dwelling older adults. JAMA Neurol 2013; 70(12): 1537-43.
       [http://dx.doi.org/10.1001/jamaneurol.2013.4258] [PMID: 24145859]

[42] Mintun MA, Larossa GN, Sheline YI, *et al.* [11C]PIB in a nondemented population: potential antecedent marker of Alzheimer disease. Neurology 2006; 67(3): 446-52.
[http://dx.doi.org/10.1212/01.wnl.0000228230.26044.a4] [PMID: 16894106]

[43] Lim AS, Yu L, Kowgier M, Schneider JA, Buchman AS, Bennett DA. Modification of the relationship of the apolipoprotein E ε4 allele to the risk of Alzheimer disease and neurofibrillary tangle density by sleep. JAMA Neurol 2013; 70(12): 1544-51. b
[http://dx.doi.org/10.1001/jamaneurol.2013.4215] [PMID: 24145819]

[44] Yaffe K, Laffan AM, Harrison SL, *et al.* Sleep-disordered breathing, hypoxia, and risk of mild cognitive impairment and dementia in older women. JAMA 2011; 306(6): 613-9.
[http://dx.doi.org/10.1001/jama.2011.1115] [PMID: 21828324]

[45] Chang WP, Liu ME, Chang WC, *et al.* Sleep apnea and the risk of dementia: a population-based 5-year follow-up study in Taiwan. PLoS One 2013; 8(10)e78655
[http://dx.doi.org/10.1371/journal.pone.0078655] [PMID: 24205289]

[46] Butler M, Nelson VA, Davila H, *et al.* Over-the-Counter Supplement Interventions to Prevent Cognitive Decline, Mild Cognitive Impairment, and Clinical Alzheimer-Type Dementia: A Systematic Review. Ann Intern Med 2018; 168(1): 52-62.
[http://dx.doi.org/10.7326/M17-1530] [PMID: 29255909]

[47] Hrehorovská M, Burda J, Domoráková I, Mechírová E. Effect of Tanakan on postischemic activity of protein synthesis machinery in the rat brain. Gen Physiol Biophys 2004; 23(4): 457-65.
[PMID: 15815080]

[48] Koh P-O. Gingko biloba extract (EGb 761) prevents cerebral ischemia-induced p70S6 kinase and S6 phosphorylation. Am J Chin Med 2010; 38(4): 727-34.
[http://dx.doi.org/10.1142/S0192415X10008196] [PMID: 20626058]

[49] Saleem S, Zhuang H, Biswal S, Christen Y, Doré S. Ginkgo biloba extract neuroprotective action is dependent on heme oxygenase 1 in ischemic reperfusion brain injury. Stroke 2008; 39(12): 3389-96.
[http://dx.doi.org/10.1161/STROKEAHA.108.523480] [PMID: 18845796]

[50] Spinnewyn B, Blavet N, Clostre F. Effects of Ginkgo biloba extract on a cerebral ischemia model in gerbils Rökan. Springer 1988; pp. 143-52.

[51] Rocher M-N, Carré D, Spinnewyn B, *et al.* Long-term treatment with standardized Ginkgo biloba extract (EGb 761) attenuates cognitive deficits and hippocampal neuron loss in a gerbil model of vascular dementia. Fitoterapia 2011; 82(7): 1075-80.
[http://dx.doi.org/10.1016/j.fitote.2011.07.001] [PMID: 21820038]

[52] Li WZ, Wu WY, Huang H, Wu YY, Yin YY. Protective effect of bilobalide on learning and memory impairment in rats with vascular dementia. Mol Med Rep 2013; 8(3): 935-41.
[http://dx.doi.org/10.3892/mmr.2013.1573] [PMID: 23835946]

[53] Schneider LS. Ginkgo biloba extract and preventing Alzheimer disease. JAMA 2008; 300(19): 2306-8.
[http://dx.doi.org/10.1001/jama.2008.675] [PMID: 19017919]

[54] Ihl R, Tribanek M, Bachinskaya N, Group GS. Efficacy and tolerability of a once daily formulation of Ginkgo biloba extract EGb 761® in Alzheimer's disease and vascular dementia: results from a randomised controlled trial. Pharmacopsychiatry 2012; 45(2): 41-6.
[http://dx.doi.org/10.1055/s-0031-1291217] [PMID: 22086747]

[55] Gauthier S, Schlaefke S. Efficacy and tolerability of Ginkgo biloba extract EGb 761® in dementia: a systematic review and meta-analysis of randomized placebo-controlled trials. Clin Interv Aging 2014; 9: 2065-77.
[http://dx.doi.org/10.2147/CIA.S72728] [PMID: 25506211]

[56] Tan M-S, Yu J-T, Tan C-C, *et al.* Efficacy and adverse effects of ginkgo biloba for cognitive impairment and dementia: a systematic review and meta-analysis. J Alzheimers Dis 2015; 43(2): 589-603.

[http://dx.doi.org/10.3233/JAD-140837] [PMID: 25114079]

[57]   Chan P-C, Xia Q, Fu PP. Ginkgo biloba leave extract: biological, medicinal, and toxicological effects. J Environ Sci Health C Environ Carcinog Ecotoxicol Rev 2007; 25(3): 211-44.
[http://dx.doi.org/10.1080/10590500701569414] [PMID: 17763047]

[58]   Wang J, Chen W, Wang Y. A ginkgo biloba extract promotes proliferation of endogenous neural stem cells in vascular dementia rats. Neural Regen Res 2013; 8(18): 1655-62.
[PMID: 25206462]

[59]   Zhang L, Wang Y. Effects of EGb761 on hippocamal synaptic plasticity of vascular dementia rats 2008.

[60]   Yao Z-X, Han Z, Drieu K, Papadopoulos V. Ginkgo biloba extract (Egb 761) inhibits β-amyloid production by lowering free cholesterol levels. J Nutr Biochem 2004; 15(12): 749-56.
[http://dx.doi.org/10.1016/j.jnutbio.2004.06.008] [PMID: 15607648]

[61]   Wang Y, Huang LQ, Tang XC, Zhang HY. Retrospect and prospect of active principles from Chinese herbs in the treatment of dementia. Acta Pharmacol Sin 2010; 31(6): 649-64.
[http://dx.doi.org/10.1038/aps.2010.46] [PMID: 20523337]

[62]   Ringman JM, Frautschy SA, Cole GM, Masterman DL, Cummings JL. A potential role of the curry spice curcumin in Alzheimer's disease. Curr Alzheimer Res 2005; 2(2): 131-6.
[http://dx.doi.org/10.2174/1567205053585882] [PMID: 15974909]

[63]   Yang F, Lim GP, Begum AN, *et al.* Curcumin inhibits formation of amyloid β oligomers and fibrils, binds plaques, and reduces amyloid *in vivo.* J Biol Chem 2005; 280(7): 5892-901.
[http://dx.doi.org/10.1074/jbc.M404751200] [PMID: 15590663]

[64]   Ringman JM, Frautschy SA, Teng E, *et al.* Oral curcumin for Alzheimer's disease: tolerability and efficacy in a 24-week randomized, double blind, placebo-controlled study. Alzheimers Res Ther 2012; 4(5): 43.
[http://dx.doi.org/10.1186/alzrt146] [PMID: 23107780]

[65]   Small GW, Siddarth P, Li Z, *et al.* Memory and brain amyloid and tau effects of a bioavailable form of curcumin in non-demented adults: a double-blind, placebo-controlled 18-month trial. Am J Geriatr Psychiatry 2018; 26(3): 266-77.
[http://dx.doi.org/10.1016/j.jagp.2017.10.010] [PMID: 29246725]

[66]   Chu S, Gu J, Feng L, *et al.* Ginsenoside Rg5 improves cognitive dysfunction and beta-amyloid deposition in STZ-induced memory impaired rats *via* attenuating neuroinflammatory responses. Int Immunopharmacol 2014; 19(2): 317-26.
[http://dx.doi.org/10.1016/j.intimp.2014.01.018] [PMID: 24503167]

[67]   Liu J-X, Cong WH, Xu L, Wang JN. Effect of combination of extracts of ginseng and ginkgo biloba on acetylcholine in amyloid beta-protein-treated rats determined by an improved HPLC. Acta Pharmacol Sin 2004; 25(9): 1118-23.
[PMID: 15339385]

[68]   Yang H, Zhang J, Breyer RM, Chen C. Altered hippocampal long-term synaptic plasticity in mice deficient in the PGE2 EP2 receptor. J Neurochem 2009; 108(1): 295-304.
[http://dx.doi.org/10.1111/j.1471-4159.2008.05766.x] [PMID: 19012750]

[69]   Shi J, Zhang S, Tang M, *et al.* The 1239G/C polymorphism in exon 5 of BACE1 gene may be associated with sporadic Alzheimer's disease in Chinese Hans. Am J Med Genet B Neuropsychiatr Genet 2004; 124B(1): 54-7.
[http://dx.doi.org/10.1002/ajmg.b.20087] [PMID: 14681914]

[70]   Sun Y, Ke J, Ma N, Chen Z, Wang C, Cui X. [Effects of root rot on saponin content in Panax notoginseng]. Zhong Yao Cai 2004; 27(2): 79-80.
[PMID: 22454992]

[71]   Choi KT. Botanical characteristics, pharmacological effects and medicinal components of Korean

Panax ginseng C A Meyer. Acta Pharmacol Sin 2008; 29(9): 1109-18.
[http://dx.doi.org/10.1111/j.1745-7254.2008.00869.x] [PMID: 18718180]

[72]   Heo JH, Lee ST, Chu K, *et al.* An open-label trial of Korean red ginseng as an adjuvant treatment for cognitive impairment in patients with Alzheimer's disease. Eur J Neurol 2008; 15(8): 865-8.
[http://dx.doi.org/10.1111/j.1468-1331.2008.02157.x] [PMID: 18684311]

[73]   Lee S-T, Chu K, Sim J-Y, Heo J-H, Kim M. Panax ginseng enhances cognitive performance in Alzheimer disease. Alzheimer Dis Assoc Disord 2008; 22(3): 222-6.
[http://dx.doi.org/10.1097/WAD.0b013e31816c92e6] [PMID: 18580589]

[74]   Heo J-H, Lee S-T, Oh MJ, *et al.* Improvement of cognitive deficit in Alzheimer's disease patients by long term treatment with korean red ginseng. J Ginseng Res 2011; 35(4): 457-61.
[http://dx.doi.org/10.5142/jgr.2011.35.4.457] [PMID: 23717092]

[75]   Jinzhou T, Aihua Z, Jing S, Eds. Ginseng may improve memory in stroke dementia patients. American Stroke Association Meeting Report.

[76]   Gui Q, Yang Y, Ying S, Zhang M. Xueshuantong improves cerebral blood perfusion in elderly patients with lacunar infarction. Neural Regen Res 2013; 8(9): 792-801.
[PMID: 25206726]

[77]   Zhao H-W, Li X-Y, Ginkgolide A. Ginkgolide A, B, and huperzine A inhibit nitric oxide-induced neurotoxicity. Int Immunopharmacol 2002; 2(11): 1551-6.
[http://dx.doi.org/10.1016/S1567-5769(02)00093-0] [PMID: 12433056]

[78]   Wang Z, Ren G, Zhao Y. A double-blind study of huperzine A and piracetam in patients with age-associated memory impairment and dementia Herbal Medicines for Nonpsychiatric Diseases Tokyo. Seiwa Choten Publishers 1999; pp. 39-50.

[79]   Zucker M. Huperzine-A: the newest brain nutrient. Let's Live 1999; pp. 47-8.

[80]   Li J, Wu HM, Zhou RL, Liu GJ, Dong BR. Huperzine A for Alzheimer's disease 2008.
[http://dx.doi.org/10.1002/14651858.CD005592.pub2]

[81]   Fu L-M, Li J-T. A systematic review of single chinese herbs for Alzheimer's disease treatment 2011.
[http://dx.doi.org/10.1093/ecam/nep136]

[82]   Yang G, Wang Y, Tian J, Liu J-P. Huperzine A for Alzheimer's disease: a systematic review and meta-analysis of randomized clinical trials. PLoS One 2013; 8(9)e74916
[http://dx.doi.org/10.1371/journal.pone.0074916] [PMID: 24086396]

[83]   Tian J, Shi J, Zhang X, Wang Y. Herbal therapy: a new pathway for the treatment of Alzheimer's disease. Alzheimers Res Ther 2010; 2(5): 30.
[http://dx.doi.org/10.1186/alzrt54] [PMID: 21067555]

[84]   Rafii MS, Walsh S, Little JT, *et al.* A phase II trial of huperzine A in mild to moderate Alzheimer disease. Neurology 2011; 76(16): 1389-94.
[http://dx.doi.org/10.1212/WNL.0b013e318216eb7b] [PMID: 21502597]

[85]   Bhattacharya SK, Bhattacharya A, Kumar A, Ghosal S. Antioxidant activity of Bacopa monniera in rat frontal cortex, striatum and hippocampus. Phytother Res 2000; 14(3): 174-9.
[http://dx.doi.org/10.1002/(SICI)1099-1573(200005)14:3<174::AID-PTR624>3.0.CO;2-O] [PMID: 10815010]

[86]   Russo A, Izzo AA, Borrelli F, Renis M, Vanella A. Free radical scavenging capacity and protective effect of Bacopa monniera L. on DNA damage. Phytother Res 2003; 17(8): 870-5.
[http://dx.doi.org/10.1002/ptr.1061] [PMID: 13680815]

[87]   Kamkaew N, Norman Scholfield C, Ingkaninan K, Taepavarapruk N, Chootip K. Bacopa monnieri increases cerebral blood flow in rat independent of blood pressure. Phytother Res 2013; 27(1): 135-8.
[http://dx.doi.org/10.1002/ptr.4685] [PMID: 22447676]

[88]   Russo A, Borrelli F. Bacopa monniera, a reputed nootropic plant: an overview. Phytomedicine 2005;

12(4): 305-17.
[http://dx.doi.org/10.1016/j.phymed.2003.12.008] [PMID: 15898709]

[89]   Stough C, Scholey A, Cropley V, *et al.* Examining the cognitive effects of a special extract of Bacopa monniera (CDRI08: Keenmnd): a review of ten years of research at Swinburne University. J Pharm Pharm Sci 2013; 16(2): 254-8.
[http://dx.doi.org/10.18433/J35G6M] [PMID: 23958194]

[90]   Hosseinzadeh H, Sadeghnia HR. Safranal, a constituent of Crocus sativus (saffron), attenuated cerebral ischemia induced oxidative damage in rat hippocampus. J Pharm Pharm Sci 2005; 8(3): 394-9.
[PMID: 16401389]

[91]   Jessie SW, Krishnakantha TP. Inhibition of human platelet aggregation and membrane lipid peroxidation by food spice, saffron. Mol Cell Biochem 2005; 278(1-2): 59-63.
[http://dx.doi.org/10.1007/s11010-005-5155-9] [PMID: 16180089]

[92]   Abe K, Saito H. Effects of saffron extract and its constituent crocin on learning behaviour and long-term potentiation. Phytother Res 2000; 14(3): 149-52.
[http://dx.doi.org/10.1002/(SICI)1099-1573(200005)14:3<149::AID-PTR665>3.0.CO;2-5] [PMID: 10815004]

[93]   Hosseinzadeh H, Ziaei T. Effects of Crocus sativus stigma extract and its constituents, crocin and safranal, on intact memory and scopolamine-induced learning deficits in rats performing the Morris water maze task. Faslnamah-i Giyahan-i Daruyi 2006; 3(19): 40-50.

[94]   Finley JW, Gao S. A perspective on Crocus sativus L.(Saffron) constituent crocin: a potent water-soluble antioxidant and potential therapy for Alzheimer's disease. J Agric Food Chem 2017; 65(5): 1005-20.
[http://dx.doi.org/10.1021/acs.jafc.6b04398] [PMID: 28098452]

[95]   Hajiaghaee R, Akhondzadeh S. Herbal medicine in the treatment of Alzheimer's disease. Faslnamah-i Giyahan-i Daruyi 2012; 1(41): 1-7.

[96]   Howes M-JR, Fang R, Houghton PJ. Effect of Chinese herbal medicine on Alzheimer's disease International review of neurobiology 135. Elsevier 2017; pp. 29-56.
[http://dx.doi.org/10.1016/bs.irn.2017.02.003]

[97]   Rahimi R, Irannejad S, Noroozian M. Avicenna's pharmacological approach to memory enhancement. Neurol Sci 2017; 38(7): 1147-57.
[http://dx.doi.org/10.1007/s10072-017-2835-7] [PMID: 28176148]

[98]   Tsolaki M, Karathanasi E, Lazarou I, *et al.* Efficacy and safety of Crocus sativus L. in patients with mild cognitive impairment: one year single-blind randomized, with parallel groups, clinical trial. J Alzheimers Dis 2016; 54(1): 129-33.
[http://dx.doi.org/10.3233/JAD-160304] [PMID: 27472878]

[99]   Akhondzadeh S, Sabet MS, Harirchian MH, *et al.* Saffron in the treatment of patients with mild to moderate Alzheimer's disease: a 16-week, randomized and placebo-controlled trial. J Clin Pharm Ther 2010; 35(5): 581-8.
[http://dx.doi.org/10.1111/j.1365-2710.2009.01133.x] [PMID: 20831681]

[100] Akhondzadeh S, Shafiee Sabet M, Harirchian MH, *et al.* A 22-week, multicenter, randomized, double-blind controlled trial of Crocus sativus in the treatment of mild-to-moderate Alzheimer's disease. Psychopharmacology (Berl) 2010; 207(4): 637-43.
[http://dx.doi.org/10.1007/s00213-009-1706-1] [PMID: 19838862]

[101] Farokhnia M, Shafiee Sabet M, Iranpour N, *et al.* Comparing the efficacy and safety of Crocus sativus L. with memantine in patients with moderate to severe Alzheimer's disease: a double-blind randomized clinical trial. Hum Psychopharmacol 2014; 29(4): 351-9.
[http://dx.doi.org/10.1002/hup.2412] [PMID: 25163440]

[102] Mandel SA, Amit T, Kalfon L, Reznichenko L, Youdim MB. Targeting multiple neurodegenerative

diseases etiologies with multimodal-acting green tea catechins. J Nutr 2008; 138(8): 1578S-83S.
[http://dx.doi.org/10.1093/jn/138.8.1578S] [PMID: 18641210]

[103] Sharangi A. Medicinal and therapeutic potentialities of tea (Camellia sinensis L.)–A review. Food Res Int 2009; 42(5-6): 529-35.
[http://dx.doi.org/10.1016/j.foodres.2009.01.007]

[104] Choi Y-T, Jung C-H, Lee S-R, *et al.* The green tea polyphenol (-)-epigallocatechin gallate attenuates β-amyloid-induced neurotoxicity in cultured hippocampal neurons. Life Sci 2001; 70(5): 603-14.
[http://dx.doi.org/10.1016/S0024-3205(01)01438-2] [PMID: 11811904]

[105] Rezai-Zadeh K, Shytle D, Sun N, *et al.* Green tea epigallocatechin-3-gallate (EGCG) modulates amyloid precursor protein cleavage and reduces cerebral amyloidosis in Alzheimer transgenic mice. J Neurosci 2005; 25(38): 8807-14.
[http://dx.doi.org/10.1523/JNEUROSCI.1521-05.2005] [PMID: 16177050]

[106] Kuriyama S, Hozawa A, Ohmori K, *et al.* Green tea consumption and cognitive function: a cross-sectional study from the Tsurugaya Project 1. Am J Clin Nutr 2006; 83(2): 355-61.
[http://dx.doi.org/10.1093/ajcn/83.2.355] [PMID: 16469995]

[107] Noguchi-Shinohara M, Yuki S, Dohmoto C, *et al.* Consumption of green tea, but not black tea or coffee, is associated with reduced risk of cognitive decline. PLoS One 2014; 9(5)e96013
[http://dx.doi.org/10.1371/journal.pone.0096013] [PMID: 24828424]

[108] Wagner H, Ulrich-Merzenich G. Synergy research: approaching a new generation of phytopharmaceuticals. Phytomedicine 2009; 16(2-3): 97-110.
[http://dx.doi.org/10.1016/j.phymed.2008.12.018] [PMID: 19211237]

[109] Zhou X, Seto SW, Chang D, *et al.* Synergistic effects of Chinese herbal medicine: a comprehensive review of methodology and current research. Front Pharmacol 2016; 7: 201.
[http://dx.doi.org/10.3389/fphar.2016.00201] [PMID: 27462269]

[110] Iwasaki K, Kobayashi S, Chimura Y, *et al.* A randomized, double-blind, placebo-controlled clinical trial of the Chinese herbal medicine "ba wei di huang wan" in the treatment of dementia. J Am Geriatr Soc 2004; 52(9): 1518-21.
[http://dx.doi.org/10.1111/j.1532-5415.2004.52415.x] [PMID: 15341554]

[111] Nagata K, Yokoyama E, Yamazaki T, *et al.* Effects of yokukansan on behavioral and psychological symptoms of vascular dementia: an open-label trial. Phytomedicine 2012; 19(6): 524-8.
[http://dx.doi.org/10.1016/j.phymed.2012.02.008] [PMID: 22421528]

[112] Man SC, Chan KW, Lu J-H, Durairajan SSK, Liu L-F, Li M. Systematic review on the efficacy and safety of herbal medicines for vascular dementia 2012.
[http://dx.doi.org/10.1155/2012/426215]

[113] Gong D, Xu J, Fan Y. Meta-analysis of clinical trials of oral Chinese herbal prescriptions for treatment of vascular dementia based on mini mental state examination scores. Eur J Integr Med 2015; 7(2): 108-17.
[http://dx.doi.org/10.1016/j.eujim.2014.11.002]

[114] Cong W-H, Liu J-X, Xu L. 2007.

[115] Xu L, Cong W, Wei C, Liu J. Effects of Weinaokang (SLT) on dysmnesia in mice models. Pharmacology and Clinics of Chinese Materia Medica 2007; 6: 60-2.

[116] Xu L, Liu JX, Cong WH, Wei CE. [Effects of Weinaokang capsule on intracephalic cholinergic system and capability of scavenging free radicas in chronic cerebral hypoperfusion rats]. Zhongguo Zhongyao Zazhi 2008; 33(5): 531-4.
[PMID: 18536376]

[117] Cong W-h, Yang B, Xu L, Dong X-x, Sheng L-s, Hou J-c, *et al.* 2012.

[118] Seto SW, Chang D, Ko WM, *et al.* Sailuotong prevents hydrogen peroxide (H2O2)-induced injury in

EA. hy926 cells. Int J Mol Sci 2017; 18(1): 95.
[http://dx.doi.org/10.3390/ijms18010095] [PMID: 28067784]

[119] Li T, Liu H-m, Lu Y, Jia Z-q, Xu L, Gao R, *et al.* A phase I tolerance and safety study of Sailuotong capsule. Zhongguo Xin Yao Zazhi 2012; (1): 16.

[120] Steiner GZ, Yeung A, Liu J-X, *et al.* The effect of Sailuotong (SLT) on neurocognitive and cardiovascular function in healthy adults: a randomised, double-blind, placebo controlled crossover pilot trial. BMC Complement Altern Med 2016; 16(1): 15.
[http://dx.doi.org/10.1186/s12906-016-0989-0] [PMID: 26762282]

[121] Chang DH-T, Colagiuri B, Luo R. Chinese medicine used to treat dementia. Advances in Natural Medicines, Nutraceuticals and Neurocognition 2013; pp. 205-23.

[122] Liu J, Chang D, Chan D, Liu J, Bensoussan A.

[123] Jia J, Wei C, Chen S, *et al.* Efficacy and safety of the compound Chinese medicine SaiLuoTong in vascular dementia: A randomized clinical trial. Alzheimers Dement (N Y) 2018; 4: 108-17.
[http://dx.doi.org/10.1016/j.trci.2018.02.004] [PMID: 29955654]

[124] Yabe T, Yamada H. Kami-Untan-To enhances choline acetyltransferase and Nerve growth factor mRNA levels in brain cultured cells. Phytomedicine 1997; 3(4): 361-7.
[http://dx.doi.org/10.1016/S0944-7113(97)80010-4] [PMID: 23195195]

[125] Yabe T, Toriizuka K, Yamada H. Kami-untan-to (KUT) improves cholinergic deficits in aged rats. Phytomedicine 1996; 2(3): 253-8.
[http://dx.doi.org/10.1016/S0944-7113(96)80051-1] [PMID: 23194625]

[126] Nakagawasai O, Yamadera F, Iwasaki K, *et al.* Effect of kami-untan-to on the impairment of learning and memory induced by thiamine-deficient feeding in mice. Neuroscience 2004; 125(1): 233-41.
[http://dx.doi.org/10.1016/j.neuroscience.2003.10.051] [PMID: 15051162]

[127] Arai H, Suzuki T, Sasaki H, Hanawa T, Toriizuka K, Yamada H. [A new interventional strategy for Alzheimer's disease by Japanese herbal medicine]. Nippon Ronen Igakkai Zasshi 2000; 37(3): 212-5.
[http://dx.doi.org/10.3143/geriatrics.37.212] [PMID: 10879069]

[128] Maruyama M, Tomita N, Iwasaki K, *et al.* Benefits of combining donepezil plus traditional Japanese herbal medicine on cognition and brain perfusion in Alzheimer's disease: a 12-week observer-blind, donepezil monotherapy controlled trial. J Am Geriatr Soc 2006; 54(5): 869-71.
[http://dx.doi.org/10.1111/j.1532-5415.2006.00722.x] [PMID: 16696770]

[129] Pandey A, Bani S, Dutt P, Kumar Satti N, Avtar Suri K, Nabi Qazi G. Multifunctional neuroprotective effect of Withanone, a compound from Withania somnifera roots in alleviating cognitive dysfunction. Cytokine 2018; 102: 211-21.
[http://dx.doi.org/10.1016/j.cyto.2017.10.019] [PMID: 29108796]

[130] Kuboyama T, Tohda C, Zhao J, Nakamura N, Hattori M, Komatsu K. Axon- or dendrite-predominant outgrowth induced by constituents from Ashwagandha. Neuroreport 2002; 13(14): 1715-20.
[http://dx.doi.org/10.1097/00001756-200210070-00005] [PMID: 12395110]

[131] Bhattacharya SK, Kumar A, Ghosal S. Effects of glycowithanolides from Withania somnifera on an animal model of Alzheimer's disease and perturbed central cholinergic markers of cognition in rats. Phytother Res 1995; 9(2): 110-3.
[http://dx.doi.org/10.1002/ptr.2650090206]

[132] Choudhary D, Bhattacharyya S, Bose S. Efficacy and safety of Ashwagandha (Withania somnifera (L.) Dunal) root extract in improving memory and cognitive functions. J Diet Suppl 2017; 14(6): 599-612.
[http://dx.doi.org/10.1080/19390211.2017.1284970] [PMID: 28471731]

[133] Park CH, Kim S-H, Choi W, *et al.* Novel anticholinesterase and antiamnesic activities of dehydroevodiamine, a constituent of Evodia rutaecarpa. Planta Med 1996; 62(5): 405-9.
[http://dx.doi.org/10.1055/s-2006-957926] [PMID: 8923803]

[134]   Park CH, Lee YJ, Lee SH, *et al.* Dehydroevodiamine.HCl prevents impairment of learning and memory and neuronal loss in rat models of cognitive disturbance. J Neurochem 2000; 74(1): 244-53.
[http://dx.doi.org/10.1046/j.1471-4159.2000.0740244.x] [PMID: 10617126]

[135]   Nishiyama N, Chu P-J, Saito H. An herbal prescription, S-113m, consisting of biota, ginseng and schizandra, improves learning performance in senescence accelerated mouse. Biol Pharm Bull 1996; 19(3): 388-93.
[http://dx.doi.org/10.1248/bpb.19.388] [PMID: 8924907]

[136]   Bhattacharya SK, Kumar A. Effect of Trasina, an ayurvedic herbal formulation, on experimental models of Alzheimer's disease and central cholinergic markers in rats. J Altern Complement Med 1997; 3(4): 327-36.
[http://dx.doi.org/10.1089/acm.1997.3.327] [PMID: 9449054]

[137]   Decker MW, Majchrzak MJ, Arnerić SP. Effects of lobeline, a nicotinic receptor agonist, on learning and memory. Pharmacol Biochem Behav 1993; 45(3): 571-6.
[http://dx.doi.org/10.1016/0091-3057(93)90508-Q] [PMID: 8332618]

[138]   Nabavi SF, Braidy N, Orhan IE, Badiee A, Daglia M, Nabavi SM. Rhodiola rosea L. and Alzheimer's disease: from farm to pharmacy. Phytother Res 2016; 30(4): 532-9.
[http://dx.doi.org/10.1002/ptr.5569] [PMID: 27059687]

[139]   Petkov VD, Stancheva SL, Tocuschieva L, Petkov VV. Changes in brain biogenic monoamines induced by the nootropic drugs adafenoxate and meclofenoxate and by citicholine (experiments on rats). Gen Pharmacol 1990; 21(1): 71-5.
[http://dx.doi.org/10.1016/0306-3623(90)90598-G] [PMID: 2105261]

[140]   Spasov AA, Wikman GK, Mandrikov VB, Mironova IA, Neumoin VV. A double-blind, placebo-controlled pilot study of the stimulating and adaptogenic effect of Rhodiola rosea SHR-5 extract on the fatigue of students caused by stress during an examination period with a repeated low-dose regimen. Phytomedicine 2000; 7(2): 85-9.
[http://dx.doi.org/10.1016/S0944-7113(00)80078-1] [PMID: 10839209]

[141]   Ahmed FAN. 2015.

[142]   Saratikov A, Krasnov E. Rhodiola rosea is a valuable medicinal plant (Golden Root). Tomsk, Russia: Tomsk State University Press 1987.

[143]   Akhondzadeh S, Noroozian M, Mohammadi M, Ohadinia S, Jamshidi AH, Khani M. Melissa officinalis extract in the treatment of patients with mild to moderate Alzheimer's disease: a double blind, randomised, placebo controlled trial. J Neurol Neurosurg Psychiatry 2003; 74(7): 863-6.
[http://dx.doi.org/10.1136/jnnp.74.7.863] [PMID: 12810768]

[144]   Akhondzadeh S, Noroozian M, Mohammadi M, Ohadinia S, Jamshidi AH, Khani M. Salvia officinalis extract in the treatment of patients with mild to moderate Alzheimer's disease: a double blind, randomized and placebo-controlled trial. J Clin Pharm Ther 2003; 28(1): 53-9.
[http://dx.doi.org/10.1046/j.1365-2710.2003.00463.x] [PMID: 12605619]

[145]   Jeong HS, Park J-S, Yang Y, Na S-H, Chung Y-A, Song I-U. Cerebral Perfusion Changes after Acetyl-L-Carnitine Treatment in Early Alzheimer's Disease Using Single Photon Emission Computed Tomography. Dement Neurocognitive Disord 2017; 16(1): 26-31.
[http://dx.doi.org/10.12779/dnd.2017.16.1.26] [PMID: 30906367]

[146]   Lolic MM, Fiskum G, Rosenthal RE. Neuroprotective effects of acetyl-L-carnitine after stroke in rats. Ann Emerg Med 1997; 29(6): 758-65.
[http://dx.doi.org/10.1016/S0196-0644(97)70197-5] [PMID: 9174521]

[147]   Bella R, Biondi R, Raffaele R, Pennisi G. Effect of acetyl-L-carnitine on geriatric patients suffering from dysthymic disorders. Int J Clin Pharmacol Res 1990; 10(6): 355-60.
[PMID: 2099360]

[148]   Pettegrew JW, Levine J, McClure RJ. Acetyl-L-carnitine physical-chemical, metabolic, and

therapeutic properties: relevance for its mode of action in Alzheimer's disease and geriatric depression. Mol Psychiatry 2000; 5(6): 616-32.
[http://dx.doi.org/10.1038/sj.mp.4000805] [PMID: 11126392]

[149] Calvani M, Carta A, Caruso G, Benedetti N, Iannuccelli M. Action of acetyl-L-carnitine in neurodegeneration and Alzheimer's disease. Ann N Y Acad Sci 1992; 663(1): 483-6.
[http://dx.doi.org/10.1111/j.1749-6632.1992.tb38710.x] [PMID: 1482095]

[150] Arrigo A, Casale R, Buonocore M, Ciano C. Effects of acetyl-L-carnitine on reaction times in patients with cerebrovascular insufficiency. Int J Clin Pharmacol Res 1990; 10(1-2): 133-7.
[PMID: 2387660]

[151] Thal LJ, Carta A, Clarke WR, *et al.* A 1-year multicenter placebo-controlled study of acetyl--carnitine in patients with Alzheimer's disease. Neurology 1996; 47(3): 705-11.
[http://dx.doi.org/10.1212/WNL.47.3.705] [PMID: 8797468]

[152] Hudson SA, Tabet N. Acetyl-l-carnitine for dementia 2003.
[http://dx.doi.org/10.1002/14651858.CD003158]

[153] Tempesta E, Troncon R, Janiri L, *et al.* Role of acetyl-L-carnitine in the treatment of cognitive deficit in chronic alcoholism. Int J Clin Pharmacol Res 1990; 10(1-2): 101-7.
[PMID: 2201652]

[154] Maczurek A, Hager K, Kenklies M, *et al.* Lipoic acid as an anti-inflammatory and neuroprotective treatment for Alzheimer's disease. Adv Drug Deliv Rev 2008; 60(13-14): 1463-70.
[http://dx.doi.org/10.1016/j.addr.2008.04.015] [PMID: 18655815]

[155] Hager K, Marahrens A, Kenklies M, Riederer P, Münch G. Alpha-lipoic acid as a new treatment option for Alzheimer [corrected] type dementia. Arch Gerontol Geriatr 2001; 32(3): 275-82.
[http://dx.doi.org/10.1016/S0167-4943(01)00104-2] [PMID: 11395173]

[156] Hager K, Kenklies M, McAfoose J, Engel J, Münch G. 2007.

[157] Dai Q, Borenstein AR, Wu Y, Jackson JC, Larson EB. Fruit and vegetable juices and Alzheimer's disease: the Kame Project. Am J Med 2006; 119(9): 751-9.
[http://dx.doi.org/10.1016/j.amjmed.2006.03.045] [PMID: 16945610]

[158] Ng T-P, Chiam P-C, Lee T, Chua H-C, Lim L, Kua E-H. Curry consumption and cognitive function in the elderly. Am J Epidemiol 2006; 164(9): 898-906.
[http://dx.doi.org/10.1093/aje/kwj267] [PMID: 16870699]

[159] Tully AM, Roche HM, Doyle R, *et al.* Low serum cholesteryl ester-docosahexaenoic acid levels in Alzheimer's disease: a case-control study. Br J Nutr 2003; 89(4): 483-9.
[http://dx.doi.org/10.1079/BJN2002804] [PMID: 12654166]

[160] Zhao BL, Li XJ, He RG, Cheng SJ, Xin WJ. Scavenging effect of extracts of green tea and natural antioxidants on active oxygen radicals. Cell Biophys 1989; 14(2): 175-85.
[http://dx.doi.org/10.1007/BF02797132] [PMID: 2472207]

[161] Sharman MJ, Gyengesi E, Liang H, *et al.* Assessment of diets containing curcumin, epigallocatechin-3-gallate, docosahexaenoic acid and α-lipoic acid on amyloid load and inflammation in a male transgenic mouse model of Alzheimer's disease: Are combinations more effective? Neurobiol Dis 2019; 124: 505-19.
[http://dx.doi.org/10.1016/j.nbd.2018.11.026] [PMID: 30610916]

[162] Hassing L, Wahlin A, Winblad B, Bäckman L. Further evidence on the effects of vitamin B12 and folate levels on episodic memory functioning: a population-based study of healthy very old adults. Biol Psychiatry 1999; 45(11): 1472-80.
[http://dx.doi.org/10.1016/S0006-3223(98)00234-0] [PMID: 10356630]

[163] Zhang D-M, Ye J-X, Mu J-S, Cui X-P. Efficacy of vitamin B supplementation on cognition in elderly patients with cognitive-related diseases: a systematic review and meta-analysis. J Geriatr Psychiatry Neurol 2017; 30(1): 50-9.

[http://dx.doi.org/10.1177/0891988716673466] [PMID: 28248558]

[164] Robinson N, Grabowski P, Rehman I. Alzheimer's disease pathogenesis: Is there a role for folate? Mech Ageing Dev 2018; 174: 86-94.
[http://dx.doi.org/10.1016/j.mad.2017.10.001] [PMID: 29037490]

[165] Morris MC, Evans DA, Schneider JA, Tangney CC, Bienias JL, Aggarwal NT. Dietary folate and vitamins B-12 and B-6 not associated with incident Alzheimer's disease. J Alzheimers Dis 2006; 9(4): 435-43.
[http://dx.doi.org/10.3233/JAD-2006-9410] [PMID: 16917153]

[166] Malouf R, Sastre AA. Vitamin B12 for cognition 2003.
[http://dx.doi.org/10.1002/14651858.CD004394]

[167] Gibson GE, Hirsch JA, Fonzetti P, Jordan BD, Cirio RT, Elder J. Vitamin B1 (thiamine) and dementia. Ann N Y Acad Sci 2016; 1367(1): 21-30.
[http://dx.doi.org/10.1111/nyas.13031] [PMID: 26971083]

[168] Blass JP, Sheu K-F, Cooper AJ, Jung EH, Gibson GE. Thiamin and Alzheimer's disease. J Nutr Sci Vitaminol (Tokyo) 1992; 38(Spec No): 401-4.
[http://dx.doi.org/10.3177/jnsv.38.Special_401] [PMID: 1297775]

[169] Mimori Y, Katsuoka H, Nakamura S. Thiamine therapy in Alzheimer's disease. Metab Brain Dis 1996; 11(1): 89-94.
[http://dx.doi.org/10.1007/BF02080934] [PMID: 8815393]

[170] Meador K, Loring D, Nichols M, et al. Preliminary findings of high-dose thiamine in dementia of Alzheimer's type. J Geriatr Psychiatry Neurol 1993; 6(4): 222-9.
[http://dx.doi.org/10.1177/089198879300600408] [PMID: 8251051]

[171] Li M-M, Yu J-T, Wang H-F, et al. Efficacy of vitamins B supplementation on mild cognitive impairment and Alzheimer's disease: a systematic review and meta-analysis. Curr Alzheimer Res 2014; 11(9): 844-52.
[PMID: 25274113]

[172] Sun Y, Lu C-J, Chien K-L, Chen S-T, Chen R-C. Efficacy of multivitamin supplementation containing vitamins B6 and B12 and folic acid as adjunctive treatment with a cholinesterase inhibitor in Alzheimer's disease: a 26-week, randomized, double-blind, placebo-controlled study in Taiwanese patients. Clin Ther 2007; 29(10): 2204-14.
[http://dx.doi.org/10.1016/j.clinthera.2007.10.012] [PMID: 18042476]

[173] Gasperi V, Sibilano M, Savini I, Catani MV. Niacin in the central nervous system: an update of biological aspects and clinical applications. Int J Mol Sci 2019; 20(4): 974.
[http://dx.doi.org/10.3390/ijms20040974] [PMID: 30813414]

[174] Morris MC, Evans DA, Bienias JL, et al. Dietary niacin and the risk of incident Alzheimer's disease and of cognitive decline. J Neurol Neurosurg Psychiatry 2004; 75(8): 1093-9.
[http://dx.doi.org/10.1136/jnnp.2003.025858] [PMID: 15258207]

[175] Choudhry F, Howlett DR, Richardson JC, Francis PT, Williams RJ. Pro-oxidant diet enhances β/γ secretase-mediated APP processing in APP/PS1 transgenic mice. Neurobiol Aging 2012; 33(5): 960-8.
[http://dx.doi.org/10.1016/j.neurobiolaging.2010.07.008] [PMID: 20724034]

[176] Engelhart MJ, Geerlings MI, Ruitenberg A, et al. Dietary intake of antioxidants and risk of Alzheimer disease. JAMA 2002; 287(24): 3223-9.
[http://dx.doi.org/10.1001/jama.287.24.3223] [PMID: 12076218]

[177] Farina N, Llewellyn D, Isaac MGEKN, Tabet N. 2017.

[178] Dysken M, Guarino P, Vertrees J.

[179] Zandi PP, Anthony JC, Khachaturian AS, et al. Reduced risk of Alzheimer disease in users of antioxidant vitamin supplements: the Cache County Study. Arch Neurol 2004; 61(1): 82-8.

[http://dx.doi.org/10.1001/archneur.61.1.82] [PMID: 14732624]

[180] Jiménez-Rubio G, Herrera-Pérez JJ, Hernández-Hernández OT, Martínez-Mota L. Relationship between androgen deficiency and memory impairment in aging and Alzheimer’s disease. Actas Esp Psiquiatr 2017; 45(5): 227-47.
[PMID: 29044447]

[181] Friess E, Trachsel L, Guldner J, Schier T, Steiger A, Holsboer F. DHEA administration increases rapid eye movement sleep and EEG power in the sigma frequency range. Am J Physiol 1995; 268(1 Pt 1): E107-13.
[PMID: 7840167]

[182] Nguyen T-V, Wu M, Lew J, *et al.* Dehydroepiandrosterone impacts working memory by shaping cortico-hippocampal structural covariance during development. Psychoneuroendocrinology 2017; 86: 110-21.
[http://dx.doi.org/10.1016/j.psyneuen.2017.09.013] [PMID: 28946055]

[183] Evans JG, Malouf R, Huppert FA, Van Niekerk JK. 2006.

[184] Pan X, Wu X, Kaminga AC, Wen SW, Liu A. Dehydroepiandrosterone and dehydroepiandrosterone sulfate in Alzheimer's disease: a systematic review and meta-analysis. Front Aging Neurosci 2019; 11: 61.
[http://dx.doi.org/10.3389/fnagi.2019.00061] [PMID: 30983988]

[185] Wolkowitz OM, Kramer JH, Reus VI, *et al.* DHEA treatment of Alzheimer's disease: a randomized, double-blind, placebo-controlled study. Neurology 2003; 60(7): 1071-6.
[http://dx.doi.org/10.1212/01.WNL.0000052994.54660.58] [PMID: 12682308]

[186] Azuma T, Nagai Y, Saito T, Funauchi M, Matsubara T, Sakoda S. The effect of dehydroepiandrosterone sulfate administration to patients with multi-infarct dementia. J Neurol Sci 1999; 162(1): 69-73.
[http://dx.doi.org/10.1016/S0022-510X(98)00295-0] [PMID: 10064172]

[187] Lv W, Du N, Liu Y, *et al.* Low testosterone level and risk of Alzheimer's disease in the elderly men: a systematic review and meta-analysis. Mol Neurobiol 2016; 53(4): 2679-84.
[http://dx.doi.org/10.1007/s12035-015-9315-y] [PMID: 26154489]

[188] Ford AH, Yeap BB, Flicker L, *et al.* Sex hormones and incident dementia in older men: The health in men study. Psychoneuroendocrinology 2018; 98: 139-47.
[http://dx.doi.org/10.1016/j.psyneuen.2018.08.013] [PMID: 30144781]

[189] Lu PH, Masterman DA, Mulnard R, *et al.* Effects of testosterone on cognition and mood in male patients with mild Alzheimer disease and healthy elderly men. Arch Neurol 2006; 63(2): 177-85.
[http://dx.doi.org/10.1001/archneur.63.2.nct50002] [PMID: 16344336]

[190] Baker LD, Sambamurti K, Craft S, *et al.* 17beta-estradiol reduces plasma Abeta40 for HRT-naïve postmenopausal women with Alzheimer disease: a preliminary study. Am J Geriatr Psychiatry 2003; 11(2): 239-44.
[http://dx.doi.org/10.1176/appi.ajgp.11.2.239] [PMID: 12611754]

[191] Hogervorst E, Yaffe K, Richards M, Huppert FA. Hormone replacement therapy to maintain cognitive function in women with dementia. Cochrane Database Syst Rev 2009; (1): CD003799
[http://dx.doi.org/10.1002/14651858.CD003799.pub2] [PMID: 19160224]

[192] Hogervorst E, Yaffe K, Richards M, Huppert F. Hormone replacement therapy for cognitive function in postmenopausal women. Cochrane Database Syst Rev 2002; (3): CD003122
[http://dx.doi.org/10.1002/14651858.CD003122] [PMID: 12137675]

[193] Shumaker SA, Legault C, Kuller L, *et al.* Conjugated equine estrogens and incidence of probable dementia and mild cognitive impairment in postmenopausal women: Women's Health Initiative Memory Study. JAMA 2004; 291(24): 2947-58.
[http://dx.doi.org/10.1001/jama.291.24.2947] [PMID: 15213206]

[194] Hashimoto M, Tanabe Y, Fujii Y, Kikuta T, Shibata H, Shido O. Chronic administration of docosahexaenoic acid ameliorates the impairment of spatial cognition learning ability in amyloid β-infused rats. J Nutr 2005; 135(3): 549-55.
[http://dx.doi.org/10.1093/jn/135.3.549] [PMID: 15735092]

[195] Florent S, Malaplate-Armand C, Youssef I, *et al.* Docosahexaenoic acid prevents neuronal apoptosis induced by soluble amyloid-β oligomers. J Neurochem 2006; 96(2): 385-95.
[http://dx.doi.org/10.1111/j.1471-4159.2005.03541.x] [PMID: 16300635]

[196] Fotuhi M, Mohassel P, Yaffe K. Fish consumption, long-chain omega-3 fatty acids and risk of cognitive decline or Alzheimer disease: a complex association. Nat Clin Pract Neurol 2009; 5(3): 140-52.
[PMID: 19262590]

[197] Zhang Y, Chen J, Qiu J, Li Y, Wang J, Jiao J. Intakes of fish and polyunsaturated fatty acids and mild-to-severe cognitive impairment risks: a dose-response meta-analysis of 21 cohort studies. Am J Clin Nutr 2016; 103(2): 330-40.
[http://dx.doi.org/10.3945/ajcn.115.124081] [PMID: 26718417]

[198] Burckhardt M, Herke M, Wustmann T, Watzke S, Langer G, Fink A. Omega-3 fatty acids for the treatment of dementia. Cochrane Database Syst Rev 2016; 4CD009002
[http://dx.doi.org/10.1002/14651858.CD009002.pub3] [PMID: 27063583]

[199] Moré MI, Freitas U, Rutenberg D. Positive effects of soy lecithin-derived phosphatidylserine plus phosphatidic acid on memory, cognition, daily functioning, and mood in elderly patients with Alzheimer's disease and dementia. Adv Ther 2014; 31(12): 1247-62.
[http://dx.doi.org/10.1007/s12325-014-0165-1] [PMID: 25414047]

[200] Cenacchi T, Bertoldin T, Farina C, Fiori MG, Crepaldi G. Cognitive decline in the elderly: a double-blind, placebo-controlled multicenter study on efficacy of phosphatidylserine administration. Aging (Milano) 1993; 5(2): 123-33.
[http://dx.doi.org/10.1007/BF03324139] [PMID: 8323999]

[201] Olivera-Pueyo J, Pelegrín-Valero C. Dietary supplements for cognitive impairment. Actas Esp Psiquiatr 2017; 45 (Suppl.): 37-47.
[PMID: 29171642]

[202] Fioravanti M, Flicker L. Nicergoline for dementia and other age associated forms of cognitive impairment 2001.
[http://dx.doi.org/10.1002/14651858.CD003159]

[203] Fioravanti M, Yanagi M. Cytidinediphosphocholine (CDP choline) for cognitive and behavioural disturbances associated with chronic cerebral disorders in the elderly. Cochrane Database Syst Rev 2004; (2): CD000269
[http://dx.doi.org/10.1002/14651858.CD000269.pub2] [PMID: 15106147]

[204] Mitka M. News about neuroprotectants for the treatment of stroke. JAMA 2002; 287(10): 1253-4.
[http://dx.doi.org/10.1001/jama.287.10.1253-JMN0313-2-1] [PMID: 11886300]

[205] Secades JJ, Lorenzo JL. Citicoline: pharmacological and clinical review, 2006 update. Methods Find Exp Clin Pharmacol 2006; 28 (Suppl. B): 1-56.
[PMID: 17171187]

[206] Spiers PA, Hochanadel G. Citicoline for traumatic brain injury: report of two cases, including my own. J Int Neuropsychol Soc 1999; 5(3): 260-4.
[http://dx.doi.org/10.1017/S1355617799533092] [PMID: 10217926]

[207] Jimbo D, Kimura Y, Taniguchi M, Inoue M, Urakami K. Effect of aromatherapy on patients with Alzheimer's disease. Psychogeriatrics 2009; 9(4): 173-9.
[http://dx.doi.org/10.1111/j.1479-8301.2009.00299.x] [PMID: 20377818]

[208] Holmes C, Hopkins V, Hensford C, MacLaughlin V, Wilkinson D, Rosenvinge H. Lavender oil as a

treatment for agitated behaviour in severe dementia: a placebo controlled study. Int J Geriatr Psychiatry 2002; 17(4): 305-8.
[http://dx.doi.org/10.1002/gps.593] [PMID: 11994882]

[209] Ballard CG, O'Brien JT, Reichelt K, Perry EK. Aromatherapy as a safe and effective treatment for the management of agitation in severe dementia: the results of a double-blind, placebo-controlled trial with Melissa. J Clin Psychiatry 2002; 63(7): 553-8.
[http://dx.doi.org/10.4088/JCP.v63n0703] [PMID: 12143909]

[210] Thorgrimsen L, Spector A, Wiles A, Orrell M. Aroma therapy for dementia 2003.

[211] Benny A, Thomas J. Essential Oils as Treatment Strategy for Alzheimer's Disease: Current and Future Perspectives. Planta Med 2019; 85(3): 239-48.
[http://dx.doi.org/10.1055/a-0758-0188] [PMID: 30360002]

[212] Snow LA, Hovanec L, Brandt J. A controlled trial of aromatherapy for agitation in nursing home patients with dementia. J Altern Complement Med 2004; 10(3): 431-7.
[http://dx.doi.org/10.1089/1075553041323696] [PMID: 15253846]

[213] Crichton RR, Dexter D, Ward RJ. Metal based neurodegenerative diseases—from molecular mechanisms to therapeutic strategies. Coord Chem Rev 2008; 252(10-11): 1189-99.
[http://dx.doi.org/10.1016/j.ccr.2007.10.019]

[214] Brewer GJ. Alzheimer's disease causation by copper toxicity and treatment with zinc. Front Aging Neurosci 2014; 6: 92.
[http://dx.doi.org/10.3389/fnagi.2014.00092] [PMID: 24860501]

[215] Brewer GJ, Kaur S. Zinc deficiency and zinc therapy efficacy with reduction of serum free copper in Alzheimer's disease 2013.
[http://dx.doi.org/10.1155/2013/586365]

[216] Brewer GJ. Copper toxicity in Alzheimer's disease: cognitive loss from ingestion of inorganic copper. J Trace Elem Med Biol 2012; 26(2-3): 89-92.
[http://dx.doi.org/10.1016/j.jtemb.2012.04.019] [PMID: 22673823]

[217] Brewer GJ. Copper excess, zinc deficiency, and cognition loss in Alzheimer's disease. Biofactors 2012; 38(2): 107-13.
[http://dx.doi.org/10.1002/biof.1005] [PMID: 22438177]

[218] Crapper McLachlan DR, Dalton AJ, Kruck TP, *et al.* Intramuscular desferrioxamine in patients with Alzheimer's disease. Lancet 1991; 337(8753): 1304-8.
[http://dx.doi.org/10.1016/0140-6736(91)92978-B] [PMID: 1674295]

[219] Sampson EL, Jenagaratnam L, McShane R. Metal protein attenuating compounds for the treatment of Alzheimer's dementia 2014.
[http://dx.doi.org/10.1002/14651858.CD005380.pub5]

[220] Jiang C, Li G, Huang P, Liu Z, Zhao B. The gut microbiota and Alzheimer's disease. J Alzheimers Dis 2017; 58(1): 1-15.
[http://dx.doi.org/10.3233/JAD-161141] [PMID: 28372330]

[221] Liang S, Wang T, Hu X, *et al.* Administration of Lactobacillus helveticus NS8 improves behavioral, cognitive, and biochemical aberrations caused by chronic restraint stress. Neuroscience 2015; 310: 561-77.
[http://dx.doi.org/10.1016/j.neuroscience.2015.09.033] [PMID: 26408987]

[222] Möhle L, Mattei D, Heimesaat MM, *et al.* Ly6Chi monocytes provide a link between antibiotic-induced changes in gut microbiota and adult hippocampal neurogenesis. Cell Rep 2016; 15(9): 1945-56.
[http://dx.doi.org/10.1016/j.celrep.2016.04.074] [PMID: 27210745]

[223] Akbari E, Asemi Z, Daneshvar Kakhaki R, *et al.* Effect of probiotic supplementation on cognitive function and metabolic status in Alzheimer's disease: a randomized, double-blind and controlled trial.

Front Aging Neurosci 2016; 8: 256.
[http://dx.doi.org/10.3389/fnagi.2016.00256] [PMID: 27891089]

[224] Lin Y-C, Wang Y-P. Status of Noninvasive Brain Stimulation in the Therapy of Alzheimer's Disease. Chin Med J (Engl) 2018; 131(24): 2899-903.
[http://dx.doi.org/10.4103/0366-6999.247217] [PMID: 30539900]

[225] Cameron MH, Lonergan E, Lee H. 2003.

[226] Gonsalvez I, Baror R, Fried P, Santarnecchi E, Pascual-Leone A. Therapeutic noninvasive brain stimulation in Alzheimer's disease. Curr Alzheimer Res 2017; 14(4): 362-76.
[PMID: 27697061]

[227] McDermott B, Porter E, Hughes D, *et al.* Gamma band neural stimulation in humans and the promise of a new modality to prevent and treat Alzheimer's disease. J Alzheimers Dis 2018; 65(2): 363-92.
[http://dx.doi.org/10.3233/JAD-180391] [PMID: 30040729]

[228] Berman MH, Halper JP, Nichols TW, Jarrett H, Lundy A, Huang JH. Photobiomodulation with near infrared light helmet in a pilot, placebo controlled clinical trial in dementia patients testing memory and cognition. J Neurol Neurosci 2017; 8(1): 176.
[http://dx.doi.org/10.21767/2171-6625.1000176] [PMID: 28593105]

[229] Ragneskog H, Bråne G, Karlsson I, Kihlgren M. Influence of dinner music on food intake and symptoms common in dementia. Scand J Caring Sci 1996; 10(1): 11-7.
[http://dx.doi.org/10.1111/j.1471-6712.1996.tb00304.x] [PMID: 8715781]

[230] Suzuki M, Tatsumi A, Otsuka T, *et al.* Physical and psychological effects of 6-week tactile massage on elderly patients with severe dementia. Am J Alzheimers Dis Other Demen 2010; 25(8): 680-6.
[http://dx.doi.org/10.1177/1533317510386215] [PMID: 21131675]

[231] Kumar AM, Tims F, Cruess DG, *et al.* Music therapy increases serum melatonin levels in patients with Alzheimer's disease. Altern Ther Health Med 1999; 5(6): 49-57.
[PMID: 10550905]

[232] Koger SM, Chapin K, Brotons M. Is music therapy an effective intervention for dementia? A meta-analytic review of literature. J Music Ther 1999; 36(1): 2-15.
[http://dx.doi.org/10.1093/jmt/36.1.2] [PMID: 10519841]

[233] Abraha I, Rimland JM, Trotta FM, *et al.* Systematic review of systematic reviews of non-pharmacological interventions to treat behavioural disturbances in older patients with dementia. The SENATOR-OnTop series. BMJ Open 2017; 7(3)e012759
[http://dx.doi.org/10.1136/bmjopen-2016-012759] [PMID: 28302633]

[234] Ueda T, Suzukamo Y, Sato M, Izumi S. Effects of music therapy on behavioral and psychological symptoms of dementia: a systematic review and meta-analysis. Ageing Res Rev 2013; 12(2): 628-41.
[http://dx.doi.org/10.1016/j.arr.2013.02.003] [PMID: 23511664]

[235] Scogin F, Bienias JL. A three-year follow-up of older adult participants in a memory-skills training program. Psychol Aging 1988; 3(4): 334-7.
[http://dx.doi.org/10.1037/0882-7974.3.4.334] [PMID: 3268276]

[236] O'Hara R, Brooks JO III, Friedman L, Schröder CM, Morgan KS, Kraemer HC. Long-term effects of mnemonic training in community-dwelling older adults. J Psychiatr Res 2007; 41(7): 585-90.
[http://dx.doi.org/10.1016/j.jpsychires.2006.04.010] [PMID: 16780878]

[237] Hill NT, Mowszowski L, Naismith SL, Chadwick VL, Valenzuela M, *et al.* 2017.Computerized Cognitive Training in Older Adults With Mild Cognitive Impairment or Dementia: A Systematic Review and Meta-Analysis
[http://dx.doi.org/10.1176/appi.ajp.2016.16030360]

[238] Liang JH, Xu Y, Lin L, Jia RX, Zhang HB, Hang L. Comparison of multiple interventions for older adults with Alzheimer disease or mild cognitive impairment: A PRISMA-compliant network meta-analysis. Medicine (Baltimore) 2018; 97(20)e10744

[http://dx.doi.org/10.1097/MD.0000000000010744] [PMID: 29768349]

[239] Rebok GW, Ball K, Guey LT, *et al.* Ten-year effects of the advanced cognitive training for independent and vital elderly cognitive training trial on cognition and everyday functioning in older adults. J Am Geriatr Soc 2014; 62(1): 16-24.
[http://dx.doi.org/10.1111/jgs.12607] [PMID: 24417410]

[240] McCurry SM, Ancoli-Israel S. Sleep dysfunction in Alzheimer's disease and other dementias. Curr Treat Options Neurol 2003; 5(3): 261-72.
[http://dx.doi.org/10.1007/s11940-003-0017-9] [PMID: 12670415]

[241] Satlin A, Volicer L, Ross V, Herz L, Campbell S. Bright light treatment of behavioral and sleep disturbances in patients with Alzheimer's disease. Am J Psychiatry 1992; 149(8): 1028-32.
[http://dx.doi.org/10.1176/ajp.149.8.1028] [PMID: 1353313]

[242] Ancoli-Israel S, Martin JL, Gehrman P, *et al.* Effect of light on agitation in institutionalized patients with severe Alzheimer disease. Am J Geriatr Psychiatry 2003; 11(2): 194-203.
[http://dx.doi.org/10.1097/00019442-200303000-00010] [PMID: 12611749]

[243] Haffmans P, Lucius S, Sival R. Bright light therapy and melatonin in motor restlessness in dementia. Eur Neuropsychopharmacol 1998; (8): S273.
[http://dx.doi.org/10.1016/S0924-977X(98)80512-9]

[244] Forbes D, Morgan D, Bangma J, Peacock S, Pelletier N, Adamson J. Light therapy for managing sleep, behavior, and mood disturbances in dementia

[245] Mitolo M, Tonon C, La Morgia C, Testa C, Carelli V, Lodi R. Effects of light treatment on sleep, cognition, mood, and behavior in Alzheimer's Disease: A Systematic Review. Dement Geriatr Cogn Disord 2018; 46(5-6): 371-84.
[http://dx.doi.org/10.1159/000494921] [PMID: 30537760]

[246] Jiang Y, Abiri R, Zhao X. Tuning up the old brain with new tricks: attention training *via* neurofeedback. Front Aging Neurosci 2017; 9: 52.
[http://dx.doi.org/10.3389/fnagi.2017.00052] [PMID: 28348527]

[247] Rozelle GR, Budzynski TH. Neurotherapy for stroke rehabilitation: a single case study. Biofeedback Self Regul 1995; 20(3): 211-28.
[http://dx.doi.org/10.1007/BF01474514] [PMID: 7495916]

[248] Glanz M, Klawansky S, Chalmers T. Biofeedback therapy in stroke rehabilitation: a review. J R Soc Med 1997; 90(1): 33-9.
[http://dx.doi.org/10.1177/014107689709000110] [PMID: 9059379]

[249] Thornton KE, Carmody DP. Efficacy of traumatic brain injury rehabilitation: interventions of QEEG-guided biofeedback, computers, strategies, and medications. Appl Psychophysiol Biofeedback 2008; 33(2): 101-24.
[http://dx.doi.org/10.1007/s10484-008-9056-z] [PMID: 18551365]

[250] McCraty R, Atkinson M, Eds. Influence of afferent cardiovascular input on cognitive performance and alpha activity. Proceedings of the Annual Meeting of the Pavlovian Society.

[251] Chung JC, Lai CK. Snoezelen for dementia 2002.
[http://dx.doi.org/10.1002/14651858.CD003152]

[252] Strøm BS, Ytrehus S, Grov EK. Sensory stimulation for persons with dementia: a review of the literature. J Clin Nurs 2016; 25(13-14): 1805-34.
[http://dx.doi.org/10.1111/jocn.13169] [PMID: 27030571]

[253] Fraser J, Kerr JR. Psychophysiological effects of back massage on elderly institutionalized patients. J Adv Nurs 1993; 18(2): 238-45.
[http://dx.doi.org/10.1046/j.1365-2648.1993.18020238.x] [PMID: 8436714]

[254] Snyder M, Egan EC, Burns KR. Efficacy of hand massage in decreasing agitation behaviors associated

with care activities in persons with dementia: A simple, easily instituted method of relaxation may decrease agitation and disruptive behaviors. Geriatr Nurs (Minneap) 1995; 16(2): 60-3.
[http://dx.doi.org/10.1016/S0197-4572(05)80005-9]

[255]   Moyle W, Cooke ML, Beattie E, *et al.* Foot massage and physiological stress in people with dementia: a randomized controlled trial. J Altern Complement Med 2014; 20(4): 305-11.
[http://dx.doi.org/10.1089/acm.2013.0177] [PMID: 24047244]

[256]   Ebel S. Designing stage-specific horticultural therapy interventions for patients with Alzheimer's disease 1991.

[257]   Epstein M, Hansen V, Hazen T. Therapeutic gardens: plant centered activities meet sensory, physical and psychosocial needs. Oreg J Aging 1991; 9: 8-14.

[258]   Detweiler MB, Warf C. Dementia wander garden aids post cerebrovascular stroke restorative therapy: a case study. Altern Ther Health Med 2005; 11(4): 54-8.
[PMID: 16053122]

[259]   Gonzalez MT, Kirkevold M. Benefits of sensory garden and horticultural activities in dementia care: a modified scoping review. J Clin Nurs 2014; 23(19-20): 2698-715.
[http://dx.doi.org/10.1111/jocn.12388] [PMID: 24128125]

[260]   Whear R, Coon JT, Bethel A, Abbott R, Stein K, Garside R. What is the impact of using outdoor spaces such as gardens on the physical and mental well-being of those with dementia? A systematic review of quantitative and qualitative evidence. J Am Med Dir Assoc 2014; 15(10): 697-705.
[http://dx.doi.org/10.1016/j.jamda.2014.05.013] [PMID: 25037168]

[261]   Khalsa DS. Stress, meditation, and Alzheimer's disease prevention: where the evidence stands. J Alzheimers Dis 2015; 48(1): 1-12.
[http://dx.doi.org/10.3233/JAD-142766] [PMID: 26445019]

[262]   Gard T, Hölzel BK, Lazar SW. The potential effects of meditation on age-related cognitive decline: a systematic review. Ann N Y Acad Sci 2014; 1307: 89-103.
[http://dx.doi.org/10.1111/nyas.12348] [PMID: 24571182]

[263]   Quintana-Hernández DJ, Miró-Barrachina MT, Ibáñez-Fernández IJ, *et al.* Mindfulness in the maintenance of cognitive capacities in Alzheimer's disease: a randomized clinical trial. J Alzheimers Dis 2016; 50(1): 217-32.
[http://dx.doi.org/10.3233/JAD-143009] [PMID: 26639952]

[264]   Larouche E, Hudon C, Goulet S. Potential benefits of mindfulness-based interventions in mild cognitive impairment and Alzheimer's disease: an interdisciplinary perspective. Behav Brain Res 2015; 276: 199-212.
[http://dx.doi.org/10.1016/j.bbr.2014.05.058] [PMID: 24893317]

[265]   Luders E. Exploring age-related brain degeneration in meditation practitioners. Ann N Y Acad Sci 2014; 1307(1): 82-8.
[http://dx.doi.org/10.1111/nyas.12217] [PMID: 23924195]

[266]   Newberg AB, Wintering N, Khalsa DS, Roggenkamp H, Waldman MR. Meditation effects on cognitive function and cerebral blood flow in subjects with memory loss: a preliminary study. J Alzheimers Dis 2010; 20(2): 517-26.
[http://dx.doi.org/10.3233/JAD-2010-1391] [PMID: 20164557]

[267]   Ngandu T, Lehtisalo J, Solomon A, *et al.* A 2 year multidomain intervention of diet, exercise, cognitive training, and vascular risk monitoring *versus* control to prevent cognitive decline in at-risk elderly people (FINGER): a randomised controlled trial. Lancet 2015; 385(9984): 2255-63.
[http://dx.doi.org/10.1016/S0140-6736(15)60461-5] [PMID: 25771249]

[268]   Bredesen DE. Reversal of cognitive decline: a novel therapeutic program. Aging (Albany NY) 2014; 6(9): 707-17.
[http://dx.doi.org/10.18632/aging.100690] [PMID: 25324467]

[269]   Ashfeld M. Effect of Therapeutic touch in treating agitation of persons with Alzheimer's Disease 2011.

[270]   Wang KL, Hermann C. Pilot study to test the effectiveness of Healing Touch on agitation in people with dementia. Geriatr Nurs 2006; 27(1): 34-40.
[http://dx.doi.org/10.1016/j.gerinurse.2005.09.014] [PMID: 16483898]

[271]   Ostuni E, Pietro MJ. World Alzheimer's Congress. Washington, D.C.. 2000.

[272]   Ostuni E, Santo Pietro M, Eds. Effects of healing touch on nursing home residents in later stages of Alzheimer disease. World Alzheimer Congress.

[273]   Woods DL, Craven RF, Whitney J. The effect of therapeutic touch on behavioral symptoms of persons with dementia. Altern Ther Health Med 2005; 11(1): 66-74.
[PMID: 15712768]

[274]   Kumarappah A, Senderovich H. Therapeutic touch in the management of responsive behavior in patients with dementia. Adv Mind Body Med 2016; 30(4): 8-13.
[PMID: 27925607]

[275]   Cai F-F, Zhang H. Effect of therapeutic touch on agitated behavior in elderly patients with dementia: A review. Int J Nurs Sci 2015; 2(3): 324-8.
[http://dx.doi.org/10.1016/j.ijnss.2015.08.002]

[276]   Lam LC, Chau RC, Wong BM, Fung AW, Tam CW, Leung GT, *et al.* 2012.

[277]   Tadros G, Ormerod S, Dobson-Smyth P, *et al.* The management of behavioural and psychological symptoms of dementia in residential homes: does Tai Chi have any role for people with dementia? Dementia 2013; 12(2): 268-79.
[http://dx.doi.org/10.1177/1471301211422769] [PMID: 24336773]

[278]   Niu Y, Wan C, Zhou B, *et al.* Breath qigong improves recognition in seniors with vascular cognitive impairment. Altern Ther Health Med 2019; 25(1): 20-6.
[PMID: 30982783]

[279]   Tashiro M, Xiang X, Okamura N, Ishizaki H, Miyazaki H, Ishii K, *et al.* Three-Dimensional PET-An Approach in Psychology. J Int Soc Life Inf Sci 1996; 14(2): 282-4.

[280]   Zhao GI, Wie Q. A case of cerebral atrophy cured by qigong. First World Conference for Academic Exchange of Medical Qigong. Beijing, China. 1988.

CHAPTER 4

Overview of Analytical Methods in Alzheimer's Disease Drugs: Optical, Chromatographic, and Electrochemical Methods

Cem Erkmen, Burcin Bozal-Palabiyik and **Bengi Uslu**[*]

Department of Analytical Chemistry, Faculty of Pharmacy, Ankara University, Ankara, Turkey

Abstract: Alzheimer's disease is a neurodegenerative disorder that results in a loss of memory, cognitive problems, and personality change. The frequency of this disease is increasing rapidly due to the aging world population. Every year, more people are exposed to this age-related disorder. It is expected that the number of individuals diagnosed with Alzheimer's disease would increase to 100 million by 2050. The treatment of Alzheimer's disease costs more than $200 billion annually in the US and this cost will possibly grow to $1 trillion annually by 2050. There are two categories of drugs used for the treatment of Alzheimer's disease; while the first category attempts to treat the symptoms of the disease (such as rivastigmine, galantamine, and donepezil), the second category focuses on the specific site or physiological factor of the disease. Still, there are attempts to develop new therapies to prevent, defer, slow down the progress, or ameliorate the symptoms of this disease. Spectrophotometric, chromatographic, and electrochemical methods have proven to be sensitive and reliable for the determination of numerous compounds. In this chapter, recent advances in different analysis methods of various Alzheimer's disease drugs are summarized. Finally, this chapter aims to explore the advantages of methods as well as highlight the future perspectives of these methods in assaying Alzheimer drugs in pharmaceutical and biological samples.

Keywords: Alzheimer's disease, Amperometry, Analysis, Donepezil, Galantamine, Gas chromatography, Huperzine A, Liquid chromatography, Memantine, Potentiometry, Rivastigmine, Spectrophotometry, Spectrofluorimetry, Tacrine, Validation, Voltammetry.

INTRODUCTION

Dementia is a brain disorder resulting in significant deterioration of personal,

---
[*] **Corresponding author Bengi Uslu:** Department of Analytical Chemistry, Faculty of Pharmacy, Ankara University, Ankara, Turkey; Tel: +903122033178; Fax: +903122131081; E-mail: buslu@pharmacy.ankara.edu.tr

Atta-ur-Rahman (Ed.)
All rights reserved-© 2020 Bentham Science Publishers

social, and professional functions and it is defined as an acquired deficiency of memorial and cognitive processes. Other brain functions affected by this disease can be enumerated as orientation, calculation, learning abilities, cognition, language abilities, judgment, and reasoning. Dementia is not an outcome of normal aging processes; although there are some significant exceptions, dementia is a progressive disorder [1]. The behavioral changes emerged out of dementia are neither conscious nor are they the results of laziness or "letting go" [2].

Most dementias are categorized as neurodegenerative diseases because they emerged out of progressive degeneration and death of nerve cells [3]. Neurodegenerative disorders are the results of abnormal accumulation of non-dissolved proteins in the brain. These proteins are toxic and they have detrimental effects on selective nerve cells. They deteriorate the functions of these cells leading to the death of these cells at the end. Moreover, this abnormal protein accumulation affects synapses between nerve cells. The neural circuits can be cut because the chemical information between cells can not be transferred properly [4].

Alzheimer's disease (AD) is the most widely seen dementia type comprising almost 50-60% of all dementia cases [5, 6]. It has characteristic neuropathological and neurochemical properties. Generally, it is very insidious at the beginning and it advances slowly but consistently across years [2]. AD can also be defined as a heterogeneous disease which is a combination of environmental and genetic factors. The most significant risk factor for AD is age. The environmental risk factors are hypertension, estrogen supplements, smoking, stroke, cardiac diseases, depression, arthritis, and diabetes [6].

AD is characterized by the development of two main lesions, namely amyloid plaques and neurofibrillary tangles (NFTs) [7]. These two lesions resulted in progressive dementia that emerged out of neuronal dysfunction and death of cells at the hippocampus and medial, temporal, and parietal lobes of the brain. The defection at the hippocampus resulted in deterioration of short term memory and affects the daily routines of the patients. Later, the brain cortex and most importantly the areas responsible for linguistic and reasoning functions are influenced. In the end, other parts of the brain are eclipsed together with atrophy and functional losses [8].

The emergence of amyloid plaques is due to their pathological overproduction or as a result of β-amyloid peptide accumulation around neurons because of deteriorated clearance along the blood-brain barrier. The β-amyloid peptides accumulated at the brain are inclined to accumulate first towards soluble pathogenic β-amyloid oligomers and fibrils and then towards insoluble β-amyloid plaques. The accumulation of β-amyloid plaques at brain arterial walls resulted in cerebral amyloid angiopathy [7]. According to the amyloid cascade hypothesis, which had been proposed at the beginning of the 1990s, the amyloid accumulations at AD resulting from a series of genetic and environmental factors and the neural cell degeneration emerging out of it ultimately leads to dementia [9].

The second significant lesion of AD is neurofibrillary tangles. The NFTs are composed of a protein called tau. Unlike plaques, the number of NFTs and their anatomical location in the brain is significantly related to the strength of dementia. Moreover, it is thought that the presence of NFTs within nerve cells is a strong indication of prospective cell death. Tau protein stabilizes the microtubules necessary for the rapid transfer of microscopic components at neurons from the nerve cell. The damaging microtubules resulted in functional losses of neurons. In AD, tau protein loses its ability to support microtubule stabilization and transforms into NFTs [10].

Since the definition of the AD at the beginning of the twentieth century, the discovery of amyloid plaques and NFTs at post-mortem brains of the patients became the basis for drug development studies for this disease; however, the discovery of drugs for the treatment of AD started with the discovery of the relationship between memory disorders and cholinergic pathways in 1984 [11]. It is understood that the neurons secreting acetylcholine for a long time are the neurons that are affected by AD the most. Acetylcholine (cholinergic) pathways are critically important for normal memory functions [12]. Since the discovery of the significance of acetylcholine for the treatment of AD, several treatments for overcoming the disease have been attempted. While earlier studies attempting to increase acetylcholine levels in the brain directly were not much promising, the methods developed for indirect boosting of acetylcholine levels by reducing its breakdown have some promising results [1]. Acetylcholinesterase (AChE), which is responsible for the hydrolysis of acetylcholine, was focused on to restore the cognitive impairment through rising neurotransmitter levels in the central nervous system [11]. The changes in neurotransmission in AD happens only after a significant level of cellular death. The lack of acetylcholine is not a reason but a result of the disease; therefore the treatments targeting for increasing the level of acetylcholine in the brain are not much effective to change the underlying causes of the disease [1]. In other words, the actual treatments of AD do not target underlying causes of AD but they attempt to cure the symptoms of the disease [1, 7, 12, 13]. Although numerous drugs are entering clinical monitoring processes, treatments curing the disease effectively can not be developed yet. That is why AD has been perceived as a priority for public health today [7].

In addition to the lack of cholinesterase, another significant neurochemical change in AD is glutamatergic over-stimulation of post-synaptic N-methyl-D-aspartate (NMDA) receptors. Therefore, AD treatment approved by the US Food and Drug Agency has two categories namely, cholinesterase inhibitors and NMDA receptor antagonists [13]. For the last 40 years, only five drugs were developed (Fig. **1**); four of them are cholinesterase inhibitors (tacrine, donepezil, rivastigmine, and galantamine; 1993-2000) and an NMDA receptor antagonist (memantine, approved by European Medicines Agency in 2002 and Food and Drug Administration (FDA) in 2003) [11]. Since 70 percent of the patients have an effective relief after cholinesterase inhibitor treatment, this class of drugs turns out to be the most significant drugs for AD treatment [13].

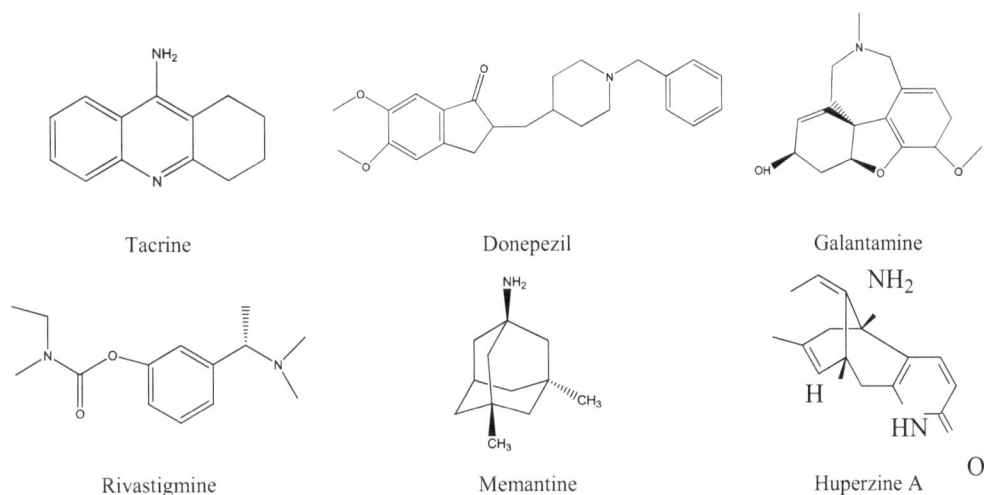

Fig. (1). Chemical structures of AD drugs.

Tacrine, the first acetylcholinesterase inhibitor approved by the FDA in 1993, is not selective for all forms of acetylcholinesterase. Moreover, hepatotoxicity and the need for frequent dosing are major limits of this drug. Both rivastigmine (approved by the FDA in 2000) and galantamine (approved by the FDA in 2001) are selectively active at the central nervous system; rivastigmine is probably mostly selective at the central nervous system and has minimal activity at the periphery. They reach maximum concentration levels just one hour after taking. Whereas, donepezil, the second approved drug by the FDA, has more peripheral effects. With a single daily dose, it has a long half-life (70 hours) in the patients. All cholinesterase inhibitors are easily absorbed along the gastrointestinal system [14].

Unlike cholinesterase inhibitors, which have been used for early or medium phases of AD, memantine is used for medium and later phases of the disease. After oral intake, memantine is absorbed well and peak concentrations are reached approximately between 3 to 7 hours. It is safe and tolerated well, it is sometimes used with donepezil [15].

Huperzine A has been utilized in Chinese traditional medicine for centuries in order to cure fever and inflammation. A while ago, its purified component was used as a prescribed drug for memory disorders that emerged due to advanced age. Although there is no effective antipyretic and anti-inflammatory properties of Huperzine A, it seems to be a significant acetylcholinesterase inhibitor. Huperzine A exerts a very significant specificity for acetylcholinesterase, and has very little efficacy on butyrylcholinesterase, unlike tacrine [2].

Ginkgo biloba is one of the oldest plants in the world and has been used for more than 3500 years in Chinese traditional medicine to treat cardiac and pulmonary diseases. Ginkgo preparations have been used in Europe since the mid-1960s for the treatment of peripheric and central vascular diseases as well as hearing and visual impairment. In 1994, German health authorities identical to the US FDA approved standard extracts of Gingko biloba as safe and effective for treating multi-infarct dementia symptoms as well as AD [2]. Gingko biloba is prescribed in Europe for memory disorders routinely and in Germany, a daily 240 mg Ginkgo substrate is approved for the treatment of AD. However, it should not be forgotten that Ginkgo might end up with gastric bleeding particularly if it is taken with aspirin [1].

These treating agents could neither prevent nor slow down neurodegenerative processes [12, 16, 17]. Therefore, during the 2010s, new revolutionary approaches based on multi-target directed ligand have been promising for AD treatment. Instead of classical "one molecule, one target" principle based drug design, this multi-target approach has been perceived as more promising. For instance, recently, a study focuses on a memantine heterodimer, namely the molecule 7-methoxytacrine, employing multi-target directed ligand theory for AD treatment [12]. Recently, AD drug discovery has focused on 'disease modifying drugs'. For this purpose, monoclonal antibodies, namely bapineuzumab, solanezumab, crenezumab, BIIB037/BART, and BAN2401, were designed and tested for the treatment of AD via making brain clear from β-amyloid [18, 19]. The other treatment options for AD are tetrahydroisoquinoline-benzimidazole hybrids as multifunctional agents [20] and chelators which can reduce the toxicity of exogenous β-amyloid [21]. There are no developed analytical methods for these compounds so these molecules are out of the chapter's scope.

Since there is no drug in the market completely treating AD and since actual drugs could only serve for ameliorating its symptoms and slowing down its progress, the detection of AD drugs at biological fluids and brain tissue is critically important for the evaluation of accurate treatment strategies.

The evaluation of pharmacokinetic parameters is very significant for the determination of the pharmacodynamic effects. These pharmacokinetic parameters are very important for ultimate dosage choice and thereby treatment. Therefore, in pharmacokinetics, drug concentration levels, bioequivalence, and therapeutic drug monitoring studies validated analytical methods are quite significant [13]. This chapter, therefore, reviews the analytical methods developed for the determination of drugs used for the treatment of AD (FDA approved drugs and Huperzine A) from pharmaceutical dosage forms, biological fluids, and tissues. In doing that, after briefly explaining chromatographic, optical, and electrochemical methods, the developed and validated methods for the determination of AD drugs since the beginning of the 2000s are covered.

ANALYSIS OF APPROVED ALZHEIMER'S DISEASE DRUGS

Chromatographic, optical, and electrochemical methods have been designed for the determination of AD drugs in different samples such as pharmaceuticals, biological samples, serum/plasma or urine, natural products, environmental samples, tissues, and others.

An important step in developing methods for the analysis of drugs from different matrices is method validation. Method validation is an essential part of the reliability of the results of the proposed methods for their routine application. The validation requirements of developed analytical methods for pharmaceutical researches have been clearly specified by several regulatory authorities such as the International Conference on Harmonization (ICH), FDA. However, the applied protocol parameters can be customized to meet the demand of the application [22 - 25].

In pharmaceutical research, validation is characterized by parameters such as selectivity, linearity, range, limit of detection (LOD), limit of quantification (LOQ), accuracy, precision (reproducibility, intermediate precision, and repeatability), and robustness [23]. In this section, these parameters will be briefly mentioned.

Selectivity is one of the primary and most significant validation parameters that informs the researchers about method reliability. It is defined as "the ability to assess unequivocally the analyte in the presence of components which may be expected to be present" by ICH guidelines. Selectivity is the study of clearly identifying/quantifying the intended analytes in the presence of other compounds [26 - 28].

The linearity (within a certain range) of an analytical method is defined as the direct proportion of analyte concentration in the sample with the obtained test results by ICH. In general, developed methods are defined linearly when the response and the concentration of the analyte has a direct proportionality. Linearity can be shown either directly on the working solutions via diluting the stock solution or separately by weighing synthetic mixtures of the test product components [23, 29, 30].

LOD, the lowest concentration to be detected for an analyte in a sample, does not need to be measured quantitatively under the specified conditions of the test. In general, any compound detected with a response at about three times the noise response level is construed to be at its LOD. LOD can be determined in three different ways (visual inspection, signal-to-noise ratio, and standard deviation of the response based on the slope of the calibration curve) [31, 32].

LOQ is defined as the lowest amount of analyte to be determined quantitatively in a sample with appropriate precision and accuracy. The LOQ value is commonly determined as 10-fold the noise response level. LOQ can also be determined by the same 3 different methods as LOD [28, 33, 34].

Accuracy evaluates the trueness of the results by evaluating the systematic and random effects on the results. It expresses how close the mean of an infinite number of results is to a reference value. In order to determine accuracy, comparison with an appropriate certified reference material and recovery of compound spiked into blank matrix procedures can be used for this purpose [35 - 37].

Precision measures the degree of closeness of the results to one another through statistical parameters that define the results. The relative standard deviation (RSD) values measured by making repeating measurements determine the precision of the method under the specified conditions [23, 38 - 41].

Robustness is a critical parameter of method validation procedure measuring the capacity of the analytical procedure to remain unaffected by small and deliberate variations. It provides the reliability of developed methods during normal usage [42 - 44].

In this section, experimental information will be given about the analysis techniques in the literature on approved drugs. In addition to experimental conditions, some important validation results are presented to readers.

Chromatographic Techniques for Analyzing Alzheimer's Disease Drugs

Separation techniques are used to separate a target analyte in a sample from other compounds and include one or more separation steps. Chromatographic methods are the most common methods preferred to achieve separation. The components of a mixture are separated based on their interactions with the mobile phase formed of the solvent and the stationary phase by a supporting material in the chromatographic method (Fig. **2**) [45 - 47].

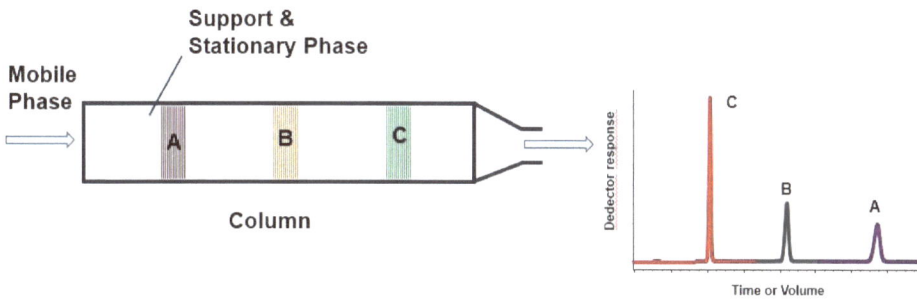

Fig. (2). Scheme of chromatographic separation.

Chromatographic techniques have already become popular in many fields of application. Researchers have adopted these techniques for their high efficiency, high resolution, speed of analysis, robustness, reliability, and availability of instrumentation [46]. These advantages also attracted the attention of those focusing on the analysis of AD drugs, as they are used in the analysis of many different classes of drugs. Literature studies made with chromatographic methods for the detection and determination of AD drugs are summarized in Table **1** according to the mobile phase, experimental conditions, retention time, and linear range, LOD, and application of the developed method.

Looking at Table **1**, it can be seen that liquid chromatography (LC) methods were the most preferred ones for AD drugs. The basic principle of LC based separation is the distribution of compounds between a liquid mobile phase and a stationary phase. In time, the development of more efficient columns increased the performance of this technique, ultimately leading to high-performance liquid chromatography (HPLC), which contributes to narrower peaks, better separations, and lower LOD values [47 - 50].

An HPLC system is typically composed of a solvent delivery system, sample injector, column, detection system, and computer [48]. Detectors are very important components in these systems. Several detectors such as ultraviolet or ultraviolet/visible detector (UV/vis), photodiode array detector (PAD), fluorescence detector, conductivity detector, refractive index detector, electrochemical detector, and mass spectrometric detector (MS), which have different features were preferred to use in HPLC [49, 51, 52].

UV detectors are frequently used detectors in LC systems. These detectors are also used for the analysis of tacrine [53], donepezil [54], memantine [55], galantamine [56 - 59], rivastigmine [60 - 62], and Huperzine A [63 - 67].

Amini and Ahmadini (2010) proposed an HPLC-UV method as a good alternative of the HPLC-MS method, which is quite expensive and still lacking in most laboratories, for the determination of rivastigmine from plasma. They obtained a very low LOD value as 0.2 ng mL^{-1} [60].

Fluorescence detectors are very specific, sensitive (0.01 ng level), and selective. They are generally used for reducing interference. Despite these advantages, there are very few compounds to be studied by these detectors, because of the lack of fluorescence characteristics [48]. These detectors were used for the determination of donepezil [68], rivastigmine [69], memantine [70 - 72], and galantamine [73 - 75]. Memantine has no fluorescence characteristic so for its determination, derivatization agents were used to detect it using fluorescence detector [70 - 72].

LC by MS is preferred for drug analysis in biological fluids for its significant specificity, sensitivity, and short analysis time [76]. MS separates gas-phase ions in the vacuum based on their mass to charge (m/z) ratio [48]. MS studies were generally used in bioequivalence studies which are made using plasma samples taken from patients or healthy volunteers after administration of the related drug. In Table **1**, there are several LC-MS/MS studies for the determination of donepezil alone [77 - 81] and in the presence of its metabolites [82 - 84]. The sensitivity of these methods is in ng mL^{-1} level.

In HPLC, the particle size of the column material is 5 µm, this size makes it pressure resistant. As a special kind of HPLC, ultra-high performance liquid chromatography (UPLC) has 2 µm particle size which allows faster analysis [85]. Lee *et al.* (2020) developed a UPLC-MS/MS analysis for the detection of donepezil and its main metabolite 6-O-desmethyl donepezil from beagle dog plasma with a very short analysis time, 2.34 and 1.42 min, respectively [82].

There are limited studies using gas chromatography (GC) for the determination of AD drugs. GC is used for studying volatile and semi-volatile analytes. However, in order to study nonvolatile polar molecules or low thermal stability compounds, derivatization is required. Especially, the coupling of GC to MS or MS/MS affords additional structural information for several compounds [86 - 88]. A GC method with an MS detector was developed by Berkov *et al.* (2011) for galantamine. Galantamine was extracted from bulbs and leaves of plants and can be detected in the range of 15–800 µg galantamine/sample [89].

There is a high-performance thin-layer chromatography (HPTLC) in the literature for rivastigmine [90]. HPTLC is based on thin-layer chromatography and preferred for better separation efficiency and detection limits. Also, HPTLC is known as high-pressure thin layer chromatography/planar chromatography or flat-bed chromatography. It is used by covering 0.1–5 mm thickness on glass or aluminum plates that are adsorbent as a stationary phase. The mobile phase is solvent or solvent mixtures. A small amount of the compounds in the mixture is dropped on the adsorbent with a capillary tube. It is immersed in the tank, which has a mobile phase in the plate, and the mobile phase is expected to move up the layer under the capillary effect, and then zones are evaluated [91].

Table 1. Summary of various published chromatographic methods for AD drugs.

Analyte	Method	Detector	Column	Mobile Phase	Analysis Conditions	Linear Range	LOD	Applications	Retention Time	Ref.
Donepezil	HPLC	UV	Phenomenex Luna Phenyl-Hexyl (150 mm × 3 mm, 3 µm)	0.02 M phosphate buffer : ACN	Temp: 30 °C IV: 35 µL FR: 0.4 mL min⁻¹ Wavelength: 210 nm	5–160 ng mL⁻¹	1.7 ng mL⁻¹	Patients serum	12.1 min	[54]
Donepezil	HPLC	Fluorescence detection	Phenyl Hypersil C18 (125 mm × 4.6 mm, 3 µm)	MeOH : 0.02 M phosphate buffer (pH 3.5) : triethyl amine (55:45:0.5, v/v/v)	IV: 15 µL Emission wavelength: 290 nm Excitation wavelength: 315 nm FR: 0.9 mL min⁻¹	5–2000 ng mL⁻¹	1.5 ng mL⁻¹	Human plasma and pharmaceuticals	-	[68]
Donepezil	HPLC	MS/MS	YMC Pack ODS-A (50 mm × 4 mm, 5 µm)	10 mM ammonium acetate (pH 5) : ACN (18:82, v/v)	Temp: ambient room temperature IV: 20 µL FR: 0.7 mL min⁻¹	0.1–100 ng mL⁻¹	-	Bioequivalence study (oral administration of 5 mg tablet to healthy human volunteers)	~0.98 min	[77]
Donepezil	HPLC	MS/MS	Betabasic-C8 (100 mm × 4.6 mm, 5 µm)	MeOH : Water : Formic acid (90:9.97:0.03, v/v/v)	Temp: 45 °C IV: 20 µL FR: 0.8 mL min⁻¹	0.15–50 ng mL⁻¹	-	Bioequivalence study (healthy human volunteers)	~1.16 min	[78]
Donepezil	HPLC	MS/MS	C18 column (75 mm × 4.6 mm, 3.5 µm)	0.1% formic acid : MeOH (70:30, v/v)	m/z: 380.2 / 91.2 and 387.3 / 98.2	50–25000 pg mL⁻¹	50 pg mL⁻¹	Human plasma	~2 min	[79]
Donepezil	HPLC	MS/MS	Gemini C18 (150 mm × 2 mm, 5 µm)	ACN : 10 mM ammonium acetate (pH 5) (70:30, v/v)	Temp: 25 °C IV: 5 µL FR: 0.2 mL min⁻¹	0.1–50 ng mL⁻¹	-	Bioequivalence study (plasma)	2.06 min	[80]
Donepezil	HPLC	MS/MS	SeQuant ZIC-HILIC (50 mm × 2 mm, 5 µm)	Formic acid : Ammonium formate : ACN (Gradient elution)	Temp: Room temperature IV: 5 µL	1–1000 ng mL⁻¹	-	3-month dose-range in rats	~5.0 min	[81]

(Table 1) cont.....

Donepezil	UPLC	MS/MS	Luna C18 (100 mm × 2.0 mm, 3 μm)	0.1% formic acid in ACN : 0.1% formic acid in distilled water (20:80, v/v)	IV: 5 μL FR: 0.65 mL min⁻¹ Detector voltage: 2.02 kV	0.02–20 ng mL⁻¹	-	Beagle dog plasma	2.34 min	[82]
6-O-desmethyl donepezil						0.02–10 ng mL⁻¹			1.42 min	
Donepezil	HPLC	MS/MS	XTerra RP (150 mm × 4.6 mm, 5 μm)	MeOH : 2 mM ammonium acetate (pH 6.2) (65:35, v/v)	Temp: 35°C IV: 10 μL FR: 0.4 mL min⁻¹	0.339–51.87 ng mL⁻¹	-	Bioequivalence study in healthy human volunteers	7.0 min	[83]
6-desmethyl donepezil						0.100–15.38 ng mL⁻¹			4.8 min	
5-desmethyl donepezil						0.103–15.76 ng mL⁻¹			5.93 min	
Donepezil N-oxide						-			-	
Donepezil	HPLC	MS/MS	Novapak C18 (150 mm × 3.9 mm, 4 μm)	0.2% formic acid in 20 mM ammonium acetate : MeOH : ACN (63:20:17, v/v/v, pH 3.43)	Temp: Ambient temperature IV: 25 μL FR: 1.0 mL min⁻¹	0.1–50 ng mL⁻¹	-	Bioequivalence study (human plasma)	3.78 min	[84]
6-O-desmethyl donepezil						0.02–10 ng mL⁻¹			2.32 min	
Donepezil	HPLC	MS/MS	Thermo Syncronis-C18 (100 mm × 4.6 mm, 5μm)	MeOH : 10 mM ammonium acetate (pH 5) (92:8, v/v)	FR: 0.7 mL min⁻¹	0.2–400 ng mL⁻¹	0.2 ng mL⁻¹	Rat plasma	2.37 min	[92]
Memantine									2.08 min	
Donepezil	HPLC UPLC	MS MS/MS	BEH C18 (50 mm × 2.1 mm, 1.7 μm)	Ammonium acetate buffer (pH 9.3) : ACN (Gradient elution)	FR: 0.4 mL min⁻¹ Source temp: 150 °C Capillary voltage: 3 kV	1–300 ng mL⁻¹	-	Human plasma	3.4 min	[93]
Galantamine						1–300 ng mL⁻¹			1.9 min	
Memantine						1–300 ng mL⁻¹			2.9 min	
Rivastigmine						0.2–50 ng mL⁻¹			3.0 min	
NAP 226-90						0.2–50 ng mL⁻¹			1.5 min	
Rivastigmine	HPLC	UV	Silica column (250 mm × 4.6 mm, 5 μm)	ACN : 50 mM sodium dihydrogen phosphate (17:83, v/v; pH 3.1 adjusted with phosphoric acid and 4 M NaOH)	Temp: 50 °C IV: 80 μL Wavelength: 200 nm	0.5–16 ng mL⁻¹	0.2 ng mL⁻¹	Capsule	~4.5 min	[60]
Rivastigmine	HPLC	UV	Kromasil KR-100 (250 mm × 4.6 mm, 5 μm)	20 mM ammonium acetate (pH 6.5) : ACN (65:35, v/v)	Temp: 25 °C IV: 100 μL Wavelength: 210 nm	50–5000 ng mL⁻¹	50 ng mL⁻¹	Urine and brain	~11.73 min	[62]
Rivastigmine	HPLC	Fluorescence detection	Inertstil, ODS-3 V, C18 (250 mm × 4.6 mm, 5 μm)	ACN : 20mM ammonium acetate buffer (pH 4.5) (26:74, v/v)	Temp: 25 °C IV: 50 μL Excitation wavelength: 220 nm Emission wavelength: 293 nm FR: 1 mL min⁻¹	10–1000 ng mL⁻¹	-	Rat plasma and brain	~7.5 min	[69]
Rivastigmine	HPLC	MS/MS	Betabasic-8 (100 mm × 4.6 mm, 5 μm)	0.1% formic acid in ACN : 0.1% formic acid in water (70:30, v/v)	Temp: 45 °C IV: 10 μL FR: 1 mL min⁻¹	0.2–20 ng mL⁻¹	-	Bioequivalence study (single oral administration of 3 mg capsule in healthy male volunteers)	~0.79 min	[94]

(Table 1) cont.....

Rivastigmine	HPTLC	TLC CAMAG scanner in the reflectance absorbance mode	HPTLC plate size: 20 cm × 10 cm	Benzene : Acetone : Ammonia (6.5:3:0.1, v/v/v)	Deuterium lamp emitting radiation: 190–350 nm	30.2–648 ng spot⁻¹	8.65 ng spot⁻¹	Capsules and human plasma	-	[90]
Rivastigmine	HPLC	PAD	Apollo C18 (150 mm × 4.6 mm, 5 µm)	20% v/v ACN in water containing 0.1% trifluoroacetic acid	Temp: 50 °C FR: 1.5 mL min⁻¹ Wavelength: 214 nm	0.5–10 µg mL⁻¹	60 ng mL⁻¹	Polymeric nanoparticle formulation matrices, drug release medium, and cellular transport medium	6.8 min	[95]
Rivastigmine	GC	MS	HP-5MS capillary column (30 m × 0.25 mm i.d. × 0.25 µm film)	Helium	Oven temp: 100 °C (1 min) to 280 °C at 20°C/min Injection and ion source temp: 250 and 230 °C FR: 1.0 mL min⁻¹	0.2 – 80 ng mL⁻¹	0.2 ng mL⁻¹	Canine plasma	-	[96]
Rivastigmine NAP 226-90	HPLC	MS/MS	Phenomenex Gemini C18 (150 mm × 2.0 mm, 5 µm)	10 mM ammonium hydroxide : MeOH (50:50, v/v) for 1 min; 10 mM ammonium hydroxide in water : MeOH (5:95, v/v) (Gradient elution)	Temp: 25 °C IV: 25 µL FR: 0.2 mL min⁻¹	0.25–50 ng mL⁻¹ 0.50–25 ng mL⁻¹	-	Patient plasma samples	~6.2 min	[97]
Rivastigmine Impurities	HPLC	UV	Xterra RP-18, (250 mm × 4.6 mm, 5 µm)	A: 10 mM dipotassium hydrogen phosphate (pH 7.6) : ACN (90:10, v/v) B: ACN : MeOH (60:40, v/v) (Gradient elution)	Temp: 40 °C FR: 1.0 mL min⁻¹ Wavelength: 210 nm	-	-	-	17.63 min ~6–17 min	[61]
Tacrine	HPLC	PAD	Waters Symmetry C18 (150 mm × 3.9 mm, 4 µm)	ACN : Buffer (17 mM sodium dodecyl sulphate and 8.3 mM sodium dihydrogen phosphate) (50:50 v/v)	FR: 1.5 mL min⁻¹ Wavelength: 247 nm	0.05–10 µg mL⁻¹	0.04 µg mL⁻¹	-	2.2 min	[53]
Galantamine	GC	MS	DB-5 MS column (30 m × 0.25 mm, 0.25 µm)	Helium	Injector temp: 280 °C FR: 0.8 mL min⁻¹ Split ratio: 1:15	15–800 µg galantamine /sample	1 µg/sample (2 µg/mL)	*Leucojum aestivum* and *Narcissus* ssp	-	[89]
Galantamine	HPLC	UV	Waters SunFire C18 (150 mm × 2.1 mm, 3.5 µm)	A: 10 mM ammonium acetate (pH 5.8) : MeOH (95:5, v/v) B: 10 mM ammonium acetate (pH 5.8): MeOH (5:95, v/v) (Gradient elution)	Temp: 30 °C Wavelength: 290 nm	2.5–60 µM	-	Capsules	16.3 min	[56]

(Table 1) cont.....

Galantamine	HPLC	UV	C18 (250 mm × 4.6 mm, 4 μm)	MeOH, ACN and ammonium formate buffer (pH 9)	Temp: 22 °C. FR: 0.7 mL min^{-1} Wavelength:212 nm	5–40 μg mL^{-1}	89 ng mL^{-1}	Tablets and artificial cerebrospinal fluid	~3.19 min	[57]
Galantamine	HPLC	UV	Supelcosil LC-18 (250 mm × 4.6 mm, 5 μm)	Ammonium carbonate : ACN (85:15 v/v)	Temp: 24 °C FR: 1.0 mL min^{-1} Wavelength: 292 nm	-	7.5 μg	*Sternbergia* species	~2.6 min	[58]
Galantamine	HPLC	UV	RP C18 ODC Spherisorb (250 mm × 4.6 mm, 5 μm)	50 mM disodium hydrogen phosphate : ACN (80:20, v/v)	Temp: 25 °C FR: 1.5 mL min^{-1} Wavelength: 280 nm	1.10^{-5}–1.10^{-3} g mL^{-1}	1.08×10^{-4} g mL^{-1}	-	~5.35 min	[59]
Galantamine	HPLC	Fluorescence detector	Synergi C18 column (inertsil, 150 mm × 4.6 mm, 5 μm)	ACN : 10 mM o□phosphoric acid (40:60, v/v)	IV: 20 μL FR: 1.2 mL min^{-1} Excitation wavelength: 375 nm Emission wavelength: 537 nm	125–2000 ng mL^{-1}	6.27 ng mL^{-1} (plasma) 70 .99 ng mL^{-1} (urine)	Human plasma and urine	16.8 min	[73]
Galantamine	HPLC	Fluorescence detection	HS F5 Supelco (150 mm × 4.6 mm, 5 μm)	A: ACN : Ammonium acetate (5 mM) (pH 6.8) B: ACN : Ammonium acetate (5 mM) (25:75, v/v) (4 min) (Gradient elution)	Temp: 30 °C FR: 1.0 mL min^{-1} IV: 4 μL Excitation wavelength: 280 nm Emission wavelength: 310 nm	0.50–63.47 nmol g^{-1}	0.04 nmol g^{-1}	Rat liver tissue	17.7 min	[74]
Epigalantamine						0.70–89.03 nmol g^{-1}	0.07 nmol g^{-1}		19.9 min	
O-desmethyl-galantamine						0.32–41.42 nmol g^{-1}	0.19 nmol g^{-1}		10.9 min	
N-desmethyl- galantamine						0.54–69.40 nmol g^{-1}	0.07 nmol g^{-1}		14.3 min	
Galantamine	HPLC	Fluorescence detection	Zorbax Eclipse XDB-C18 (75 mm × 4.6 mm, 3.5 μm)	Ammonium acetate buffer (pH 5.0) : ACN (94:6, v/v)	Temp: 20 °C FR: 2.2 mL min^{-1} Excitation wavelength: 230 nm Emission wavelength: 290 to 440 nm	-	11 ng mL^{-1}	Artificial and natural water	-	[75]
Epigalantamine							8 ng mL^{-1}			
O-Desmethyl-galantamine							9 ng mL^{-1}			
N-Desmethyl-galantamine							12 ng mL^{-1}			
Galantamine N-oxide							10 ng mL^{-1}			
Memantine	HPLC	MS/MS	Kinetex XB-C18 (100 mm × 2.1 mm, 2.6 μm)	ACN : 0.1% formic acid in a 10 mM ammonium formate solution	Temp: 40 °C IV: 2 μL	0.5–20 μg kg^{-1}	1.0 μg kg^{-1} (LOQ)	Deep-fried chicken, fried chicken, fried quail egg and grilled chicken	10 min	[98]
Memantine	HPLC	MS/MS	Zorbax SB-C18 (75 mm × 4.6 mm, 3.5 μm)	ACN : 0.1% formic acid (65:35, v/v)	Temp: 40 °C IV: 20 μL FR: 0.6 mL min^{-1}	50–50,000 ng mL^{-1}	-	Bioequivalence study (plasma concentrations in human volunteers)	1.45 min	[99]
Memantine	HPLC	Fluorescence detector	Novapak C18 (150 mm × 3.9 mm, 4 μm)	ACN : Water (73:27, v/v)	Temp: 40 °C IV: 10 μL FR: 1.2 mL min^{-1} Excitation wavelength: 260 nm Emission wavelength: 315 nm	10–100 ng mL^{-1}	3 ng mL^{-1}	Plasma and vitreous humour	~10.5 min	[70]
Memantine	HPLC	UV	Lichrospher C18 (150 mm × 6 mm, 5 μm)	MeOH : water (85:15, v/v)	Temp: 25 °C FR: 1.0 mL min^{-1} Wavelength: 256 nm	0.05–2 μg mL^{-1}	20 ng mL^{-1}	Rat plasma	19 min	[55]
Memantine	HPLC	Fluorescence detector	Diamonsil C18 (150 mm × 4.6 mm, 5 μm)	ACN : Water (67:33, v/v)	Temp: 25 °C IV: 20 μL FR: 1.0 mL min^{-1}	0.025–5.0 μg mL^{-1}	10 ng mL^{-1}	Rat plasma	~23.69 min	[71]

(Table 1) cont.....

Memantine	HPLC	Fluorescence detector	Chromolith Performance RP-18e (100 mm × 4.6 mm)	ACN : 0.025 M phosphate buffer (pH 4.6) (50:50, v/v)	Temp: 25 °C FR: 2.5 mL min⁻¹ Excitation wavelength: 335 nm Emission wavelength: 440 nm	2–80 ng mL⁻¹	-	Human plasma	~4.2 min	[72]
Memantine	GC	MS	HP- 35 fused-silica capillary column (30 m × 0.32 mm i.d., 0.25 mm film thickness)	Carrier gas: Hydrogen Auxiliary gas: Argon-methane (95:5)	Injector and detector temp: 280 °C FR: 4 mL min⁻¹ and 26 mL min⁻¹	5.0 – 100 ng mL⁻¹	-	Bioequivalence study	-	[100]
Memantine	GC	MS	DB-5 MS fused-silica capillary column (15 m × 0.25 mm, i.d., 0.25 µm film thickness)	Helium	Injector temp: 270 °C FR: 1.5 mL min⁻¹ Column temp: 100 °C (1 min), increase of 40 °C min⁻¹ to 300 °C	0.117 – 30 ng mL⁻¹	-	Human plasma	-	[101]
Huperzine A	HPLC	UV	Agilent Zorbax SB-C18 (150 mm × 4.6 mm, 5 µm)	MeOH : 1 mM l-Lysine water solution (50:50, v/v)	FR: 0.6 mL min⁻¹ Wavelength: 310 nm	1–25 µg mL⁻¹	-	*Huperzia serrata*	-	[63]
Huperzine A	HPLC	UV	Diamonsil C18 (200 mm × 4.6 mm, 5 µm)	MeOH : Ammonium acetate (80 mM, pH 6.0) (40: 60, v/v)	FR: 1.0 mL min⁻¹	5–100 µg mL⁻¹	-	*Huperzia serrata*	14.6 min	[64]
		PAD-MS/MS	-	MeOH : Ammonium acetate (80 mM, pH 6.0) (30/70, v/v)	Wavelenght: 230 and 310 nm Source voltage: 4.5 kV Capillary temp: 275 °C					
Huperzine A	HPLC	UV-MS	Waters Symmetry C18 (250 mm × 4.0 mm, 5 µm)	MeOH:10 mM ammonium acetate (pH 3.5) (Gradient elution)	Capillary temp: 250 °C FR: 1.0 mL min⁻¹ Wavelength: 308 nm	-	5 ng	Lycopodiaceae species	8.8 min	[102]
Huperzine A	HPLC	UV	Hypercosil Gold C18 (250 mm × 4.6 mm)	A: Water (include 30 mM NaPF₆) B: ACN (Gradient elution)	Temp: 24 °C FR: 0.6 mL min⁻¹	25–1252 µg mL⁻¹	14 ng mL⁻¹	*Huperzia selago* plants	~12 min	[65]
Huperzine A	HPLC	PAD	C18 column (250 mm × 4.6 mm, 5 µm)	MeOH: 0.1% (v/v) TFA (18:82, v/v) (Gradient elution)	Temp: 30 °C IV: 10 µL FR: 1.0 mL min⁻¹ Wavelength: 308 nm	-	-	*Huperzia serrata*	~21 min	[103]
Huperzine A	HPLC	Fluorescence detector	Kromasil C8 (150 mm × 4.6 mm, 5 µm)	MeOH : Water: Triethanol amine (45:55:0.05, v/v/v)	FR: 1.0 mL min⁻¹ Excitation wavelength: 310 nm Emission wavelength: 370 nm	2.5–250 ng mL⁻¹	2.5 ng mL⁻¹	Rat plasma	~7 min	[104]

(Table 1) cont.....

Huperzine A	HPLC	PAD	YMC Basic C18 (250 mm × 4.6 mm, 5 μm)	10 mM ammonium acetate (pH : 3.5) : MeOH (Gradient elution)	Temp: 30 °C IV: 20 μL FR: 1.0 mL min⁻¹ Capillary temp: 250 °C Capillary voltage: 20 V	0.8–11.6 μg mL⁻¹	-	Tablets and capsules	9.51 min	[105]
		MS							-	
Huperzine A	HPLC	MS/MS	Nucleosil C18 (50 mm × 4.6 mm, 5 μm)	ACN : MeOH : 10 mM ammonium acetate (35:40:25, v/v/v)	Temp: 500 °C FR: 0.4 mL min⁻¹ Voltage: 5000 V	0.05–20 ng mL⁻¹	0.01 ng mL⁻¹	Dog plasma	~1.6 min	[76]
Huperzine A	HPLC	UV	Shimpack CLC-ODS (150 mm × 4.6 mm, 5 μm)	MeOH : Water : Glacial acetic acid (50:48.5:1.5, v/v/v) include 0.0015 M dodecylsulfonate	Temp: 35 °C FR: 1.0 mL min⁻¹ Wavelength: 313 nm	0.5–12 ng mL⁻¹	0.5 ng mL⁻¹	Beagle dog serum	~12 min	[66]
Huperzine A	HPLC	UV	Nucleosil C18 (250 mm × 4.6 mm, 5 μm)	1 mM ammonium acetate : MeOH (70:30, v/v)	Temp: 35 °C FR: 0.7 mL min⁻¹ Wavelength: 313 nm	0.02–5.0 nmol mL⁻¹	-	Rat plasma	~14 min	[67]
Huperzine A	UPLC	MS	Zorbax Eclipse Plus C18 (50 mm × 2.1 mm, 1.8 μm)	A: 0.2% aqueous acetic acid B: 0.2% acetic acid in MeOH (Gradient elution)	Temp: 60 °C IV: 2 μL FR: 1.3 mL min⁻¹	0.01–50 μg mL⁻¹	5.1–10.1 ng mL⁻¹	*Huperzia*	-	[106]

Optical Techniques for Analyzing Alzheimer's Disease Drugs

Spectrophotometric analysis techniques are based on the production or interaction of electromagnetic radiation with the drug. Considering the spectroscopic methods developed for the analysis of the drugs used to treat AD, generally, spectrofluorimetric methods, which are based on light emission, have been preferred [107]. Spectrofluorimetry is a selective and quite sensitive technique for the determination of some materials with fluorescent characteristics [108, 109]. The reason for high sensitivity is the different wavelengths between the exciting and fluorescence radiation, while the reason for high specificity is the dependence on the excitation and the emission spectra. In fluorescence analysis, under appropriate excitation, the amount of scattered light is used as a scale for the concentration of the material in concern. It is possible to determine materials at ng level with spectrofluorimetry [109]. This method is limited by the low fluorescence or non-fluorescence of some materials. In order to analyze such materials spectrofluorimetrically, hydrolysis, treatment with heat, forming complexes with ligands, labeling with a fluorogenic agent, and chemical derivatization techniques such as changing solvent polarity are used [110].

Since memantine does not show fluorescence characteristics, the Hantzsch condensation reaction was used to analyze it using spectrofluorimetry. Hantzsch condensation reaction is a multi-component reaction that occurs between formaldehyde, a 2 β-keto ester such as acetylacetone, and a nitrogen donor such as the primary amino group to form a colored and/or fluorescent product. In this reaction, the nitrogen donor is memantine. The resulting product gave yellow fluorescence at 418 nm excitation and 484.5 nm emission wavelengths. With this method, 0.2-1 µg mL^{-1} linear range and 0.0153 µg mL^{-1} LOD value were obtained for memantine [111].

Another example is the study for galantamine, which cannot be directly determined using spectrofluorometry because of its low fluorescence intensity. Culzoni *et al.* (2010) increased the fluorescent intensity of galantamine using sodium dodecyl sulfate to determine it by the spectrofluorimetric method. Although this process can significantly increase the sensitivity of the method, it is difficult to improve selectivity in complex matrices because of spectral overlap caused by matrix interactions. Therefore, in this study, the second-order multivariate algorithm was used to increase the selectivity of the method. The developed approach made galantamine determination at the level of ng mL^{-1} without any need for separation and in the presence of uncalibrated interferences [112].

Spectrofluorimetry was also used to understand the interaction between rivastigmine tartrate and bovine serum albumin by Shamsi *et al.* (2020) [113].

In a spectrofluorometric study for rivastigmine determination, the solution of the substance was prepared in triple distilled water and relative fluorescence intensity was measured using 220 nm excitation and 289 nm emission wavelengths. This method is preferred for its simplicity, sensitivity, fast response, precision, accuracy, and cost-effectiveness; it has been used for the quantification of rivastigmine from capsules and used for *in vitro* dissolution studies [114].

UV-vis spectroscopy is based on light absorption [107]. This method is widely used for the quantitative analysis of substances absorbed in the region of 200-800 nm due to the high sensitivity of the method [108]. In the literature, there is a UV-vis spectrophotometric method combined with chemometry for the determination of donepezil and rivastigmine at the same time. The mixtures of these two compounds gave absorption spectra between 200-400 nm in 0.1 M HCl. The developed method was applied to Exelon and Doenza tablets [115].

In spectrophotometric studies, first and second derivatives of spectra are taken to increase sensitivity. This method has advantages for removing the effect of interfering substances from the spectrum. Salem *et al.* (2010) performed rivastigmine analysis using first and second derivative spectrophotometry in the presence of its major metabolite NAP 226-90. Rivastigmine analysis using first and second derivative spectrophotometry was performed in the presence of its metabolite with accuracy values of 99.97% and 100.18%, respectively [116].

There is another study for the analysis of rivastigmine in the literature which uses chemiluminescence. Chemiluminescence reveals light from electronically excited species which is produced by an exothermic chemical reaction. This method is preferred for its high precision, wide linear range, low-cost instrumentation, and ease of use. Iranifam *et al.* (2017), developed simple and sensitive CuO nanoparticles (CuO NPs) –catalyzed chemiluminescence system, hydrogen peroxide–hydrogen carbonate –CuO NPs, for rivastigmine determination. The developed method focused on the inhibitory effect of rivastigmine on the chemiluminescence system. Experimental parameters such as chemiluminescence intensity, concentrations of hydrogen carbonate, CuO NPs, and hydrogen peroxide were optimized to achieve the highest light emission eliciting from this chemiluminescence system. Finally, the linearity range was found as $1.0 \times 10^{-6} - 4.0 \times 10^{-4}$ M and LOD was found as 8.6×10^{-7} M for rivastigmine [117].

In recent years, colorimetry, which is an optical method, was developed for a drug (tacrine) used in the treatment of AD [118]. Colorimetric analysis is based on a simple chemical reaction between the analyte and specific reagents to produce a visibly observable product. The color of the product is used to indicate the presence of an analyte in the sample. The colorimetric tests are generally preferred as the first chemical method which is applied to a sample by analysts because of the simplicity and fast response time of the method. Reagents and necessary materials are cheap and easily available and also, many commercial test kits are available on the markets [119]. Huang *et al.* (2020), developed a colorimetric assay of acetylcholinesterase inhibitor tacrine based on molybdenum oxide nanoparticles as peroxidase mimetics. Based on the peroxidase-like activity of molybdenum oxide nanoparticles, H_2O_2 was determined using colorimetry for tacrine determination. Very low LOD value (0.9 nM) was obtained which showed the sensitivity of developed colorimetric method [118].

Electrochemical Techniques for Analyzing Alzheimer's Disease Drugs

Electroanalytical methods are popular methods in analytical chemistry that are frequently chosen for the analysis of drugs [120]. Electroanalytical methods not only determine very small amounts of an electroactive drug but also they inform the researcher about its physical and chemical properties [121]. These method are highly sensitive, low cost, and rapid; there is little solvent and sample requirement and these qualities make them alternative or complementary to other separation techniques [122]. One of the most important advantages of electrochemical methods is there no need for any separation or extraction procedure because these methods do not affect matrix interferences [123].

Among the electrochemical methods, it has been observed that the most commonly used method in the analysis of AD drugs is voltammetry. Voltammetry is an electrochemical method based on measuring current as a function of potential applied to a cell. This method can be classified as cyclic voltammetry (CV), differential pulse voltammetry (DPV), and square wave voltammetry (SWV) according to applied potential waveform [123].

A wide variety of working electrodes are used in voltammetry such as carbon-based electrodes, metal electrodes, and hanging mercury drop electrodes [124, 125]. Golcu and Ozkan (2006) developed DPV and SWV techniques for donepezil and obtained an LOD value of 10^{-7} M using bare glassy carbon electrode (GCE) [126]. Aparicio *et al.* (2000) determined tacrine in the presence of its 1-OH-tacrine metabolite with carbon paste electrode (CPE) from pharmaceutical formulations and urine at the level of µg/mL [127].

The accumulation of the substance on the electrode surface before measurement which is called stripping voltammetry provides more sensitive analysis [123]. Ghoneim *et al.* (2009) performed an analysis for donepezil using square-wave adsorptive cathodic stripping voltammetry at hanging mercury drop electrode and donepezil was analyzed in the level of 10^{-10} M [128]. The method using the hanging mercury drop electrode is called polarography and it is used for substances that can be reduced [125].

Various nanomaterials have been used in recent years to increase the sensitivity of voltammetric methods. Nanomaterials can increase the selectivity of the method as well as providing more sensitive determination by increasing the surface area of the electrode [129]. Nanomaterials were also used for the determination of AD drugs. As an example of these studies, an adsorptive stripping differential pulse voltammetry technique for rivastigmine determination can be given. In this study, CPE was modified with graphene nanosheets and gold nanoparticles and rivastigmine was determined at the nM level [130].

Another electrochemical method, potentiometry, was also applied for rivastigmine determination and this method measures the accumulation of a charge potential at the working electrode [131]. In this study, a molecularly imprinted polymer (MIP) modified electrode is used, which has been used frequently in recent years for sensor development. It increases the selectivity of the method because cavities are created with the same size and shape of the substance on the polymer surface. Arvand *et al.* (2012) obtained a µM level LOD value in this study [132].

Another electrochemical method, amperometry, is a method that is frequently used in biosensors where a biological material is integrated onto the electrode surface. Amperometry is a method based on the measurement of current at a constant potential [133]. There is an amperometric method in the literature for donepezil where a biosensor was developed. In this study, acetylcholinesterase enzyme was attached onto the GCE surface to make a biosensor. The method was applied to urine samples with a level of nM [134].

Electrochemical methods developed for drugs used in the treatment of AD are summarized in Table **2** by means of some experimental conditions and validation parameters.

Table 2. Electrochemical studies for the analysis of AD drugs.

Drugs	Electroanalytical Technique	Electrode	Calibration Range	LOD	Application	Ref.
Tacrine 1-OH-tacrine	DPV	CPE	-	0.06 μg mL^{-1} 0.18 μg mL^{-1}	Pharmaceuticals and human urine	[127]
Tacrine	SWV	AChE/Magnetic NPs/SPE	-	8.1 μM	-	[135]
Donepezil	DPV SWV	GCE	1×10^{-6} - 1×10^{-4} M	2.90×10^{-7} M (DPV) 2.34×10^{-7} M (DPV serum) 2.63×10^{-7} M (SWV) 1.51×10^{-7} M (SWV serum)	Tablets and serum	[126]
Donepezil	Square-wave adsorptive cathodic stripping voltammetry	Hanging mercury drop electrode	3×10^{-9} -1×10^{-7} M	9.3×10^{-10} M 1.5×10^{-9} M (serum)	Tablets and serum	[128]
Donepezil	Square wave anodic stripping voltammetry	Gold electrode	29.1–65.4 μg mL^{-1}	-	-	[136]
Donepezil	DPV	CoFe$_2$O$_4$ NPs/CPE	5.0×10^{-6} -2.0×10^{-5} M	-	Tablets and urine	[137]
Donepezil	CV	poly (SBT)/N-CNDs/CoNPs/PGE	1.5 nM–400 μM	0.5 nM	Tablets and rabbit plasma	[138]
Donepezil	Amperometry	AChE/GCE	-	0.46 nM	Urine	[134]
Rivastigmine	Potentiometrry	MIP/PVC membrane electrode	1.0×10^{-5} – 1.0×10^{-2} M	6.3×10^{-6} M	Human serum, plasma, urine, rat brain and tablets	[132]
Rivastigmine	DPV	MIP/CPE	2.0–1000 μM	0.44 μM	Capsules, serum and urine	[139]
Rivastigmine	Adsorptive stripping differential pulse voltammetry	Graphene nanosheets/AuNPs/CPE	2.0×10^{-7} -6.0×10^{-4} M	5.3×10^{-8} M	Capsules, serum and urine	[130]
Rivastigmine	SWV	β-CD/MWCNTs/GCE	10- 1500 μM	2.0 μM	Pharmaceuticals	[140]
Rivastigmine	DPV	γ-FeOOH/ N@CCNS/PGE	3.0–90.0 nM	0.99 nM	Tablets and serum	[141]

(Table 2) cont.....

Galantamine	DPV	AgNPs/poly 3MT/GCE	1.0–700 μM	0.18 μM	Human cerebrospinal fluid, Narcissus and tablets	[142]
Galantamine	DPV	Graphene ink/GCE	20–180 μM	4.1 μM	Galantamine hydrobromide injection	[143]
Memantine	CV	MnO_2-PEDOT/GCE	20–100 μM	-	-	[144]

ANALYSIS OF THE MAIN MOLECULAR TARGETS UNDER INVESTIGATION IN ALZHEIMER'S DISEASE RESEARCH

Nowadays, there are some novel researches for the treatment of AD and there are some new promising drug candidates [145]. In recent years, scientists and researchers focus on the efforts and efficiency of cholinesterase inhibitors to cure the cause of AD. Zueva *et al.* (2019) described a series of 6-methyluracil derivatives as potent reversible inhibitors of cholinesterase. Researchers found that C-35 (1,3-bis[5-(o nitrobenzylethylamino)pentyl]-6-methyluracil), which is a compound of these derivatives, reduces the rate of oligomerization of β-amyloid peptide *in vitro*. Moreover, it is found that this compound restores memory parameters in transgenic mice. Zueva *et al.* (2019) used the HPLC-MS technique to analyze C-35 concentration in brain samples. In this study, the Zorbax Eclipse XDB-C18 column (50 mm × 3.0 mm, 1.8 μm) was used as the stationary phase. Gradient elution with methanol and 1% formic acid used as the mobile phase. The flow rate was 0.4 mL min^{-1} and capillary voltage was –4500 V. The linearity range was found as 5–50 ng mL^{-1} for C-35 in brain homogenate [146].

Another prospective treatment research is related to neuropathogenesis of AD which focuses on the pathological changes in the brain due to different amounts and distribution of d-amino acids. Considering AD, concentrations of d-amino acid have significant differences between people who are healthy and AD patients. Li *et al.* (2017) developed UPLC–MS/MS method for investigating the potential correlation between d-amino acids and AD. In this method, 11 d-amino acids were simultaneously determined in different regions of the rat brain. UPLC BEH C18 (50 mm × 2.1 mm, 1.7 μm) column was used as the stationary phase. Acetonitrile and water (containing 8 mM ammonium hydrogen carbonate) were used as mobile phase with a flow rate of 0.6 mL min^{-1}. Through multiple reaction monitoring in the positive ion mode, it was possible to detect analytes. Analytes were detected by multiple reaction monitoring in the positive ion mode. The lower LOQ values were found as 0.06–10 ng mL^{-1}. The intra- and inter-day RSD values were found in the range of 3.6–12% and 5.7–12%, respectively [147].

A new development related to AD drugs occurred in 2016. A new FDA-approved drug, which is used both to treat Parkinson's disease as well as AD is pimavanserin. Ezzeldin *et al.* (2020) developed a UPLC-MS/MS method for the determination of pimavanserin in the brain and the pharmacokinetic study was performed in mice. Pimavanserin and vilazodone(used as internal standard) were extracted by the liquid-liquid extraction technique. Acquity UPLC BEH C18 was chosen as an analytical column. The mobile phase consisted of 0.1% formic acid in acetonitrile and 0.1% formic acid in 20 mM ammonium acetate buffer (70:30 v/v) at a flow rate of 0.25 mL min^{-1}. The linearity range of the method was found >0.99 over the range of 0.1- 300 ng mL^{-1} in plasma and 0.25 -300 ng g^{-1} in the brain homogenate [148].

CONCLUSION

Considering the severity of increasing AD presence in society and the staggering costs of AD treatment, developing an effective treatment for AD is a significant aim for the researchers. Within this framework, this chapter attempts to review analysis studies for AD drugs in order to evaluate pharmacokinetic parameters for the determination of pharmacodynamic effects. This review reveals that the most popular method for the detection and determination of AD drugs is HPLC due to its rapid measurement of drugs with their metabolites. Particularly, during HPLC studies, three detectors, namely UV, MS, and fluorescence detectors, are widely preferred. Among them, MS detectors are used for bioequivalence studies, which is an essential part of pharmacokinetic studies. Optical methods, on the other hand, are another category of analytical methods; spectrophotometry and spectrofluorimetry are generally chosen by researchers utilizing optical methods. Electrochemical methods are the third category, which is quite promising for informing the transformation of drugs into their metabolites after their intake. The use of nanomaterials as an electrode modifier increases sensitivity and thereby decreases LOD values. Finally, this chapter provides essential information to help the research of analytical methods in Alzheimer's disease drugs in pharmaceuticals, biological samples, and natural products.

ABBREVIATIONS

SPE	Screen printed electrode
MWCNTs	Multi-walled carbon nanotubes
NPs	Nanoparticles
β-CD	beta cyclodextrin
3MT	3-methylthiophene
ACN	Acetonitrile
CoNPs	cobalt nanoparticles

FR	Flow rate
IV	Injection volume
MeOH	Methanol
N@CCNS	N-chitosan carbon nanosheets
N-CNDs	N-doped carbon nanodots
PEDOT	poly 3,4-ethylenedioxythiophene
PGE	pencil graphite electrode
SBT	Solochrome black T
Temp	Temperature

CONSENT FOR PUBLICATION

Not applicable.

CONFLICT OF INTEREST

The authors declare no conflict of interest, financial or otherwise.

ACKNOWLEDGEMENTS

Declared none.

REFERENCES

[1]   Draper B. Dealing With Dementia: A Guide To Alzheimer's Disease and Other Dementias. Australia: Allen & Unwin 2004.

[2]   Henderson AS, Jorm AF. Definition and Epidemiology of Dementia.Dementia. Wiley 2002; pp. 1-68. [http://dx.doi.org/10.1002/0470842350.ch1]

[3]   Syarifah-Noratiqah S-B, Naina-Mohamed I, Zulfarina MS, Qodriyah HMS. Natural Polyphenols in the Treatment of Alzheimer's Disease. Curr Drug Targets 2018; 19(8): 927-37. [http://dx.doi.org/10.2174/1389450118666170328122527] [PMID: 28356027]

[4]   Taipa R, Pinho J, Melo-Pires M. Clinico-pathological correlations of the most common neurodegenerative dementias. Front Neurol 2012; 3: 68. [http://dx.doi.org/10.3389/fneur.2012.00068] [PMID: 22557993]

[5]   Jones RW. Drug Treatment in Dementia. Bath, UK: Wiley-Blackwell 2000. [http://dx.doi.org/10.1002/9780470698747]

[6]   Ridge PG, Mukherjee S, Crane PK, Kauwe JS. Alzheimer's disease: analyzing the missing heritability. PLoS One 2013; 8(11): e79771. [http://dx.doi.org/10.1371/journal.pone.0079771] [PMID: 24244562]

[7]   De Matteis L, Martín-Rapún R, de la Fuente JM. Nanotechnology in Personalized Medicine: A Promising Tool for Alzheimer's Disease Treatment. Curr Med Chem 2018; 25(35): 4602-15. [http://dx.doi.org/10.2174/0929867324666171012112026] [PMID: 29022501]

[8]   Sliwinski M. Dementia.Encyclopedia of Health Psychology. New York: Kluwer Academic 2004; pp. 75-7. [http://dx.doi.org/10.1007/978-0-387-22557-9_4]

[9]   Ricciarelli R, Fedele E. The Amyloid Cascade Hypothesis in Alzheimer's Disease: It's Time to Change Our Mind. Curr Neuropharmacol 2017; 15(6): 926-35.
[http://dx.doi.org/10.2174/1570159X15666170116143743] [PMID: 28093977]

[10]   Brion JP. Neurofibrillary tangles and Alzheimer's disease. Eur Neurol 1998; 40(3): 130-40.
[http://dx.doi.org/10.1159/000007969] [PMID: 9748670]

[11]   De Simone A, Naldi M, Tedesco D, Bartolini M, Davani L, Andrisano V. Advanced analytical methodologies in Alzheimer's disease drug discovery. J Pharm Biomed Anal 2020; 178112899
[http://dx.doi.org/10.1016/j.jpba.2019.112899] [PMID: 31606562]

[12]   Kumar K, Kumar A, Keegan RM, Deshmukh R. Recent advances in the neurobiology and neuropharmacology of Alzheimer's disease. Biomed Pharmacother 2018; 98: 297-307.
[http://dx.doi.org/10.1016/j.biopha.2017.12.053] [PMID: 29274586]

[13]   Ponnayyan Sulochana S, Sharma K, Mullangi R, Sukumaran SK. Review of the validated HPLC and LC-MS/MS methods for determination of drugs used in clinical practice for Alzheimer's disease. Biomed Chromatogr 2014; 28(11): 1431-90.
[http://dx.doi.org/10.1002/bmc.3116] [PMID: 24515838]

[14]   Jeffrey K, Grossberg G. Cholinesterase Inhibitors.Lau L-F, Brodney MA. Topics in Medicinal Chemistry. Springer Verlag Berlin Heidelberg 2008; pp. 26-51.

[15]   Sadock BJ, Sadock VA. Concise Textbook of Clinical Psychiatry. 3rd ed., Philadelphia, USA: Lippincott Williams & Wilkins 2007.

[16]   Yacoubian TA. Neurodegenerative Disorders: Why Do We Need New Therapies?Drug Discovery Approaches for the Treatment of Neurodegenerative Disorders: Alzheimer's Disease. London, UK: Academic Press 2017; pp. 1-16.
[http://dx.doi.org/10.1016/B978-0-12-802810-0.00001-5]

[17]   Van Bulck M, Sierra-Magro A, Alarcon-Gil J, Perez-Castillo A, Morales-Garcia JA. Novel approaches for the treatment of alzheimer's and parkinson's disease. Int J Mol Sci 2019; 20(3): 719.
[http://dx.doi.org/10.3390/ijms20030719] [PMID: 30743990]

[18]   Prins ND, Scheltens P. Treating Alzheimer's disease with monoclonal antibodies: current status and outlook for the future. Alzheimers Res Ther 2013; 5(6): 56.
[http://dx.doi.org/10.1186/alzrt220] [PMID: 24216217]

[19]   Satlin A, Wang J, Logovinsky V, *et al.* Design of a Bayesian adaptive phase 2 proof-of-concept trial for BAN2401, a putative disease-modifying monoclonal antibody for the treatment of Alzheimer's disease. Alzheimers Dement (N Y) 2016; 2(1): 1-12.
[http://dx.doi.org/10.1016/j.trci.2016.01.001] [PMID: 29067290]

[20]   Fang Y, Zhou H, Gu Q, Xu J. Synthesis and evaluation of tetrahydroisoquinoline-benzimidazole hybrids as multifunctional agents for the treatment of Alzheimer's disease. Eur J Med Chem 2019; 167: 133-45.
[http://dx.doi.org/10.1016/j.ejmech.2019.02.008] [PMID: 30771601]

[21]   D'Acunto CW, Kaplánek R, Gbelcová H, *et al.* Metallomics for Alzheimer's disease treatment: Use of new generation of chelators combining metal-cation binding and transport properties. Eur J Med Chem 2018; 150: 140-55.
[http://dx.doi.org/10.1016/j.ejmech.2018.02.084] [PMID: 29525434]

[22]   Van den Broeck WMM. Drug Targets, Target Identification, Validation, and Screening.The Practice of Medicinal Chemistry. San Diego, CA: Academic Press 2015; pp. 45-70.
[http://dx.doi.org/10.1016/B978-0-12-417205-0.00003-1]

[23]   Kurbanoglu S, Uslu B, Ozkan SA. Validation of Analytical Methods for the Assessment of Hazards in Food.Food Safety and Preservation. Academic Press 2018; pp. 59-90.
[http://dx.doi.org/10.1016/B978-0-12-814956-0.00004-4]

[24]   González O, Alonso RM. Validation of bioanalytical chromatographic methods for the quantification of drugs in biological fluids.Handbook of Analytical Separations. Elsevier B.V. 2020; pp. 115-34.

[25]   Issaq HJ, Veenstra TD. Analytical methods and biomarker validation.Proteomic and Metabolomic Approaches to Biomarker Discovery. London, UK: Academic Press 2020; pp. 389-93.
[http://dx.doi.org/10.1016/B978-0-12-818607-7.00022-0]

[26]   Christodoulakis G, Satchell S. The Analytics of Risk Model Validation. London, UK: Elsevier 2008.

[27]   Bruce P, Minkkinen P, Riekkola M-L. Practical Method Validation: Validation Sufficient for an Analysis Method. Mikrochim Acta 1998; 128: 93-106.
[http://dx.doi.org/10.1007/BF01242196]

[28]   Branch SK. Guidelines from the International Conference on Harmonisation (ICH). J Pharm Biomed Anal 2005; 38(5): 798-805.
[http://dx.doi.org/10.1016/j.jpba.2005.02.037] [PMID: 16076542]

[29]   Horwitz W. Evaluation of Regulations Analytical Methods Used for Regulation of Foods and Drugs. Anal Chem 1982; 54: 67A-76A.
[http://dx.doi.org/10.1021/ac00238a765]

[30]   González AG, Herrador MÁ, Asuero AG, *et al.* The Correlation Coefficient Attacks Again. Accredit Qual Assur 2006; 11: 256-8.
[http://dx.doi.org/10.1007/s00769-006-0153-5]

[31]   Hartmann C, Massart DL, McDowall RD. An analysis of the Washington Conference Report on bioanalytical method validation. J Pharm Biomed Anal 1994; 12(11): 1337-43.
[http://dx.doi.org/10.1016/0731-7085(94)00083-2] [PMID: 7849129]

[32]   Hartmann C, Smeyers-Verbeke J, Massart DL, McDowall RD. Validation of bioanalytical chromatographic methods. J Pharm Biomed Anal 1998; 17(2): 193-218.
[http://dx.doi.org/10.1016/S0731-7085(97)00198-2] [PMID: 9638572]

[33]   Danzer K, Currie LA. Guidelines for calibration in analytical chemistry-Part 1. Fundamentals and single component calibration. Pure Appl Chem 1998; 70: 993-1014.
[http://dx.doi.org/10.1351/pac199870040993]

[34]   Buick AR, Doig MV, Jeal SC, Land GS, McDowall RD. Method validation in the bioanalytical laboratory. J Pharm Biomed Anal 1990; 8(8-12): 629-37.
[http://dx.doi.org/10.1016/0731-7085(90)80093-5] [PMID: 2100599]

[35]   Rozet E, Ziemons E, Marini RD, Boulanger B, Hubert P. Quality by design compliant analytical method validation. Anal Chem 2012; 84(1): 106-12.
[http://dx.doi.org/10.1021/ac202664s] [PMID: 22107128]

[36]   Rozet E, Marini RD, Ziemons E, Boulanger B, Hubert P. Advances in validation, risk and uncertainty assessment of bioanalytical methods. J Pharm Biomed Anal 2011; 55(4): 848-58.
[http://dx.doi.org/10.1016/j.jpba.2010.12.018] [PMID: 21237607]

[37]   Rozet E, Marini RD, Ziemons E, *et al.* Total error and uncertainty: Friends or foes? TrAC -. Trends Analyt Chem 2011; 30: 797-806.
[http://dx.doi.org/10.1016/j.trac.2010.12.009]

[38]   Green JM. A Practical Guide to Analytical Method Validation. Anal Chem News Featur 1996; 305A-9A.

[39]   Thompson M, Wood R. Harmonized Guidelines For Internal Quality Control in Analytical Chemistry Laboratories. Pure Appl Chem 1995; 67: 649-66.
[http://dx.doi.org/10.1351/pac199567040649]

[40]   Rösslein M, Rezzonico S, Hedinger R, *et al.* Repeatability: Some aspects concerning the evaluation of the measurement uncertainty. Accredit Qual Assur 2007; 12: 425-34.
[http://dx.doi.org/10.1007/s00769-007-0278-1]

[41] Thompson M, Ellison SLR, Wood R. The International Harmonized Protocol for the proficiency testing of analytical chemistry laboratories: (IUPAC technical report). Pure Appl Chem 2006; 78: 145-96.
[http://dx.doi.org/10.1351/pac200678010145]

[42] Heyden YV, Massart DL, Zhu Y, Hoogmartens J, De Beer J. Ruggedness tests on the high performance liquid chromatography assay of the United States Pharmacopoeia 23 for tetracycline.HCl: comparison of different columns in an interlaboratory approach. J Pharm Biomed Anal 1996; 14(8-10): 1313-26.
[http://dx.doi.org/10.1016/S0731-7085(96)01754-2] [PMID: 8818050]

[43] Currie LA. Nomenclature in evaluation of analytical methods including detection and quanti®cation capabilities 1 (IUPAC Recommendations 1995). Anal Chim Acta 1999; 391: 105-26.
[http://dx.doi.org/10.1016/S0003-2670(99)00104-X]

[44] Cuadros-Rodríguez L, Romero R, Bosque-Sendra JM. The role of the robustness/ruggedness and inertia studies in research and development of analytical processes. Crit Rev Anal Chem 2005; 35: 57-69.
[http://dx.doi.org/10.1080/10408340590947934]

[45] Berkowitz SA. Chromatography (other than size-exclusion chromatography) and electrophoresis. Biophysical Characterization of Proteins in Developing Biopharmaceuticals. Amsterdam, Netherlands: Elsevier 2020; pp. 431-56.
[http://dx.doi.org/10.1016/B978-0-444-64173-1.00014-7]

[46] Hage DS. Chromatography. Principles and Applications of Clinical Mass Spectrometry USA: Elsevier 2018. 1-32.

[47] Barnett KL, Harrington B, Graul TW. Validation of Liquid Chromatographic Methods. 2013.
[http://dx.doi.org/10.1016/B978-0-12-415806-1.00003-6]

[48] Lozano-Sánchez J, Borrás-Linares I, Sass-Kiss A, *et al.* Chromatographic Technique: High-Performance Liquid Chromatography (HPLC).Modern Techniques for Food Authentication. New York, NY: Academic Press 2018; pp. 459-526.
[http://dx.doi.org/10.1016/B978-0-12-814264-6.00013-X]

[49] Crowley TE. High-performance liquid chromatography.Purification and Characterization of Secondary Metabolites San Diego, United States, Elsevier, 2019. 49-58.

[50] Snyder LR, Dolan JW. Milestones in the Development of Liquid Chromatography. USA, ELsevier 2013.
[http://dx.doi.org/10.1016/B978-0-12-415807-8.00001-8]

[51] Eriksson KO. Reversed Phase Chromatography.Biopharmaceutical Processing: Development, Design, and Implementation of Manufacturing Processes. Amsterdam, Netherlands: Elsevier 2018; pp. 433-9.
[http://dx.doi.org/10.1016/B978-0-08-100623-8.00022-0]

[52] Moldoveanu SC, David V. Parameters that Characterize HPLC Analysis.Essentials in Modern HPLC Separations. Amsterdam, Netherlands: Elsevier 2013; pp. 53-83.
[http://dx.doi.org/10.1016/B978-0-12-385013-3.00002-1]

[53] Szymanski P, Karpiński A, Mikiciuk-Olasik E. Synthesis, biological activity and HPLC validation of 1,2,3,4-tetrahydroacridine derivatives as acetylcholinesterase inhibitors. Eur J Med Chem 2011; 46(8): 3250-7.
[http://dx.doi.org/10.1016/j.ejmech.2011.04.038] [PMID: 21570751]

[54] Koeber R, Kluenemann HH, Waimer R, *et al.* Implementation of a cost-effective HPLC/UV-approach for medical routine quantification of donepezil in human serum. J Chromatogr B Analyt Technol Biomed Life Sci 2012; 881-882: 1-11.
[http://dx.doi.org/10.1016/j.jchromb.2011.10.027] [PMID: 22204871]

[55] Shuangjin C, Fang F, Han L, Ming M. New method for high-performance liquid chromatographic

determination of amantadine and its analogues in rat plasma. J Pharm Biomed Anal 2007; 44(5): 1100-5.
[http://dx.doi.org/10.1016/j.jpba.2007.04.021] [PMID: 17553649]

[56]   Marques LA, Maada I, de Kanter FJJ, *et al.* Stability-indicating study of the anti-Alzheimer's drug galantamine hydrobromide. J Pharm Biomed Anal 2011; 55(1): 85-92.
[http://dx.doi.org/10.1016/j.jpba.2011.01.022] [PMID: 21300511]

[57]   Lohan S, Kaur R, Bharti S, *et al.* QbD-Enabled Development and Validation of a Liquid Chromatographic Method for Estimating Galantamine Hydrobromide in Biological Fluids. Curr Pharm Anal 2017; 14: 527-40.
[http://dx.doi.org/10.2174/1573412913666170912111144]

[58]   Acikara ÖB, Yilmaz BS, Yazgan D, *et al.* Quantification of galantamine in sternbergia species by high performance liquid chromatography. Turkish J Pharm Sci 2019; 16: 32-6.
[http://dx.doi.org/10.4274/tjps.95967]

[59]   Doncheva Tsvetkova D, Petrova Obreshkova D. Validation of HPLC Method for Simultaneous Determination of Galantamine Hydrobromide/Pymadine. Int J Pharm Res Allied Sci 2018; 7: 233-47.

[60]   Amini H, Ahmadiani A. High-performance liquid chromatographic determination of rivastigmine in human plasma for application in pharmacokinetic studies. Iran J Pharm Res 2010; 9(2): 115-21.
[PMID: 24363716]

[61]   Thomas S, Shandilya S, Bharati A, Paul SK, Agarwal A, Mathela CS. Identification, characterization and quantification of new impurities by LC-ESI/MS/MS and LC-UV methods in rivastigmine tartrate active pharmaceutical ingredient. J Pharm Biomed Anal 2012; 57: 39-51.
[http://dx.doi.org/10.1016/j.jpba.2011.08.014] [PMID: 21880452]

[62]   Arumugam K, Chamallamudi MR, Gilibili RR, *et al.* Development and validation of a HPLC method for quantification of rivastigmine in rat urine and identification of a novel metabolite in urine by LC-MS/MS. Biomed Chromatogr 2011; 25(3): 353-61.
[http://dx.doi.org/10.1002/bmc.1455] [PMID: 20540167]

[63]   Pan J, Jin R, Hu X. Application of L-Amino acid in determination of Huperzine A by high performance liquid chromatography. J Chromatogr B Analyt Technol Biomed Life Sci 2006; 836(1-2): 108-10.
[http://dx.doi.org/10.1016/j.jchromb.2006.03.018] [PMID: 16574511]

[64]   Wu Q, Gu Y. Quantification of huperzine A in Huperzia serrata by HPLC-UV and identification of the major constituents in its alkaloid extracts by HPLC-DAD-MS-MS. J Pharm Biomed Anal 2006; 40(4): 993-8.
[http://dx.doi.org/10.1016/j.jpba.2005.07.047] [PMID: 16337768]

[65]   Szypuła W, Kiss A, Pietrosiuk A, *et al.* Determination of huperzine a in Huperzia selago plants from wild population and obtained in in vitro culture by high-performance liquid chromatography using a chaotropic mobile phase. Acta Chromatogr 2011; 23: 339-52.
[http://dx.doi.org/10.1556/AChrom.23.2011.2.11]

[66]   Ye J, Zeng S, Zhang W, Chen G. Ion-pair reverse-phase high performance liquid chromatography method for determination of Huperzine-A in beagle dog serum. J Chromatogr B Analyt Technol Biomed Life Sci 2005; 817(2): 187-91.
[http://dx.doi.org/10.1016/j.jchromb.2004.12.002] [PMID: 15686984]

[67]   Wei G, Xiao S, Lu R, Liu C. Simultaneous determination of ZT-1 and its metabolite Huperzine A in plasma by high-performance liquid chromatography with ultraviolet detection. J Chromatogr B Analyt Technol Biomed Life Sci 2006; 830(1): 120-5.
[http://dx.doi.org/10.1016/j.jchromb.2005.10.027] [PMID: 16288904]

[68]   Abonassif MA, Hefnawy MM, Kassem MG, Mostafa GA. Determination of donepezil hydrochloride in human plasma and pharmaceutical formulations by HPLC with fluorescence detection. Acta Pharm 2011; 61(4): 403-13.

[http://dx.doi.org/10.2478/V10007-011-0035-1] [PMID: 22202199]

[69]   Arumugam K, Chamallamudi M, Mallayasamy S, *et al.* High performance liquid chromatographic fluorescence detection method for the quantification of rivastigmine in rat plasma and brain: application to preclinical pharmacokinetic studies in rats. J Young Pharm 2011; 3(4): 315-21.
[http://dx.doi.org/10.4103/0975-1483.90244] [PMID: 22224039]

[70]   Puente B, Hernandez E, Perez S, *et al.* Determination of memantine in plasma and vitreous humour by HPLC with precolumn derivatization and fluorescence detection. J Chromatogr Sci 2011; 49(10): 745-52.
[http://dx.doi.org/10.1093/chrsci/49.10.745] [PMID: 22080801]

[71]   Xie MF, Zhou W, Tong XY, *et al.* High-performance liquid chromatographic determination of memantine hydrochloride in rat plasma using sensitive fluorometric derivatization. J Sep Sci 2011; 34(3): 241-6.
[http://dx.doi.org/10.1002/jssc.201000579] [PMID: 21268245]

[72]   Zarghi A, Shafaati A, Foroutan SM, Khoddam A, Madadian B. Sensitive and rapid HPLC method for determination of memantine in human plasma using OPA derivatization and fluorescence detection: application to pharmacokinetic studies. Sci Pharm 2010; 78(4): 847-56.
[http://dx.doi.org/10.3797/scipharm.1008-17] [PMID: 21179320]

[73]   Özdemir E, Tatar Ulu S. Highly sensitive HPLC method for the determination of galantamine in human plasma and urine through derivatization with dansyl chloride using fluorescence detector. Luminescence 2017; 32(7): 1145-9.
[http://dx.doi.org/10.1002/bio.3301] [PMID: 28430400]

[74]   Maláková J, Nobilis M, Svoboda Z, *et al.* High-performance liquid chromatographic method with UV photodiode-array, fluorescence and mass spectrometric detection for simultaneous determination of galantamine and its phase I metabolites in biological samples. J Chromatogr B Analyt Technol Biomed Life Sci 2007; 853(1-2): 265-74.
[http://dx.doi.org/10.1016/j.jchromb.2007.03.025] [PMID: 17416214]

[75]   Culzoni MJ, Aucelio RQ, Escandar GM. High-performance liquid chromatography with fast-scanning fluorescence detection and multivariate curve resolution for the efficient determination of galantamine and its main metabolites in serum. Anal Chim Acta 2012; 740: 27-35.
[http://dx.doi.org/10.1016/j.aca.2012.06.034] [PMID: 22840647]

[76]   Wang Y, Chu D, Gu J, Fawcett JP, Wu Y, Liu W. Liquid chromatographic-tandem mass spectrometric method for the quantitation of huperzine A in dog plasma. J Chromatogr B Analyt Technol Biomed Life Sci 2004; 803(2): 375-8.
[http://dx.doi.org/10.1016/j.jchromb.2004.01.013] [PMID: 15063351]

[77]   Apostolou C, Dotsikas Y, Kousoulos C, Loukas YL. Quantitative determination of donepezil in human plasma by liquid chromatography/tandem mass spectrometry employing an automated liquid-liquid extraction based on 96-well format plates. Application to a bioequivalence study. J Chromatogr B Analyt Technol Biomed Life Sci 2007; 848(2): 239-44.
[http://dx.doi.org/10.1016/j.jchromb.2006.10.037] [PMID: 17113365]

[78]   Shah HJ, Kundlik ML, Pandya A, *et al.* A rapid and specific approach for direct measurement of donepezil concentration in human plasma by LC-MS/MS employing solid-phase extraction. Biomed Chromatogr 2009; 23(2): 141-51.
[http://dx.doi.org/10.1002/bmc.1095] [PMID: 18823072]

[79]   Katakam P, Kalakuntla RR, Adiki SK, *et al.* Development and validation of a liquid chromatography mass spectrometry method for the determination of donepezil in human plasma. J Pharm Res 2013; 7: 720-6.
[http://dx.doi.org/10.1016/j.jopr.2013.08.021]

[80]   Kim KA, Lim JL, Kim C, Park JY. Pharmacokinetic comparison of orally disintegrating and conventional donepezil formulations in healthy Korean male subjects: a single-dose, randomized,

open-label, 2-sequence, 2-period crossover study. Clin Ther 2011; 33(7): 965-72.
[http://dx.doi.org/10.1016/j.clinthera.2011.06.003] [PMID: 21723605]

[81]   Meier-Davis SR, Meng M, Yuan W, *et al.* Dried blood spot analysis of donepezil in support of a GLP 3-month dose-range finding study in rats. Int J Toxicol 2012; 31(4): 337-47.
[http://dx.doi.org/10.1177/1091581812447957] [PMID: 22705881]

[82]   Lee CB, Min JS, Chae SU, *et al.* Simultaneous determination of donepezil, 6-O-desmethyl donepezil and spinosin in beagle dog plasma using liquid chromatography–tandem mass spectrometry and its application to a drug-drug interaction study. J Pharm Biomed Anal 2020; 178112919
[http://dx.doi.org/10.1016/j.jpba.2019.112919] [PMID: 31654856]

[83]   Khuroo AH, Gurule SJ, Monif T, Goswami D, Saha A, Singh SK. ESI-MS/MS stability-indicating bioanalytical method development and validation for simultaneous estimation of donepezil, 5-desmethyl donepezil and 6-desmethyl donepezil in human plasma. Biomed Chromatogr 2012; 26(5): 636-49.
[http://dx.doi.org/10.1002/bmc.1709] [PMID: 22120680]

[84]   Patel BN, Sharma N, Sanyal M, Shrivastav PS. Quantitation of donepezil and its active metabolite 6-O-desmethyl donepezil in human plasma by a selective and sensitive liquid chromatography-tandem mass spectrometric method. Anal Chim Acta 2008; 629(1-2): 145-57.
[http://dx.doi.org/10.1016/j.aca.2008.09.048] [PMID: 18940331]

[85]   Gumustas M, Kurbanoglu S, Uslu B, *et al.* UPLC versus HPLC on drug analysis: Advantageous, applications and their validation parameters. Chromatographia 2013; 76: 1365-427.
[http://dx.doi.org/10.1007/s10337-013-2477-8]

[86]   Gröger TM, Käfer U, Zimmermann R. Gas chromatography in combination with fast high-resolution time-of-flight mass spectrometry: Technical overview and perspectives for data visualization. TrAC -. Trends Analyt Chem 2020; 122115677
[http://dx.doi.org/10.1016/j.trac.2019.115677]

[87]   Zoccali M, Tranchida PQ, Mondello L. Fast gas chromatography-mass spectrometry: A review of the last decade. TrAC -. Trends Analyt Chem 2019; 118: 444-52.
[http://dx.doi.org/10.1016/j.trac.2019.06.006]

[88]   Ruiz-Matute AI, Rodríguez-Sánchez S, Sanz ML, *et al.* Chromatographic Technique: Gas Chromatography (GC).Modern Techniques for Food Authentication. New York, NY: Academic Press 2018; pp. 415-58.
[http://dx.doi.org/10.1016/B978-0-12-814264-6.00012-8]

[89]   Berkov S, Bastida J, Viladomat F, Codina C. Development and validation of a GC-MS method for rapid determination of galanthamine in Leucojum aestivum and Narcissus ssp.: a metabolomic approach. Talanta 2011; 83(5): 1455-65.
[http://dx.doi.org/10.1016/j.talanta.2010.11.029] [PMID: 21238736]

[90]   Mohamed FA, Khashaba PY, El-Wekil MM, *et al.* Spectrodensitometric determination of rivastigmine after vortex assisted magnetic solid phase extraction. Microchem J 2019; 147: 764-74.
[http://dx.doi.org/10.1016/j.microc.2019.03.085]

[91]   Shewiyo DH, Kaale E, Risha PG, Dejaegher B, Smeyers-Verbeke J, Vander Heyden Y. HPTLC methods to assay active ingredients in pharmaceutical formulations: a review of the method development and validation steps. J Pharm Biomed Anal 2012; 66: 11-23.
[http://dx.doi.org/10.1016/j.jpba.2012.03.034] [PMID: 22494517]

[92]   Bhateria M, Ramakrishna R, Pakala DB, Bhatta RS. Development of an LC-MS/MS method for simultaneous determination of memantine and donepezil in rat plasma and its application to pharmacokinetic study. J Chromatogr B Analyt Technol Biomed Life Sci 2015; 1001: 131-9.
[http://dx.doi.org/10.1016/j.jchromb.2015.07.042] [PMID: 26280281]

[93]   Noetzli M, Ansermot N, Dobrinas M, Eap CB. Simultaneous determination of antidementia drugs in human plasma: procedure transfer from HPLC-MS to UPLC-MS/MS. J Pharm Biomed Anal 2012;

64-65: 16-25.
[http://dx.doi.org/10.1016/j.jpba.2012.02.008] [PMID: 22410501]

[94] Bhatt J, Subbaiah G, Kambli S, *et al.* A rapid and sensitive liquid chromatography-tandem mass spectrometry (LC-MS/MS) method for the estimation of rivastigmine in human plasma. J Chromatogr B Analyt Technol Biomed Life Sci 2007; 852(1-2): 115-21.
[http://dx.doi.org/10.1016/j.jchromb.2007.01.003] [PMID: 17296337]

[95] Huda NH, Gauri B, Benson HAE, Chen Y. A Stability Indicating HPLC Assay Method for Analysis of Rivastigmine Hydrogen Tartrate in Dual-Ligand Nanoparticle Formulation Matrices and Cell Transport Medium. J Anal Methods Chem 2018; 20181841937
[http://dx.doi.org/10.1155/2018/1841937] [PMID: 29686925]

[96] Sha Y, Deng C, Liu Z, Huang T, Yang B, Duan G. Headspace solid-phase microextraction and capillary gas chromatographic-mass spectrometric determination of rivastigmine in canine plasma samples. J Chromatogr B Analyt Technol Biomed Life Sci 2004; 806(2): 271-6.
[http://dx.doi.org/10.1016/j.jchromb.2004.04.006] [PMID: 15171938]

[97] Frankfort SV, Ouwehand M, van Maanen MJ, Rosing H, Tulner LR, Beijnen JH. A simple and sensitive assay for the quantitative analysis of rivastigmine and its metabolite NAP 226-90 in human EDTA plasma using coupled liquid chromatography and tandem mass spectrometry. Rapid Commun Mass Spectrom 2006; 20(22): 3330-6.
[http://dx.doi.org/10.1002/rcm.2737] [PMID: 17044120]

[98] Tsuruoka Y, Nakajima T, Kanda M, *et al.* Simultaneous determination of amantadine, rimantadine, and memantine in processed products, chicken tissues, and eggs by liquid chromatography with tandem mass spectrometry. J Chromatogr B Analyt Technol Biomed Life Sci 2017; 1044-1045: 142-8.
[http://dx.doi.org/10.1016/j.jchromb.2017.01.014] [PMID: 28107701]

[99] Konda RK, Challa BR, Chandu BR, Chandrasekhar KB. Bioanalytical method development and validation of memantine in human plasma by high performance liquid chromatography with tandem mass spectrometry: application to bioequivalence study. J Anal Methods Chem 2012; 2012101249
[http://dx.doi.org/10.1155/2012/101249] [PMID: 22567548]

[100] Chládek J, Žaludek B, Sova P, *et al.* Steady-state bioequivalence studies of two memantine tablet and oral solution formulations in healthy volunteers. J Appl Biomed 2008; 6: 39-45.
[http://dx.doi.org/10.32725/jab.2008.006]

[101] Leis HJ, Fauler G, Windischhofer W. Quantitative analysis of memantine in human plasma by gas chromatography/negative ion chemical ionization/mass spectrometry. J Mass Spectrom 2002; 37(5): 477-80.
[http://dx.doi.org/10.1002/jms.303] [PMID: 12112752]

[102] Borloz A, Marston A, Hostettmann K. The determination of huperzine A in European Lycopodiaceae species by HPLC-UV-MS. Phytochem Anal 2006; 17(5): 332-6.
[http://dx.doi.org/10.1002/pca.922] [PMID: 17019934]

[103] Zhang H, Liang H, Kuang P, Yuan Q, Wang Y. Simultaneously preparative purification of Huperzine A and Huperzine B from Huperzia serrata by macroporous resin and preparative high performance liquid chromatography. J Chromatogr B Analyt Technol Biomed Life Sci 2012; 904: 65-72.
[http://dx.doi.org/10.1016/j.jchromb.2012.07.019] [PMID: 22877738]

[104] Yue P, Tao T, Zhao Y, Ren J, Chai X. Determination of Huperzine A in rat plasma by high-performance liquid chromatography with a fluorescence detector. J Pharm Biomed Anal 2007; 44(1): 309-12.
[http://dx.doi.org/10.1016/j.jpba.2007.02.019] [PMID: 17408901]

[105] Yang QP, Kou XL, Fugal KB, McLaughlin JL. Determination of huperzine A in formulated products by reversed-phase-liquid chromatography using diode array and electrospray ionization mass spectrometric detection. Phytomedicine 2003; 10(2-3): 200-5.
[http://dx.doi.org/10.1078/094471103321659942] [PMID: 12725577]

[106] Cuthbertson D, Piljac-Žegarac J, Lange BM. Validation of a microscale extraction and high-throughput UHPLC-QTOF-MS analysis method for huperzine A in Huperzia. Biomed Chromatogr 2012; 26(10): 1191-5.
[http://dx.doi.org/10.1002/bmc.2677] [PMID: 22275140]

[107] Meurens M. Spectrophotometric techniques.Food Authenticity and Traceability. Boca Raton, FL: Woodhead Publishing Limited 2003; pp. 184-96.
[http://dx.doi.org/10.1533/9781855737181.1.184]

[108] Nelson DL. Introduction to Spectroscopy.Spectroscopic Methods in Food Analysis. Boca Raton, FL: CRC Press 2018; pp. 5-33.

[109] Nahata A. Department of Pharmaceutical Sciences, Doctor Hari Singh Gour Vishwavidyalaya. Pharm Anal Acta 2011; 2: 107.

[110] Ben-Zur R, Hake H, Hassoon S, *et al.* Optical analytical methods for detection of pesticides. Rev Anal Chem 2011; 30: 123-39.
[http://dx.doi.org/10.1515/REVAC.2011.104]

[111] Atia NN, Marzouq MA, Hassan AI, Eltoukhi WE. A new spectrofluorimetric assay for quantification of Amisulpride and Memantine hydrochloride in real human plasma sample and pharmaceutical formulations via Hantzsch reaction. Spectrochim Acta A Mol Biomol Spectrosc 2020; 224117388
[http://dx.doi.org/10.1016/j.saa.2019.117388] [PMID: 31357052]

[112] Culzoni MJ, Aucelio RQ, Escandar GM. Spectrofluorimetry in organized media coupled to second-order multivariate calibration for the determination of galantamine in the presence of uncalibrated interferences. Talanta 2010; 82(1): 325-32.
[http://dx.doi.org/10.1016/j.talanta.2010.04.043] [PMID: 20685474]

[113] Shamsi A, Mohammad T, Anwar S, *et al.* Probing the interaction of Rivastigmine Tartrate, an important Alzheimer's drug, with serum albumin: Attempting treatment of Alzheimer's disease. Int J Biol Macromol 2020; 148: 533-42.
[http://dx.doi.org/10.1016/j.ijbiomac.2020.01.134] [PMID: 31954794]

[114] Kapil R, Dhawan S, Singh B. Development and validation of a spectrofluorimetric method for the estimation of rivastigmine in formulations. Indian J Pharm Sci 2009; 71(5): 585-9.
[http://dx.doi.org/10.4103/0250-474X.58179] [PMID: 20502586]

[115] Pekcan Ertokuş G, Çatalyürek KN. Spectrophotometric and Chemometric Methods for Simultaneous Determination of Alzheimer's drugs in Pharmaceutical Tablets. Int J Pharm Res Allied Sci 2017; 6: 73-9.

[116] Salem MY, El-Kosasy AM, El-Bardicy MG, Abd El-Rahman MK. Spectrophotometric and spectrodensitometric methods for the determination of rivastigmine hydrogen tartrate in presence of its degradation product. Drug Test Anal 2010; 2(5): 225-33.
[http://dx.doi.org/10.1002/dta.121] [PMID: 20468010]

[117] Iranifam M, Hendekhale NR. CuO nanoparticles-catalyzed hydrogen peroxide–sodium hydrogen carbonate chemiluminescence system used for quenchometric determination of atorvastatin, rivastigmine and topiramate. Sens Actuators B Chem 2017; 243: 532-41.
[http://dx.doi.org/10.1016/j.snb.2016.12.013]

[118] Huang L, Li Z, Guo L. Colorimetric assay of acetylcholinesterase inhibitor tacrine based on MoO_2 nanoparticles as peroxidase mimetics. Spectrochim Acta A Mol Biomol Spectrosc 2020; 224117412
[http://dx.doi.org/10.1016/j.saa.2019.117412] [PMID: 31357051]

[119] Choodum A. NicDaeid N. Quantitative Colorimetric Assays for Methamphetamine.Neuropathology of Drug Addictions and Substance Misuse. London, UK: Academic Press 2016; pp. 349-59.
[http://dx.doi.org/10.1016/B978-0-12-800212-4.00034-0]

[120] Bozal-Palabiyik B, Uslu B. Voltammetric Investigation and Determination of Piribedil in Pharmaceutical Dosage Forms Using Carbon-Based Electrodes. Curr Pharm Anal 2017; 13: 91-8.

[http://dx.doi.org/10.2174/1573412912666160422155747]

[121] Ozkan S. Principles and Techniques of Electroanalytical Stripping Methods for Pharmaceutically Active Compounds in Dosage Forms and Biological Samples. Curr Pharm Anal 2009; 5: 127-43.
[http://dx.doi.org/10.2174/1573412097881728 70]

[122] Uslu B, Ozkan SA. Electroanalytical Methods for the Determination of Pharmaceuticals: A Review of Recent Trends and Developments. Anal Lett 2011; 44: 2644-702.
[http://dx.doi.org/10.1080/00032719.2011.553010]

[123] Özkan SA, Uslu B, Aboul-Enein HY. Analysis of pharmaceuticals and biological fluids using modern electroanalytical techniques. Crit Rev Anal Chem 2003; 33: 155-81.
[http://dx.doi.org/10.1080/713609162]

[124] Uslu B, Ozkan SA. Electroanalytical application of carbon based electrodes to the pharmaceuticals. Anal Lett 2007; 40: 817-53.
[http://dx.doi.org/10.1080/00032710701242121]

[125] Uslu B, Ozkan SA. Solid electrodes in electroanalytical chemistry: present applications and prospects for high throughput screening of drug compounds. Comb Chem High Throughput Screen 2007; 10(7): 495-513.
[http://dx.doi.org/10.2174/138620707782152425] [PMID: 17979634]

[126] Golcu A, Ozkan SA. Electroanalytical determination of donepezil HCl in tablets and human serum by differential pulse and osteryoung square wave voltammetry at a glassy carbon electrode. Pharmazie 2006; 61(9): 760-5.
[PMID: 17020151]

[127] Aparicio I, Callejón M, Jiménez JC, Bello MA, Guiraúm A. Electrochemical oxidation at carbon paste electrode of tacrine and 1-hydroxytacrine and differential pulse voltammetric determination of tacrine in pharmaceuticals and human urine. Analyst (Lond) 2000; 125(11): 2016-9.
[http://dx.doi.org/10.1039/b005874m] [PMID: 11193091]

[128] Ghoneim EM, El-Attar MA, Ghoneim MM. Determination of Donepezil Hydrochloride in Pharmaceutical Formulation and Human Serum by Square-Wave Adsorptive Cathodic Stripping Voltammetry. Chem Anal (Warsaw) 2009; 54: 389-402.

[129] Bozal-Palabiyik B, Dogan-Topal B, Ozkan SA, *et al.* New Trends in Electrochemical Sensors Modified with Carbon Nanotubes and Graphene for Pharmaceutical Analysis. 2018.
[http://dx.doi.org/10.2174/9781681085746118020009]

[130] Kalambate PK, Biradar MR, Karna SP, *et al.* Adsorptive stripping differential pulse voltammetry determination of rivastigmine at graphene nanosheet-gold nanoparticle/carbon paste electrode. J Electroanal Chem (Lausanne Switz) 2015; 757: 150-8.
[http://dx.doi.org/10.1016/j.jelechem.2015.09.027]

[131] Su L, Jia W, Hou C, Lei Y. Microbial biosensors: a review. Biosens Bioelectron 2011; 26(5): 1788-99.
[http://dx.doi.org/10.1016/j.bios.2010.09.005] [PMID: 20951023]

[132] Arvand M, Fallahi P. Man-Tailored Biomimetic Sensor of Molecularly Imprinted Materials for the Potentiometric Measurement of Rivastigmine in Tablets and Biological Fluids and Employing the Taguchi Optimization Methodology to Optimize the MIP-Based Membranes. Electroanalysis 2012; 24: 1852-63.
[http://dx.doi.org/10.1002/elan.201200247]

[133] Parsajoo C, Kauffmann J-M. Electrochemical Biosensors for Drug Analysis.Monographs in Electrochemistry: Electroanalysis in Biomedical and Pharmaceutical Sciences. New York: Springer-Verlag 2015; pp. 144-86.

[134] Ivanov A, Davletshina R, Sharafieva I, Evtugyn G. Electrochemical biosensor based on polyelectrolyte complexes for the determination of reversible inhibitors of acetylcholinesterase. Talanta 2019; 194: 723-30.

[http://dx.doi.org/10.1016/j.talanta.2018.10.100] [PMID: 30609597]

[135] Kostelnik A, Cegan A, Pohanka M. Electrochemical Determination of Activity of Acetylcholinesterase Immobilized on Magnetic Particles. Int J Electrochem Sci 2016; 11: 4840-9.
[http://dx.doi.org/10.20964/2016.06.39]

[136] Mladenović AR, Mijin DŽ, Drmanić SŽ, *et al.* Electrochemical oxidation of donepezil and its voltammetric determination at gold electrode. Electroanalysis 2014; 26: 893-7.
[http://dx.doi.org/10.1002/elan.201400034]

[137] Ramadan NK, Derar AR, Mohamed TA, *et al.* Cobalt Ferrite Nanoparticles Modified Carbon Paste Miniaturized Electrode with Enhanced Sensitivity for Electrochemical Sensing of Donepezil Hydrochloride. Anal Bioanal Electrochem 2018; 10: 1259-72.

[138] Mohamed FA, Khashaba PY, Shahin RY, *et al.* Tunable ternary nanocomposite prepared by electrodeposition for biosensing of centrally acting reversible acetyl cholinesterase inhibitor donepezil hydrochloride in real samples. Colloids Surf A Physicochem Eng Asp 2019; 567: 76-85.
[http://dx.doi.org/10.1016/j.colsurfa.2019.01.033]

[139] Arvand M, Fallahi P. Voltammetric determination of rivastigmine in pharmaceutical and biological samples using molecularly imprinted polymer modified carbon paste electrode. Sens Actuators B Chem 2013; 188: 797-805.
[http://dx.doi.org/10.1016/j.snb.2013.07.092]

[140] Kılıçyaldır B, Avan AA, Güçlü K, *et al.* Electrochemical Determination of Rivastigmine Hydrogen Tartrate at β-Cyclodextrin/Multi-Walled Carbon Nanotubes Modified Electrode. Curr Pharm Anal 2019; 15: 211-6.
[http://dx.doi.org/10.2174/1573412913666171115162250]

[141] Mohamed FA, Khashaba PY, Shahin RY, *et al.* A determination approach for rivastigmine by lepidocrocite nanoparticles supported on N-chitosan carbon nanosheets/Anti-fouling PAS: Application to biosensing. J Electrochem Soc 2019; 166: H41-6.
[http://dx.doi.org/10.1149/2.0961902jes]

[142] Arvand M, Habibi MF, Hemmati S. A novel one-step electrochemical preparation of silver nanoparticles/poly(3-methylthiophene) nanocomposite for detection of galantamine in human cerebrospinal fluid and narcissus. J Electroanal Chem (Lausanne Switz) 2017; 785: 220-8.
[http://dx.doi.org/10.1016/j.jelechem.2016.12.048]

[143] Shi H, Zheng Y, Wu M, *et al.* Graphene ink modified glassy carbon electrode as electrochemical sensor for galantamine determination. Int J Electrochem Sci 2019; 14: 1546-55.
[http://dx.doi.org/10.20964/2019.02.56]

[144] Wu Z-Y, Thiagarajan S, Chen S-M, *et al.* Electrochemical Preparation and Characterization of MnO2-PEDOT Hybrid Film and its Application in Electrocatalytic Oxidation of Memantine Hydrochloride. Int J Electrochem Sci 2012; 7: 1230-41.

[145] Cummings J, Lee G, Ritter A, Sabbagh M, Zhong K. Alzheimer's disease drug development pipeline: 2019. Alzheimers Dement (N Y) 2019; 5: 272-93.
[http://dx.doi.org/10.1016/j.trci.2019.05.008] [PMID: 31334330]

[146] Zueva I, Dias J, Lushchekina S, *et al.* New evidence for dual binding site inhibitors of acetylcholinesterase as improved drugs for treatment of Alzheimer's disease. Neuropharmacology 2019; 155: 131-41.
[http://dx.doi.org/10.1016/j.neuropharm.2019.05.025] [PMID: 31132435]

[147] Li Z, Xing Y, Guo X, Cui Y. Development of an UPLC-MS/MS method for simultaneous quantitation of 11 d-amino acids in different regions of rat brain: Application to a study on the associations of d-amino acid concentration changes and Alzheimer's disease. J Chromatogr B Analyt Technol Biomed Life Sci 2017; 1058: 40-6.
[http://dx.doi.org/10.1016/j.jchromb.2017.05.011] [PMID: 28531844]

[148]   Ezzeldin E, Iqbal M, Asiri YA, Ali AA, El-Nahhas T. A rapid, simple and highly sensitive UPLC-MS/MS method for quantitation of pimavanserin in plasma and tissues: Application to pharmacokinetics and brain uptake studies in mice. J Chromatogr B Analyt Technol Biomed Life Sci 2020; 1143122015
[http://dx.doi.org/10.1016/j.jchromb.2020.122015] [PMID: 32174544]

CHAPTER 5

Targeting Alzheimer's Disorders Through Nanomedicine

Mahima Kaushik[1,*] and **Dhriti Jha**[2]

[1] *Cluster Innovation Centre, University of Delhi, Delhi, India*

[2] *Bhaskaracharya College of Applied Sciences, University of Delhi, Delhi, India*

Abstract: Neurodegenerative diseases have been known to exist in the human population for a long time, yet most of them do not have a cure even until now. With the recent surge of nanotechnology and its applications to physiology as nanomedicine, a new hope for a better future has been offered. Nanomedicine has opened up many areas of further research that can lead us to a probable cure for these diseases. This review particularly focuses on Alzheimer's disease (AD), which is a progressive neurodegenerative disorder. AD primarily affects the cerebral cortex in the forebrain and the hippocampus in the midbrain. According to the World Alzheimer's report, there are 46.8 million AD patients worldwide as of 2016 with a high probability of this statistic becoming double in the next two decades. For understanding this crucial situation, studying the pathology of AD, and recent advancements in the treatment of AD, specifically through nanomedicine becomes extremely important. Through this review, we intend to explore all the relevant aspects in relation to nanotechnological advancements for the treatment of Alzheimer's disease.

Keywords: Alzheimer's disease, Drug Delivery System, Lipid-based Nanoparticles, Neurodegenerative diseases, Nanoparticles, Nanomedicine, Neuroinflammation, Polymeric Nanoparticles.

INTRODUCTION

Neurodegenerative diseases (NDs) are defined as a set of diseases that cause structural modifications to the neurons present in the human nervous system resulting in malfunctioning of the neuronal cells. This ultimately leads to a loss of cognitive, sensory and motor abilities. Though NDs have been prevalent in the human population for a long time, the drugs currently in clinical use are only for disease management and play a little-to-no role in disease cure.

* **Corresponding author Mahima Kaushik:** Cluster Innovation Centre, University of Delhi, Delhi, India; Tel: 91-011-276667020; Fax: 91-011-27666706; E-mails: mkaushik@cic.du.ac.in, kaushikmahima2011@gmail.com

Atta-ur-Rahman (Ed.)
All rights reserved-© 2020 Bentham Science Publishers

Some of the most popular NDs include Alzheimer's disease (AD), Amyotrophic Lateral Sclerosis (ALS), Parkinson's disease (PD) and Frontotemporal dementia (FTD). Most of the NDs are caused due to misfolding of essential messenger proteins such as amyloid plaques, transactive response DNA-binding protein 43 (TDP-43), dopamine neurotransmitters and serotonin. AD is caused by an irreversible neuronal loss and vascular toxicity resulting due to extracellular deposition of Aβ peptide, in the form of senile plaques, and neurofibrillary tangles of phosphorylated tau protein. In PD, there is loss of neurons involved in the dopaminergic pathway and proteinaceous Lewy bodies, which are α-synuclein aggregates present in the central nervous systems (CNS) and peripheral nervous system (PNS) [1, 2]. Both AD and PD are characterized by progressive memory loss, depleting decision -making ability and, degradation of cognitive and sensory perceptions [3 - 6]. A group of disorders showing rigidity, tremors and stature instability are categorized under a clinical syndrome termed as Parkinsonism [7], including PD as one such kind of disorder. ALS is a type of motor neuron degenerative disorder, where there is damage to motor neurons of both the CNS and the PNS. Here, the nervous connections between both cortex-brainstem and spinal cord-muscles are severely affected [8]. Although, aggregation of TDP-43 is thought to be a major cause of ALS, yet the disease pathway has not been clearly understood until now [8]. FTD is another type of neurodegenerative disorder, which is often linked to ALS due to their genetic and neuropathological common links [9]. A patient of FTD shows signs of forgetfulness and altered social behaviour. Language dysfunctionality is also observed in some kinds of dementia [9].

The recent surge in nanomedicine has given new hope as a prospective cure for NDs. The field of nanomedicine involves the use of nanotechnology for designing particles, with sizes ranging in the order of 10^{-9} m (nanometer), which can potentially be used as drugs and in drug delivery systems to inhibit such misfolding and agglomeration. Most diseases occur due to malfunctioning of some biological aspect involved, at a nanoscale [10]. Through nanoparticle-based drugs and delivery systems, there is an attempt to utilise the similarity in size shared between the nanoparticle and the biological molecules, and target the factors that are ultimately resulting in the disease. Due to their small size, nanomaterials display a high surface area to volume ratio. This translates to higher reactivity as the particles offer more surface area for the ligand to interact with, making it a favourable option for transporting drugs to their targets [11]. It is the dynamic and versatile nature of nanoparticle-based drug designing that gives it an edge over traditional medicine. Still, the extent of nanotoxicity caused and the unknown long-term effects of nanoparticle therapy can pose as disadvantages of this approach. Thus, clinical application of nanotechnology as medicine in NDs require exhaustive studies in appropriate models [12].

Major types of nanoparticles (NPs) commonly associated with biomedical research include liposomes, albumin-based, polymeric NPs, microemulsions, metallic NPs like gold, nano-emulsions, iron oxide and quantum dots [13]. Amongst these, polymeric NPs, lipid NPs, micro- and nano-emulsions are majorly involved in designing nanomedical therapies.

This review particularly focuses on AD, which is a progressive neurodegenerative disorder, primarily affecting the cerebral cortex in the forebrain and the Hippocampus in the midbrain [14, 15]. 46.8 million AD patients were reported worldwide in 2016 by world Alzheimer's report, with an expected double in number in the next two decades [16]. AD can be sporadic or familial in nature. The cases of familial AD are far lower than patients of acquired AD (measured 11.3% in the Asian population in 2016) [17]. The genes that are majorly observed as carriers of mutation leading to AD are *APP, PSEN1 and PSEN2* [5, 18]. Although AD is the most common of all the NDs, researchers do not yet completely understand its complex pathophysiology. Through this review, we attempt to explore the major pathways of the disease, the current treatment used for it and the prospective solutions that nanomedicine-directed research has to offer in future.

AD PATHOLOGY AND CURRENT TREATMENT TRENDS

There are many factors that need to be considered in order to find a cure for a disease. A detailed understanding of the disease pathway can serve as a starting point for exploring potential treatments. Understanding AD and its pathology have provided an insightful approach towards designing treatments and drugs for it. AD results majorly due to inter-cell neurofibrillary tangles and intracellular neuritic plaques [19 - 21]. Paired helical filaments (PHF) form the tangles and amyloid aggregation results in these plaques. Two major pathways (Fig. 1) have been suggested as probable disease pathologies for AD: Beta-amyloid (Aβ) cascade and Tau pathology. Both Aβ and Tau protein are associated with normal neuronal metabolism. Aβ is processed through a complex enzyme using amyloid precursor protein (APP) and is cleaved into peptides of various sizes [22, 23]. Modification in the proteolytic processing by APP results in aggregation of Aβ in the neuronal synapses, leading to a decline in the signaling efficiency. Even though inter-neuron Aβ deposition is identified as a crucial factor in AD, yet Aβ accumulation may not be the stand-alone cause of AD. It has been reported that even with the presence of Aβ deposits, the subject still did not suffer from AD.

Tau is a major microtubule-associated protein (MAP) present in six molecular isoforms in a mature neuronal cell in the human nervous system. It is majorly responsible for stabilizing the microtubular network in the neuronal cell.

Hyperphosphorylation of Tau causes it to mis -fold and aggregate in the form of PHFs (Fig. **2**). Along with conformational changes in all isoforms of Tau, truncation of tau also occurs following its hyperphosphorylation [24]. This mechanism has been closely related to neurodegenerative disorders, which in this case is AD [25].

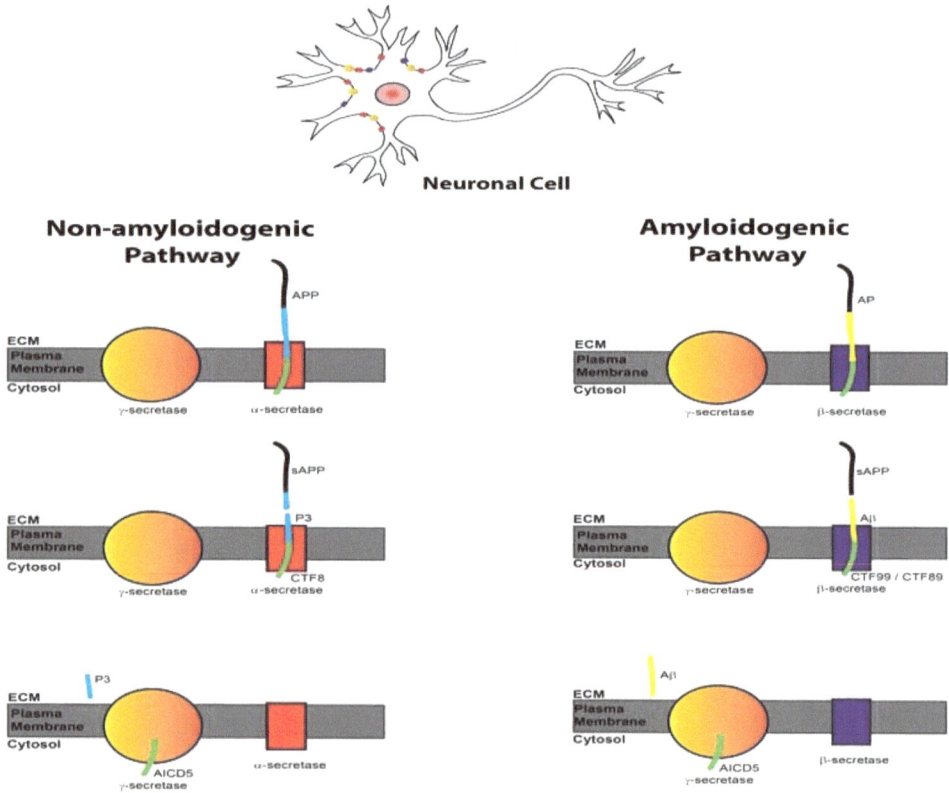

Fig. (1). Two most popular theories explaining Alzheimer's disease pathology: Non-Amyloidogenic Pathway (Left) and Amyloidogenic Pathway (Right).

In both Aβ cascade and Tau pathology, there is an increased amount of protein production. This leads to an imbalance, where the product synthesized is more than the amount required. This increased production may be due to a defect in the precursor for the proteins involved or the hormone regulation pathway for the control of these protein precursors may be affected. In the case of familial AD, it might be due to mutation of the gene factors (*APP, PSEN1 and PSEN2*) coding for involved protein precursors and hormones.

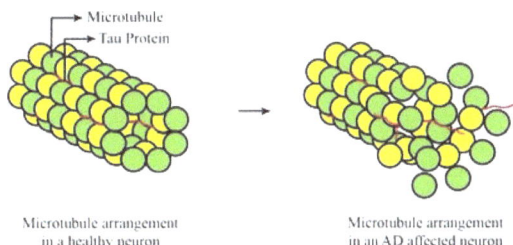

Fig. (2). Misfolding of the microtubule-associated protein, Tau causing aggregates in the form of paired helical filaments (PHFs).

Other hypotheses that have been put forward in pursuit of explaining AD's physiopathology include Vascular hypothesis, oxidative stress, disturbed calcium homeostasis, metallic imbalance and neurotransmitter imbalance [26].

Among these, Neurotransmitters involved in AD and their pathology are the most distinguished. All the currently approved drugs used for the treatment of AD are based on the regulation of neurotransmitters. As per the fact sheet published by the National Institute of Aging, the current five drugs that have been FDA approved and are in clinical use for the treatment of AD, are based on two major neurotransmitters (Acetylcholine and Galantamine) involved in signal transduction in the brain. The malfunctioning of these neurotransmitters is majorly associated with AD. Table **1** summarizes these drugs, their route of administration and their working principle. These drugs differ in their mechanism of action based on the pathway that they regulate. They fall into two major categories: Cholinesterase inhibitors and N-methyl-D-aspartate (NMDA) receptor antagonists.

Table 1. Drugs used for the treatment of AD.

Cholinesterase Inhibitors		
Drugs	**Route of Administration**	**Working Principle**
Razadyne□ (galantamine)	Tablets (Oral)	● Prevents breakdown of Acetylcholine ● Causes release of Acetylcholine by stimulating nicotine receptors
Exelon□ (rivastigmine)	Transdermal patch	● Prevents breakdown of acetylcholine and butyrylcholine
Aricept□ (donepezil)	Tablets (Oral)	● Prevents breakdown of acetylcholine
NMDA Receptor Antagonists		

(Table 1) cont.....

Cholinesterase Inhibitors		
Drugs	**Route of Administration**	**Working Principle**
Namenda □ (memantine)	Tablets (Oral)	• Blocks NMDA receptors to prevent brain cell damage due to excess of Glutamate
Namzaric □ (memantine and donepezil)	Tablets (Oral)	• Blocks NMDA receptors to prevent brain cell damage due to excess of Glutamate • Prevents breakdown of Acetylcholine

Cholinesterase pathway explains how the decreased cholinergic transmission significantly contributes to beta- amyloid pathology and enhances tau hyperphosphorylation. Treatment in the form of acetylcholinesterase helps to maintain functioning of brain cells by inhibiting the breakdown of acetylcholine by cholinesterase. Drugs that follow this principle are Razadyne® (galantamine), Exelon® (rivastigmine) and Aricept® (donepezil).

Drugs that act as N-methyl -D-aspartate (NMDA) receptor antagonists function by regulating the activity of glutamate. In the brain, glutamate is sensed by the neurons via NMDA receptors. These receptors play a major role in cell signaling in the CNS. Insufficient NMDR signaling is involved in NDs and ultimately leads to neuronal cell death [27]. NMDA receptors are present in two groups- synaptic or extra synaptic. Studies have drawn a relation between extra synaptic NMDA receptors and ecotoxicity, ultimately leading to neuronal death [28 - 31]. Namenda® (memantine) and Namzaric® (memantine and donepezil) are NMDA receptor antagonists currently used for AD treatment.

FACTORS INFLUENCING THE TARGETING OF AD

The Brain, along with the spinal cord, constitutes the central nervous system (CNS). Since it is the higher functional center of the human body that is responsible for the regulation of all *in-vivo* mechanisms, it possesses a carefully integrated and well-oriented defense mechanism, consisting of the Blood- Brain Barrier (BBB) and the immune system. BBB, together with various other enzymes, restricts the entry of many macro-particles, nanoparticles, and drugs for maintenance of the internal homeostasis of the brain [32 - 34]. Thus, there is a huge challenge in targeted drug delivery to the brain. There are other crucial problems that are encountered, when looking for a cure for AD. These include Aβ oligomerization, reactive gliosis (majorly due to oxidative stress) and neuroinflammation [35, 36]. Even with our current understanding of the disease, the drug delivery pathway through the BBB is a major challenge.

Hence, while designing a drug to deal with the primary cause, prevent it from recurring and to counter the ill- effects of AD, there is a focus on the existent cause-effect continuum. Here, we discuss the three major aspects that can primarily help in the regulation of neurodegeneration in AD. These are Neuroinflammation, Neurotoxicity and Drug Delivery System (DDS)/ pathway.

Neuroinflammation involves the dysfunction, degeneration and ultimate cell death of neurons in the CNS [37]. This is a characteristic feature for most NDs. This process affects both the neurons and the neuroglia leading to breakdown of the neuronal circuitry. It is yet to be proved, whether neuroinflammation is a cause or a consequence of NDs, nonetheless it is considered a characteristic feature observed in most NDs [38]. Supporting cells involved in neuroinflammation include phagocytic microglia, neuron- supportive astrocytes and neurotransmission-enhancers oligodendrocytes. Among these, microglia are the deciding factors responsible for neuroinflammation. These cells phagocytose pathogens and cellular debris in the CNS. In case of excessive stress, induced due to factors such as protein aggregation in AD, they release excess amounts of pro-inflammatory factors. These factors include reactive oxygen species, chemokines, cytokines and various growth factors. All these components have the ability to harm the neuronal cells, when accumulated in more than required amounts, particularly in sensitive areas of the brain. In a normal human CNS, astrocytes are responsible for supporting and linking the neurons to the blood. In diseased conditions, these astrocytes lead to over-secretion of inflammatory molecules causing neurodegeneration. In AD patients, abnormalities associated with myelin sheath have also been observed. This sheds light on the improper functioning of oligodendrocytes. All three cells have been observed to have specific stimuli, which leads to their altered working but Aβ deposits seem to be the common, crucial factor that links them all. These cells tend to accumulate around these protein deposits, leading to aggregation of their secretions in plaque-rich areas resulting in inflammation in these regions [39]. These neuroinflammatory responses cause neurotoxicity, which results in dysfunctionality and apoptosis.

So, while designing a drug, there is a focus on preventing this protein aggregation in the cranial nervous tissue, as it seems to be the major reason behind neuroinflammation. In case of AD, the protein is beta-amyloid (Aβ). It has been seen that fibrillization of Aβ is dependent on factors concerning both the properties of the solution and the environment surrounding it [40]. Particularly, the presence of nanoparticles has the ability to either enhance or inhibit Aβ fibrillation [40]. The nanoparticle therapy targets this property of the Aβ fibrils by destabilizing the fibrils by modifying their surrounding environment, solubilizing the oligomers, countering the effects of Aβ deposition on the brain cells and preventing the formation of Aβ fibrils all together by adsorbing the monomers present in the solution.

TARGETING AD THROUGH NANOMEDICINE

Nanotechnology holds an edge over all the other available current fields being utilized for AD treatment, as here the focus is on the manipulation and modification of drug delivery networks and their associated systems. Currently, there is no available cure for AD. The treatment is used to manage the symptoms of the disease but not the root cause itself because it's still unclear. There are various proposed hypotheses but the variability in symptomatic expression is quite high, leading to a lack of a concrete pathway for AD progression. A large number of nanoparticle-based therapeutic approaches have been tried throughout the world for the treatment of Alzheimer's Disease (Fig. **3**) and some of these have been discussed in the following section.

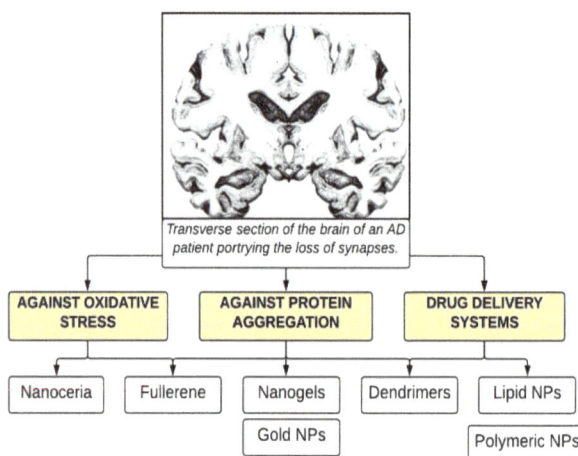

Transverse section of the brain of an AD patient portrying the loss of synapses.

| AGAINST OXIDATIVE STRESS | AGAINST PROTEIN AGGREGATION | DRUG DELIVERY SYSTEMS |

| Nanoceria | Fullerene | Nanogels | Dendrimers | Lipid NPs |
| | | Gold NPs | | Polymeric NPs |

Fig. (3). Different Nanoparticle-based Therapeutic approaches for the treatment of Alzheimer's Disease.

Gold nanoparticles have been shown to act as a multifunctional Aβ-deposit inhibitor. In a temporary arrangement with polyoxometalates (POMs) and a peptide component, these have the ability to act as Aβ peroxidase inhibitor, Aβ fibril inhibitor and dissociating agent against Aβ fibrils [41]. Gold NPs can also prevent further Aβ deposits by thermoregulation of the surrounding region [42]. *Palmal et al* reported that use of curcumin-functionalized gold NPs can also inhibit Aβ deposition and can disintegrate the fibrils [43]. Gold NPs have also been reported to recognize Aβ deposits by acting as β-sheet breakers, when conjugated with peptides THRPPMWSPVWP (THR) and CLPFFD [44]. The size, versatile nature, and the high reactivity offered by nanoparticles due to their high surface-to-volume ratio are some of the advantages for extensive applications of nanoparticles. The number of gold nanoparticles that need to be introduced in neurons is unclear till now. Also, some amount of neurotoxicity is associated with gold nanoparticles, as they are non-biodegradable. Neurotoxicity, Lysosomal dysfunction, and autophagy dysfunction are some of the disadvantages associated with some nanoparticles. Although, the long-term effects of these therapies is yet to be explored.

Nanogels are another type of NPs used to target protein aggregation. These types of NPs are nanoscale analogs for natural proteins that assist in the process of protein formation. These nanogels are used in treatment to inhibit fibril formation. Their high stability and responsiveness to the surrounding environment make them a suitable medium for nano-medicinal purposes. Pullulan- bearing Cholesterol nanogels (CHP) are being explored as a potential treatment for AD due to their inhibitory nature towards amyloid aggregation and the resultant neurotoxicity due to it [45]. They can also act as complements to the antibody immunotherapy given in case of AD and various other neurological disorders [46]. Interestingly, Nanogels have been shown as neural drug uptake enhancers in various *in vivo* studies [47]. Recently, therapeutic approaches to engineer stable, compatible delivery systems for proper insulin delivery to the brain have also been acknowledged as probable solutions. Insulin is a major hormone, essential for cellular growth and survival. Dysfunction of its pathway can lead to AD [48]. *Picone et al* synthesized a carboxylated poly(N-vinyl pyrrolidone) nanogel system as a delivery pathway for insulin to the brain [49]. This insulin nanogel conjugate has been reported to be neuroprotective, protecting the neurons against the ill-effects of Aβ deposits [49].

Dendrimers are multilayered nanoparticles, made up of a single/ multiple polymer. The structural components include a core, an intermediary layer and an external surface through which they interact to the target. Due to their target versatility, high biocompatibility and reduced biotoxicity, these NPs have been investigated as potential macromolecules for drug delivery and inhibition of Aβ plaque deposition. A quite recent review highlights lot of possibilities of the usage of nanoparticles-dendrimers as a class of well-defined branched polymers for the early diagnosis and therapy of AD [50]. Poly(amidoamine) (PAMAM), an ionic dendrimer, whose charge varies depending on the surface functional group [51]. PAMAM has been reported to have enhanced drug delivery through the BBB and increased drug uptake [52, 53]. *Posadas et al* observed the anti- inflammatory effect of phosphorus bearing dendrimers in both mouse and human macrophages [54]. They reported a concentration dependent inhibition of neuroinflammation inducers nitric oxide synthase and nitrite production by PAMAM. Maltose decorated poly(propyleneimine) dendrimer is another type of low toxicity, nanosized alternative which, at high ratios, causes clumping of amyloid fibrils, thus hindering the fibril formation and the corresponding symptoms [55]. Despite many advantages, further investigations are still required to explore the potential of dendrimers as a type of nanomedicine.

Other than targeting the protein aggregates, some nano -therapies have been directed towards reduction and removal of oxidative stress. This stress is majorly due to the accumulation of reactive oxygen species (ROS) and reactive nitrogen species (RNS). In normal physiological conditions, ROS and RNS assist as molecular signals indicative of stress [56]. The irony is that overproduction of these "stress-indicating" molecules leads to more oxidative and nitrosative stress, which ultimately results in diseases like AD and ALS. Due to the high oxygen requirement of the brain, in most NDs, oxidative stress becomes a major risk factor [56]. Thus, a homeostatic balance between ROS and antioxidants is required for its optimal functioning. Due to the requirement of such critical balance, ROS ceasing nanoparticle -based nanomedicines need to be strategically designed to ensure excessive ROS removal without the disruption of this "redox homeostasis". Currently, Cerium-, Platinum-based nanoparticles, Fullerenes and metal chelators are being researched as potential oxidative stress focused therapies for NDs.

Fullerenes act as scavengers for ROS and free radicals. Due to their ability to easily accept free electrons, various derivatives of fullerene have shown potential for application in nanomedicine. Among them, derivatives with hydroxyl groups are most promising. Due to their enhanced hydrophilic properties, these are more compatible with polar solvents and are thus, considered as better-suited for neuroprotection [57]. Carboxyfullerene ($C_{60}(C(COOH)_2)_2$) and Fullerenol ($C_{60}(OH)_{22}$) are derived from fullerene and have been particularly researched as potential therapies due to their binary nature of being both an antitoxin and a protector for the neuronal cells. Fullerenol, a polyhydroxy fullerene derivative, has been shown as an anti-Aβ oligomerization molecule, as decreasing agents for neural toxicity and as mediators of selective anti-acetylcholinesterase activity [58, 59].

Cerium nanoparticles (ceria or Nanoceria) are used in another way of nanomedicinal approach. They react with ROS species such as superoxide radicals ($O_2{}^-$) and convert them into oxygen (O_2). Since $O_2{}^-$ converts to hydrogen peroxide (H_2O_2) by the action of enzyme superoxide dismutase (SOD), these NPs prevent H_2O_2 formation by mimicking the action of SOD. Moreover, they also catalyze degradation of H_2O_2, accompanied by generation of Ce^{3+} from Ce^{4+} [60]. Currently, research trends for these NPs are directed towards designing methodologies to enhance superoxide- scavenging ability, tissue-specificity and longevity of their activity of nanoceria [61]. Platinum-based nanoparticles are also ROS scavenging NPs that catalyze the conversion of $O_2{}^-$ and H_2O_2. Although their direct application as a treatment for NDs need further research and investigation.

Characteristic Aβ plaques in AD are made of peptides that have the ability to bind to various metals, particularly Copper (Cu^{2+}), Zinc (Zn^{2+}) and Iron (Fe^{3+}). This results in accumulation of these metals at the site of protein deposit and leads to a metallic imbalance. This may also be one of the reasons behind the disturbance of oxidative homeostasis observed in AD [62]. Metal chelators are being explored as future therapy for restoring this ionic imbalance. Due to their strong oxidizing nature, administration of conventional metal chelators is toxic and harmful. Instead, nanoparticle delivery systems are being used to deliver these metal chelators to their target site with reduced tissue damage and without disturbing body's intrinsic functioning. An efficient delivery system is designed by conjugating an appropriate metal chelator with a biocompatible, non- toxic NP which has high target specificity, *in vivo* stability, reduces metal chelators toxicity, while retaining its original efficiency [63]. Currently, the metal chelation therapies that are used for AD patients include Desferrioxamine (DFO), iodochlorhydroxyquin (clioquinol) and ethylene diamine tetra acetic acid (EDTA), but their toxic nature results in many harmful side-effects [63]. Efficient NP delivery systems have been designed for delivery of DFO, an FDA approved iron chelator, using carboxyl-functionalized NPs [63]. Another study designed a delivery system, where NPs made of 1,2-Dioleoyl-snglycero-3-pho-pho-ethanolamine-N-[4 -(p-maleimidophenyl) butyramide] (MPB -PE) and pyridyldithio-propionyl phospho-ethanolamine (PDP-PE) are used for delivery of the copper chelator d-penicillamine to the brain. It was reported that this metal chelator was transported through the BBB passively, without changing the structural integrity of the BBB [64].

DRUG DELIVERY SYSTEMS AND NANOMEDICINE

Nanoparticle mediated drug delivery systems (DDS) have become one of the hot topics for research in recent years. A wide range of nano-carriers, such as emulsions, lipid-based NPs, polymeric NPs, metal -based carriers and carbon nanotubes *etc*. have been quite extensively explored. Among the various types of NPs, lipid -based NPs and polymer-based NPs (Fig. **4**) are considered ideal for designing DDS due to their high biocompatibility, target specificity and ability to enhance drug solubility and stability [65].

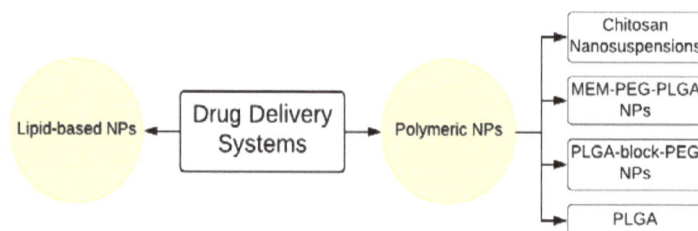

Fig. (4). Nanoparticles used for designing drug delivery systems in AD.

Liposomes are a type of lipid-based NPs, constituted by a singular or a double layer of lipid and an internal aqueous pocket. This aqueous pocket can encapsulate hydrophilic molecules. They have majorly been studied for drug delivery and transport. Liposomes can be used to target certain organelles in the cytoplasm. The level of toxicity/ side effects of liposomes are still unclear but since it has been shown that the reticuloendothelial system (RES) is very quick to detect and cause degradation of liposomes, it is an indication of a reduction in the resultant side effects. These NPs have been used to effectively deliver both drugs as well as peptides that act as antibodies across the BBB [65 - 67]. PEGylated liposomes targeting Glutathione (GSH- PEG) were used to deliver a single domain Aβ plaque-binding llama antibody fragment (VHH-pa2H) to the brain. It was able to cross the BBB and bind to the amyloid plaques with high affinity [68]. Drugs that down regulate certain gene expressions have also been delivered using liposomes. Rivastigmine, a cholinesterase inhibitor, when administered in the form of a liposome made of egg PC(L-α-phosphatidylcholine), Chol and dihexadecyl phosphate (DCP) in an AlCl3-induced rat model of AD, downregulates AChE, IL1B, and BACE1 gene expression. As a result, a decrease in the amyloidogenic processes is observed [69]. Liposomes can also be used as multifunctional NPs. *Bana et al* formulated liposomes made of sphingomyelin and Chol, functionalized with phosphatidic acid (PA) and mApoE. Ionic interactions of PA and mApoE with beta-amyloid plaques resulted in their disintegration [70]. Solid Lipid NPs (SLNs) are another type of lipid-based nanoparticles that are stable nanocarriers. Due to their small size (40 -200nm) and hydrophobic lipid core, they can be used to deliver drugs across the BBB [64]. SLNs, functionalized with monoclonal antibody (OX26 mAb) and Cetylpalmitate, were observed to effectively deliver active drug grape seed extract and Resveratrol in human brain-like endothelial cells [71]. The drugs were able to suppress fibrillar formation, with grape extract being more effective in doing so than resveratrol [71]. Researchers have also tried tackling the issue of neuronal cell death associated with AD and most other NDs. Heparinized cationic SLNs loaded with nerve growth factor (NGF-loaded HCSLNs) induced differentiation in neuron-like induced pluripotent mouse stem cells (iPSCs) [72]

PLGA (poly- d, l-lactide-co -glycolide) is one of the best polymeric nanoparticles to be used in drug delivery systems due to minimal system toxicity offered by it. It hydrolyzes inside the body and results in metabolite monomers, lactic acid and glycolic acid. These monomers are familiar to the human body and hence, they are dealt with efficiency [73]. *Md S et al* [74] prepared donepezil-loaded PLGA NPs for sustained release and efficient brain targeting by diffusion-evaporation technique. The size of the drug loaded NPs was found to be 89.67 ± 6.43 nm with a 1:1 drug polymer ratio. Donepezil exhibited a biphasic release pattern- an initial burst release followed by sustained release. Uptake of donepezil via loaded NP may help in an improved AD treatment although more clinical analysis and evaluation is yet to be performed.

Baysal et al prepared donepezil loaded poly (lactic-co-glycolic acid)-block-poly (ethylene glycol) nanoparticles (PLGA-block-PEG NPs) by double emulsion method [75]. They found that the drug loaded NPs could easily cross the BBB and showed controlled release profile in the *in vitro* BBB model setup. Moreover, these particles were reported to have a destabilizing effect on amyloid-beta fibril formation [75]. These NPs, thus, may be able to target the characteristic causative amyloid plaques but this hypothesis is yet to be tested.

PLGA NPs have also been used as carrier for drugs Galantamine and Rivastigmine Tartrate. Novel galantamine-loaded NPs were prepared using nano-emulsion templating for the first time by *Fornaguera et al* [76]. With an encapsulation efficiency higher than 90% by wt and a sustained drug release profile comparable to aqueous and micellar solutions, these drug-loaded NPs portrayed a promising profile to that of an advanced DDS for the treatment of NDs [76]. Rivastigmine Tartrate loaded biopolymeric NPs PLGA and PBCA demonstrated faster regain of memory loss in amnesic mice as compared to RT solution [77]. These NPs were prepared by nanoprecipitation and emulsion polymerization techniques [77]. Although they portrayed the suitability of these NPs in sustained drug delivery to the brain, their performance *in vivo* is yet to be tested.

Recently, Memantine-loaded PLGA NPs were prepared by double emulsion method for oral administration to make memantine (MEM) treatment more effective [78]. It was found MEM followed a slower release profile from the NPs against the free drug solution, allowing the reduction of drug administration frequency *in vivo* [78]. MEM-PEG-PLGA NPs also demonstrated a decrease in memory impairment in behavioral tests performed on transgenic mice. MEM-PEG-PLGA NPs have already been found to reduce beta-amyloid plaques and inflammation caused by them in AD [78].

Chitosan is another biodegradable biopolymer, which consists of glucosamine and N-acetylglucosamine as copolymers. It is prepared by the partial N -deacetylation of crustacean derived natural biopolymer chitin [79]. Due to its low cost, biodegradability and lack of toxicity, it is a very attractive substance for diverse applications in pharmaceutical and medical fields. Chitosan and its derivatives have been used both as a drug and as drug carriers because of their high biocompatibility [79]. *Hanafy et al.* worked with galantamine/ chitosan complex NPs to assess their pharmacological, toxicological and neuronal localization using male Wistar rats through nasal administration [80]. These particles were found to exhibit no negative alterations to the pharmacological efficiency of galantamine hydrobromide (GH). Also, no signs of toxicity or histopathological manifestations were observed [80].

Rivastigmine loaded chitosan NPs have also been investigated. These were prepared by ionic gelation method and administered through the intranasal pathway, and were shown to have higher drug transport efficiency and direct transport percentage [81]. It has been shown that coating of rivastigmine loaded chitosan NPs by polysorbate-80 increased the concentration of rivastigmine 3.82 folds in mice brain in comparison to free drug rivastigmine with intravenous administration [82].

Another interesting approach for the treatment of AD combines the potential of RNA therapy and the efficiency of nanoparticles designed to be used as a delivery system. This approach has been used to introduce protein that inhibit or decrease amyloid plaque formation. *Lin et al* developed a cationic polymer-based PEGylated nanocarrier, which formed polyplex nanomicelles, for the delivery of NEP-expressing mRNA [83]. In contrast, another class of RNA used to target a

specific protein coding mRNA have been introduced in the brain. These are called small interfering RNAs (siRNAs) [84]. To enable an effective gene silencing effect, siRNAs are released away from the lysosomes. *Liu et al* (2013) [85] used polyethylene glycol-polyethyleneimine (PEG-PEI) complexes as a delivery system for delivering ROCK-II-siRNA. Rho -associated kinase (ROCK-II) is a kinase involved in Aβ formation. PEG-PEI/ ROCK-II-siRNA complexes were able to suppress the expression of ROCK-II mRNA *in vitro* in C17.2 neural stem cells [85]. *Farr et al* demonstrated an improvement in learning and memory in an AD mouse after treatment with specially designed antisense oligonucleotides that bind to a specific mRNA of a gene [86].

Over the past few years, NPs have been extensively explored as drug delivery systems. A wide range of compounds and treatments that can be delivered using them have also been expanded. Some of the quite recent nano-drug delivery applications in the field of AD therapeutics, and their future prospects on potential molecular targets have been reviewed [87]. In the current pharmacotherapies for AD different drug molecules that have been successfully encapsulated in nanoparticles (NPs) and their route of administration has also been elaborated in much detail [88]. Usually the industrial perspectives of these delivery systems are ignored, which needs special attention for getting these therapeutics to the market [89]. History, epidemiology, pathogenesis and therapeutics of neurological diseases and their possible roles as neurotheranostics in personalized medicines have been quite extensively reviewed [90].

Interestingly, traditional medicines are also one of the fields that has had an interesting collaboration with nanotechnology [91]. Medicinal plant extracts have long been used for the treatment of different ailments. So, it only made sense to combine this old traditional approach with the very promising approach of nanomedicine. Curcumin, Coumarin, Quercetin, Ginsenoside, Catechins, Choline, Magnolol and honokiol are only some examples of the compounds extracted from plants and used for AD treatment [91]. This interface provides an insightful, fresh perspective to view treatment of AD and other NDs and will be the field to watch out for the future.

OUTLOOK

Over the years, many researchers have attempted to understand the disease pathology of various NDs, particularly AD. Currently, Beta-amyloid (Aβ) Cascade and Tau pathology are the major pathways accepted for AD but there are still various aspects of AD that these pathologies fail to explain. With a limited knowledge about the cause and a lack of practically applicable solutions paired with the ambiguous nature of neuronal sensitivity, the process of drug design has become more complex. There are still very few FDA approved drugs for AD treatment and hence there is an urgent need to work on this area.

Nanotechnology brings new hope in the form of nanomedicine as a probable key to the mystery of NDs. Although most of the research is still in the preliminary stage and most drugs have failed to reach the stage of clinical trials, still this seems to be a field to look out for. Particularly, the drug delivery systems designed using nanoparticles are quite promising and if they perform well in clinical trials, they will be able to deal with the critical issues of improper drug delivery and high toxicity that we face with AD therapy today. Furthermore, the use of these drug delivery systems in gene therapy and delivery of medicinal plant extracts are fields worth exploring, as they hold exciting perspectives for future research.

With the increasing number of AD patients throughout the world, there is a need for immediate solutions to this problem. Nanomedicine may help us unlock solutions relevant not just for NDs but for many other disorders that are still scarcely understood.

CONFLICT OF INTEREST

The authors confirm that this chapter contents have no conflict of interest.

CONSENT FOR PUBLICATION

Not applicable.

ACKNOWLEDGEMENTS

Author would like to acknowledge the University of Delhi and Department of Science and Technology, Gov. of India for DU-DST Purse Grant.

REFERENCES

[1]   Srikanth M, Kessler JA. Nanotechnology-novel therapeutics for CNS disorders. Nat Rev Neurol 2012; 8(6): 307-18.
        [http://dx.doi.org/10.1038/nrneurol.2012.76] [PMID: 22526003]

[2]   Giráldez-Pérez R, Antolín-Vallespín M, Muñoz M, Sánchez-Capelo A. Models of α-synuclein aggregation in Parkinson's disease. Acta Neuropathol Commun 2014; 2: 176.
        [http://dx.doi.org/10.1186/s40478-014-0176-9] [PMID: 25497491]

[3]   Desai AK, Grossberg GT. Diagnosis and treatment of Alzheimer's disease. Neurology 2005; 64(12) (Suppl. 3): S34-9.
        [http://dx.doi.org/10.1212/WNL.64.12_suppl_3.S34] [PMID: 15994222]

[4]   Citron M. Alzheimer's disease: strategies for disease modification. Nat Rev Drug Discov 2010; 9(5): 387-98.
        [http://dx.doi.org/10.1038/nrd2896] [PMID: 20431570]

[5]   Hardy J, Cullen K. Amyloid at the blood vessel wall. Nat Med 2006; 12(7): 756-7.
        [http://dx.doi.org/10.1038/nm0706-756] [PMID: 16829930]

[6]   Nussbaum RL, Ellis CE. Alzheimer's disease and Parkinson's disease. N Engl J Med 2003; 348(14): 1356-64.
        [http://dx.doi.org/10.1056/NEJM2003ra020003] [PMID: 12672864]

[7]   Dickson DW. Parkinson's disease and parkinsonism: neuropathology. Cold Spring Harb Perspect Med 2012; 2(8)a009258
        [http://dx.doi.org/10.1101/cshperspect.a009258] [PMID: 22908195]

[8]   Hardiman O, Al-Chalabi A, Chio A, *et al.* Amyotrophic lateral sclerosis. Nat Rev Dis Primers 2017; 3: 17071.
        [http://dx.doi.org/10.1038/nrdp.2017.71] [PMID: 28980624]

[9]   Ferrari R, Kapogiannis D, Huey ED, Momeni P. FTD and ALS: a tale of two diseases. Curr Alzheimer Res 2011; 8(3): 273-94.
        [http://dx.doi.org/10.2174/156720511795563700] [PMID: 21222600]

[10]   Kim BYS, Rutka JT, Chan WCW. Nanomedicine. N Engl J Med 2010; 363(25): 2434-43.

[http://dx.doi.org/10.1056/NEJMra0912273] [PMID: 21158659]

[11]   Soares S, Sousa J, Pais A, Vitorino C. Nanomedicine: Principles, Properties, and Regulatory Issues. Front Chem 2018; 6: 360.
[http://dx.doi.org/10.3389/fchem.2018.00360] [PMID: 30177965]

[12]   Catalan-Figueroa J, Palma-Florez S, Alvarez G, Fritz HF, Jara MO, Morales JO. Nanomedicine and nanotoxicology: the pros and cons for neurodegeneration and brain cancer. Nanomedicine (Lond) 2016; 11(2): 171-87.
[http://dx.doi.org/10.2217/nnm.15.189] [PMID: 26653284]

[13]   Wang EC, Wang AZ. Nanoparticles and their applications in cell and molecular biology. Integr Biol 2014; 6(1): 9-26.
[http://dx.doi.org/10.1039/c3ib40165k] [PMID: 24104563]

[14]   Kumar A, Singh A, Ekavali . A review on Alzheimer's disease pathophysiology and its management: an update. Pharmacol Rep 2015; 67(2): 195-203.
[http://dx.doi.org/10.1016/j.pharep.2014.09.004] [PMID: 25712639]

[15]   Masters CL, Bateman R, Blennow K, Rowe CC, Sperling RA, Cummings JL. Alzheimer's disease. Nat Rev Dis Primers 2015; 1: 15056.
[http://dx.doi.org/10.1038/nrdp.2015.56] [PMID: 27188934]

[16]   Ali G-C, Guerchet M, Wu Y-T, Prince M, Prina M. Chapter 2: The global prevalence of dementia The Global Impact of Dementia An analysis of prevalence, incidence, cost and trends. London: Alzheimer's Disease International (ADI); 2015; pp. 10-29.

[17]   Shea YF, Chu LW, Chan AO, Ha J, Li Y, Song YQ. A systematic review of familial Alzheimer's disease: Differences in presentation of clinical features among three mutated genes and potential ethnic differences. J Formos Med Assoc 2016; 115(2): 67-75.
[http://dx.doi.org/10.1016/j.jfma.2015.08.004] [PMID: 26337232]

[18]   Guerreiro R, Hardy J. Genetics of Alzheimer's disease. Neurotherapeutics 2014; 11(4): 732-7.
[http://dx.doi.org/10.1007/s13311-014-0295-9] [PMID: 25113539]

[19]   Brion JP, Passareiro H, Nunez J, Flament Durand J. Mise en evidence immunologique de la proteine tau au nivea u des lesions de degenerescence neurofibrillaire de la maladie d'Alzheimer. Arch Biol (Bruxelles) 1985; 95: 229-35.

[20]   Grundke-Iqbal I, Iqbal K, Tung YC, Quinlan M, Wisniewski HM, Binder LI. Abnormal phosphorylation of the microtubule-associated protein tau (tau) in Alzheimer cytoskeletal pathology. Proc Natl Acad Sci USA 1986; 83(13): 4913-7.
[http://dx.doi.org/10.1073/pnas.83.13.4913] [PMID: 3088567]

[21]   Agdeppa ED, Kepe V, Liu J, *et al.* Binding characteristics of radiofluorinated 6-dialkylamino-2-naphthylethylidene derivatives as positron emission tomography imaging probes for beta-amyloid plaques in Alzheimer's disease. J Neurosci 2001; 21(24): RC189.
[http://dx.doi.org/10.1523/JNEUROSCI.21-24-j0004.2001] [PMID: 11734604]

[22]   Golde TE, Eckman CB, Younkin SG. Biochemical detection of Abeta isoforms: implications for pathogenesis, diagnosis, and treatment of Alzheimer's disease. Biochim Biophys Acta 2000; 1502(1): 172-87.
[http://dx.doi.org/10.1016/S0925-4439(00)00043-0] [PMID: 10899442]

[23]   Selkoe DJ. Alzheimer's disease: genes, proteins, and therapy. Physiol Rev 2001; 81(2): 741-66.
[http://dx.doi.org/10.1152/physrev.2001.81.2.741] [PMID: 11274343]

[24]   Iqbal K, Liu F, Gong CX, Grundke-Iqbal I. Tau in Alzheimer disease and related tauopathies. Curr Alzheimer Res 2010; 7(8): 656-64.
[http://dx.doi.org/10.2174/156720510793611592] [PMID: 20678074]

[25]   Hansson O, Grothe MJ, Strandberg TO, *et al.* Tau Pathology Distribution in Alzheimer's disease Corresponds Differentially to Cognition-Relevant Functional Brain Networks. Front Neurosci 2017;

11: 167.
[http://dx.doi.org/10.3389/fnins.2017.00167] [PMID: 28408865]

[26] Mohandas E, Rajmohan V, Raghunath B. Neurobiology of Alzheimer's disease. Indian J Psychiatry 2009; 51(1): 55-61.
[http://dx.doi.org/10.4103/0019-5545.44908] [PMID: 19742193]

[27] Wang R, Reddy PH. Role of glutamate and NMDA receptors in Alzheimer's disease. J Alzheimers Dis 2017; 57(4): 1041-8.
[http://dx.doi.org/10.3233/JAD-160763] [PMID: 27662322]

[28] Sattler R, Xiong Z, Lu WY, MacDonald JF, Tymianski M. Distinct roles of synaptic and extrasynaptic NMDA receptors in excitotoxicity. J Neurosci 2000; 20(1): 22-33.
[http://dx.doi.org/10.1523/JNEUROSCI.20-01-00022.2000] [PMID: 10627577]

[29] Hardingham GE, Fukunaga Y, Bading H. Extrasynaptic NMDARs oppose synaptic NMDARs by triggering CREB shut-off and cell death pathways. Nat Neurosci 2002; 5(5): 405-14.
[http://dx.doi.org/10.1038/nn835] [PMID: 11953750]

[30] Léveillé F, El Gaamouch F, Gouix E, *et al.* Neuronal viability is controlled by a functional relation between synaptic and extrasynaptic NMDA receptors. FASEB J 2008; 22(12): 4258-71.
[http://dx.doi.org/10.1096/fj.08-107268] [PMID: 18711223]

[31] Stanika RI, Pivovarova NB, Brantner CA, Watts CA, Winters CA, Andrews SB. Coupling diverse routes of calcium entry to mitochondrial dysfunction and glutamate excitotoxicity. Proc Natl Acad Sci USA 2009; 106(24): 9854-9.
[http://dx.doi.org/10.1073/pnas.0903546106] [PMID: 19482936]

[32] Abbott NJ, Romero IA. Transporting therapeutics across the blood-brain barrier. Mol Med Today 1996; 2(3): 106-13.
[http://dx.doi.org/10.1016/1357-4310(96)88720-X] [PMID: 8796867]

[33] Alam MI, Beg S, Samad A, *et al.* Strategy for effective brain drug delivery. Eur J Pharm Sci 2010; 40(5): 385-403.
[http://dx.doi.org/10.1016/j.ejps.2010.05.003] [PMID: 20497904]

[34] Davis SS. Biomedical applications of nanotechnology--implications for drug targeting and gene therapy. Trends Biotechnol 1997; 15(6): 217-24.
[http://dx.doi.org/10.1016/S0167-7799(97)01036-6] [PMID: 9183864]

[35] Nazem A, Mansoori GA. Nanotechnology solutions for Alzheimer's disease: advances in research tools, diagnostic methods and therapeutic agents. J Alzheimers Dis 2008; 13(2): 199-223.
[http://dx.doi.org/10.3233/JAD-2008-13210] [PMID: 18376062]

[36] Du X, Wang X, Geng M. Alzheimer's disease hypothesis and related therapies. Transl Neurodegener 2018; 7: 2.
[http://dx.doi.org/10.1186/s40035-018-0107-y] [PMID: 29423193]

[37] Hooker JM. Neuroinflammation: Brain on Fire? ACS Chem Neurosci 2016; 7(4): 415-5.
[http://dx.doi.org/10.1021/acschemneuro.6b00104] [PMID: 27094163]

[38] Wyss-Coray T, Rogers J. Inflammation in Alzheimer disease-a brief review of the basic science and clinical literature. Cold Spring Harb Perspect Med 2012; 2(1)a006346
[http://dx.doi.org/10.1101/cshperspect.a006346] [PMID: 22315714]

[39] Raskin J, Cummings J, Hardy J, Schuh K, Dean RA. Neurobiology of Alzheimer's Disease: Integrated Molecular, Physiological, Anatomical, Biomarker, and Cognitive Dimensions. Curr Alzheimer Res 2015; 12(8): 712-22.
[http://dx.doi.org/10.2174/1567205012666150701103107] [PMID: 26412218]

[40] Cabaleiro-Lago C, Szczepankiewicz O, Linse S. The effect of nanoparticles on amyloid aggregation depends on the protein stability and intrinsic aggregation rate. Langmuir 2012; 28(3): 1852-7.
[http://dx.doi.org/10.1021/la203078w] [PMID: 22168533]

[41]   Gao N, Sun H, Dong K, Ren J, Qu X. Gold-nanoparticle-based multifunctional amyloid-β inhibitor against Alzheimer's disease. Chemistry 2015; 21(2): 829-35.
[http://dx.doi.org/10.1002/chem.201404562] [PMID: 25376633]

[42]   Kogan MJ, Bastus NG, Amigo R, *et al.* Nanoparticle-mediated local and remote manipulation of protein aggregation. Nano Lett 2006; 6(1): 110-5.
[http://dx.doi.org/10.1021/nl0516862] [PMID: 16402797]

[43]   Palmal S, Maity AR, Singh BK, Basu S, Jana NR, Jana NR. Inhibition of amyloid fibril growth and dissolution of amyloid fibrils by curcumin-gold nanoparticles. Chemistry 2014; 20(20): 6184-91.
[http://dx.doi.org/10.1002/chem.201400079] [PMID: 24691975]

[44]   Prades R, Guerrero S, Araya E, *et al.* Delivery of gold nanoparticles to the brain by conjugation with a peptide that recognizes the transferrin receptor. Biomaterials 2012; 33(29): 7194-205.
[http://dx.doi.org/10.1016/j.biomaterials.2012.06.063] [PMID: 22795856]

[45]   Ikeda K, Okada T, Sawada S, Akiyoshi K, Matsuzaki K. Inhibition of the formation of amyloid beta-protein fibrils using biocompatible nanogels as artificial chaperones. FEBS Lett 2006; 580(28-29): 6587-95.
[http://dx.doi.org/10.1016/j.febslet.2006.11.009] [PMID: 17125770]

[46]   Boridy S, Takahashi H, Akiyoshi K, Maysinger D. The binding of pullulan modified cholesteryl nanogels to Abeta oligomers and their suppression of cytotoxicity. Biomaterials 2009; 30(29): 5583-91.
[http://dx.doi.org/10.1016/j.biomaterials.2009.06.010] [PMID: 19577802]

[47]   Panahi Y, Mohammadhosseini M, Abadi AJ, Akbarzadeh A, Mellatyar H. An Update on Biomedical Application of Nanotechnology for Alzheimer's Disease Diagnosis and Therapy. Drug Res (Stuttg) 2016; 66(11): 580-6.
[http://dx.doi.org/10.1055/s-0042-112811] [PMID: 27701713]

[48]   Pardeshi R, Bolshette N, Gadhave K, *et al.* Insulin signaling: An opportunistic target to minify the risk of Alzheimer's disease. Psychoneuroendocrinology 2017; 83: 159-71.
[http://dx.doi.org/10.1016/j.psyneuen.2017.05.004] [PMID: 28624654]

[49]   Picone P, Ditta LA, Sabatino MA, *et al.* Ionizing radiation-engineered nanogels as insulin nanocarriers for the development of a new strategy for the treatment of Alzheimer's disease. Biomaterials 2016; 80: 179-94.
[http://dx.doi.org/10.1016/j.biomaterials.2015.11.057] [PMID: 26708643]

[50]   Aliev G, Ashraf GM, Tarasov VV, *et al.* Alzheimer's disease – future therapy based on dendrimers. Curr Neuropharmacol 2019; 17(3): 288-94.
[http://dx.doi.org/10.2174/1570159X16666180918164623] [PMID: 30227819]

[51]   Xu L, Zhang H, Wu Y. Dendrimer advances for the central nervous system delivery of therapeutics. ACS Chem Neurosci 2014; 5(1): 2-13.
[http://dx.doi.org/10.1021/cn400182z] [PMID: 24274162]

[52]   Cunha S, Amaral MH, Lobo JM, Silva AC. Therapeutic Strategies for Alzheimer's and Parkinson's Diseases by Means of Drug Delivery Systems. Curr Med Chem 2016; 23(31): 3618-31.
[http://dx.doi.org/10.2174/0929867323666160824162401] [PMID: 27554805]

[53]   Nyitrai G, Héja L, Jablonkai I, Pál I, Visy J, Kardos J. Polyamidoamine dendrimer impairs mitochondrial oxidation in brain tissue. J Nanobiotechnology 2013; 11: 9.
[http://dx.doi.org/10.1186/1477-3155-11-9] [PMID: 23556550]

[54]   Posadas I, Romero-Castillo L, El Brahmi N, *et al.* Neutral high-generation phosphorus dendrimers inhibit macrophage-mediated inflammatory response *in vitro* and *in vivo*. Proc Natl Acad Sci USA 2017; 114(37): E7660-9.
[http://dx.doi.org/10.1073/pnas.1704858114] [PMID: 28847956]

[55]   Klementieva O, Benseny-Cases N, Gella A, Appelhans D, Voit B, Cladera J. Dense shell

glycodendrimers as potential nontoxic anti-amyloidogenic agents in Alzheimer's disease. Amyloid-dendrimer aggregates morphology and cell toxicity. Biomacromolecules 2011; 12(11): 3903-9.
[http://dx.doi.org/10.1021/bm2008636] [PMID: 21936579]

[56] Di Meo S, Reed TT, Venditti P, Victor VM. Role of ROS and RNS Sources in Physiological and Pathological Conditions. Oxid Med Cell Longev 2016; 20161245049
[http://dx.doi.org/10.1155/2016/1245049] [PMID: 27478531]

[57] Grebowski J, Kazmierska P, Krokosz A. Fullerenols as a new therapeutic approach in nanomedicine. BioMed Res Int 2013; 2013751913
[http://dx.doi.org/10.1155/2013/751913] [PMID: 24222914]

[58] Huang HM, Ou HC, Hsieh SJ, Chiang LY. Blockage of amyloid beta peptide-induced cytosolic free calcium by fullerenol-1, carboxylate C60 in PC12 cells. Life Sci 2000; 66(16): 1525-33.
[http://dx.doi.org/10.1016/S0024-3205(00)00470-7] [PMID: 10794500]

[59] Xie L, Luo Y, Lin D, Xi W, Yang X, Wei G. The molecular mechanism of fullerene-inhibited aggregation of Alzheimer's β-amyloid peptide fragment. Nanoscale 2014; 6(16): 9752-62.
[http://dx.doi.org/10.1039/C4NR01005A] [PMID: 25004796]

[60] Nash KM, Ahmed S. Nanomedicine in the ROS-mediated pathophysiology: Applications and clinical advances. Nanomedicine (Lond) 2015; 11(8): 2033-40.
[http://dx.doi.org/10.1016/j.nano.2015.07.003] [PMID: 26255114]

[61] Rajeshkumar S, Naik P. Synthesis and biomedical applications of Cerium oxide nanoparticles - A Review. Biotechnol Rep (Amst) 2017; 17: 1-5.
[http://dx.doi.org/10.1016/j.btre.2017.11.008] [PMID: 29234605]

[62] Jellinger KA. The relevance of metals in the pathophysiology of neurodegeneration, pathological considerations. Int Rev Neurobiol 2013; 110: 1-47.
[http://dx.doi.org/10.1016/B978-0-12-410502-7.00002-8] [PMID: 24209432]

[63] Liu G, Men P, Kudo W, Perry G, Smith MA. Nanoparticle-chelator conjugates as inhibitors of amyloid-β aggregation and neurotoxicity: a novel therapeutic approach for Alzheimer disease. Neurosci Lett 2009; 455(3): 187-90.
[http://dx.doi.org/10.1016/j.neulet.2009.03.064] [PMID: 19429118]

[64] Cui Z, Lockman PR, Atwood CS, *et al.* Novel D-penicillamine carrying nanoparticles for metal chelation therapy in Alzheimer's and other CNS diseases. Eur J Pharm Biopharm 2005; 59(2): 263-72.
[http://dx.doi.org/10.1016/j.ejpb.2004.07.009] [PMID: 15661498]

[65] Masserini M. Nanoparticles for brain drug delivery. ISRN Biochem 2013; 2013238428
[http://dx.doi.org/10.1155/2013/238428] [PMID: 25937958]

[66] Wong KH, Riaz MK, Xie Y, *et al.* Review of Current Strategies for Delivering Alzheimer's Disease Drugs across the Blood-Brain Barrier. Int J Mol Sci 2019; 20(2): 381.
[http://dx.doi.org/10.3390/ijms20020381] [PMID: 30658419]

[67] J SJ, Jimena CF, Dalet FE, Guadalupe TJ, Antonio SM. Scope of Lipid Nanoparticles in Neuroscience: Impact on the Treatment of Neurodegenerative Diseases. Curr Pharm Des 2017; 23(21): 3120-33.
[http://dx.doi.org/10.2174/1381612823666170301123504] [PMID: 28260513]

[68] Tosi G, Pederzoli F, Belletti D, *et al.* Nanomedicine in Alzheimer's disease: Amyloid beta targeting strategy Prog Brain Res 2019; 245: 57-88.

[69] Cardoso AM, Guedes JR, Cardoso AL, *et al.* Recent Trends in Nanotechnology Toward CNS Diseases: Lipid-Based Nanoparticles and Exosomes for Targeted Therapeutic Delivery. Int Rev Neurobiol 2016; 130: 1-40.
[http://dx.doi.org/10.1016/bs.irn.2016.05.002] [PMID: 27678173]

[70] Bana L, Minniti S, Salvati E, *et al.* Liposomes bi-functionalized with phosphatidic acid and an ApoE-derived peptide affect Aβ aggregation features and cross the blood-brain-barrier: implications for therapy of Alzheimer disease. Nanomedicine (Lond) 2014; 10(7): 1583-90.

[http://dx.doi.org/10.1016/j.nano.2013.12.001] [PMID: 24333591]

[71] Loureiro JA, Andrade S, Duarte A, *et al.* Resveratrol and Grape Extract-loaded Solid Lipid Nanoparticles for the Treatment of Alzheimer's Disease. Molecules 2017; 22(2)E277 [http://dx.doi.org/10.3390/molecules22020277] [PMID: 28208831]

[72] Kuo YC, Rajesh R. Nerve growth factor-loaded heparinized cationic solid lipid nanoparticles for regulating membrane charge of induced pluripotent stem cells during differentiation. Mater Sci Eng C Mater Biol Appl 2017; 177: 680-9. [http://dx.doi.org/10.1016/j.msec.2017.03.303]

[73] Kumari A, Yadav SK, Yadav SC. Biodegradable polymeric nanoparticles based drug delivery systems. Colloids Surf B Biointerfaces 2010; 75(1): 1-18. [http://dx.doi.org/10.1016/j.colsurfb.2009.09.001] [PMID: 19782542]

[74] Md S, Ali M, Baboota S, *et al.* Preparation, characterization, *in vivo* biodistribution and pharmacokinetic studies of donepezil-loaded PLGA nanoparticles for brain targeting. Drug Dev Ind Pharm 2013; 40: 278-87.

[75] Baysal I, Ucar G, Gultekinoglu M, Ulubayram K, Yabanoglu-Ciftci S. Donepezil loaded PLGA--PEG nanoparticles: their ability to induce destabilization of amyloid fibrils and to cross blood brain barrier in vitro. J Neural Transm (Vienna) 2017; 124(1): 33-45. [http://dx.doi.org/10.1007/s00702-016-1527-4] [PMID: 26911385]

[76] Fornaguera C, Feiner-Gracia N, Calderó G, García-Celma MJ, Solans C. Galantamine-loaded PLGA nanoparticles, from nano-emulsion templating, as novel advanced drug delivery systems to treat neurodegenerative diseases. Nanoscale 2015; 7(28): 12076-84. [http://dx.doi.org/10.1039/C5NR03474D] [PMID: 26118655]

[77] Joshi SA, Chavhan SS, Sawant KK. Rivastigmine-loaded PLGA and PBCA nanoparticles: preparation, optimization, characterization, *in vitro* and pharmacodynamic studies. Eur J Pharm Biopharm 2010; 76(2): 189-99. [http://dx.doi.org/10.1016/j.ejpb.2010.07.007] [PMID: 20637869]

[78] Sánchez-López E, Ettcheto M, Egea MA, *et al.* Memantine loaded PLGA PEGylated nanoparticles for Alzheimer's disease: *in vitro* and *in vivo* characterization. J Nanobiotechnology 2018; 16(1): 32. [http://dx.doi.org/10.1186/s12951-018-0356-z] [PMID: 29587747]

[79] Ouyang Q-Q, Zhao S, Li SD, Song C. Application of Chitosan, Chitooligosaccharide, and Their Derivatives in the Treatment of Alzheimer's Disease. Mar Drugs 2017; 15(11): 322. [http://dx.doi.org/10.3390/md15110322] [PMID: 29112116]

[80] Hanafy AS, Farid RM, Helmy MW, ElGamal SS. Pharmacological, toxicological and neuronal localization assessment of galantamine/chitosan complex nanoparticles in rats: future potential contribution in Alzheimer's disease management. Drug Deliv 2016; 23(8): 3111-22. [http://dx.doi.org/10.3109/10717544.2016.1153748] [PMID: 26942549]

[81] Fazil M, Md S, Haque S, *et al.* Development and evaluation of rivastigmine loaded chitosan nanoparticles for brain targeting. Eur J Pharm Sci 2012; 47(1): 6-15. [http://dx.doi.org/10.1016/j.ejps.2012.04.013] [PMID: 22561106]

[82] Khemariya RP, Khemariya PS. New-fangled approach in the management of Alzheimer by formulation of polysorbate 80 coated chitosan nanoparticles of rivastigmine for brain delivery and their *in vivo* evaluation. International Journal of Current Research in Medical Sciences 2016; 2(2): 18-29.http://s-o-i.org/1.15/ijcrms-2016-2-2-3

[83] Lin CY, Perche F, Ikegami M, Uchida S, Kataoka K, Itaka K. Messenger RNA-based therapeutics for brain diseases: An animal study for augmenting clearance of beta-amyloid by intracerebral administration of neprilysin mRNA loaded in polyplex nanomicelles. J Control Release 2016; 235: 268-75.

[84] Lares MR, Rossi JJ, Ouellet DL. RNAi and small interfering RNAs in human disease therapeutic

applications. Trends Biotechnol 2010; 28(11): 570-9.
[http://dx.doi.org/10.1016/j.tibtech.2010.07.009] [PMID: 20833440]

[85]   Liu Y, Liu Z, Wang Y, *et al.* Investigation of the performance of PEG-PEI/ROCK-II-siRNA complexes for Alzheimer's disease in vitro. Brain Res 2013; 1490: 43-51.
[http://dx.doi.org/10.1016/j.brainres.2012.10.039] [PMID: 23103413]

[86]   Farr SA, Erickson MA, Niehoff ML, Banks WA, Morley JE. Central and peripheral administration of antisense oligonucleotide targeting amyloid-β protein precursor improves learning and memory and reduces neuroinflammatory cytokines in Tg2576 (AβPPswe) mice. J Alzheimers Dis 2014; 40(4): 1005-16.
[http://dx.doi.org/10.3233/JAD-131883] [PMID: 24577464]

[87]   Karthivashan G, Ganesan P, Park S-Y, Kim JS, Choi D-K. Therapeutic strategies and nano-drug delivery applications in management of ageing Alzheimer's disease. Drug Deliv 2018; 25(1): 307-20.
[http://dx.doi.org/10.1080/10717544.2018.1428243] [PMID: 29350055]

[88]   Sahni JK, Doggui S, Ali J, Baboota S, Dao L, Ramassamy C. Neurotherapeutic applications of nanoparticles in Alzheimer's disease. J Control Release 2011; 152(2): 208-31.
[http://dx.doi.org/10.1016/j.jconrel.2010.11.033] [PMID: 21134407]

[89]   Wen MM, El-Salamouni NS, El-Refaie WM, *et al.* Nanotechnology-based drug delivery systems for Alzheimer's disease management: Technical, industrial, and clinical challenges. J Control Release 2017; 245: 95-107.
[http://dx.doi.org/10.1016/j.jconrel.2016.11.025] [PMID: 27889394]

[90]   Kevadiya BD, Ottemann BM, Thomas MB, *et al.* Neurotheranostics as personalized medicines 2018.

[91]   Ovais M, Zia N, Ahmad I, *et al.* Phyto-Therapeutic and Nanomedicinal Approaches to Cure Alzheimer's Disease: Present Status and Future Opportunities. Front Aging Neurosci 2018; 10: 284.
[http://dx.doi.org/10.3389/fnagi.2018.00284] [PMID: 30405389]

CHAPTER 6

Current Challenges in Alzheimer's Disease

Umar Mushtaq[1,*], **Amrina Shafi**[2] and **Firdous A. Khanday**[2]

[1] Department of Biotechnology, School of life Science, Central University of Kashmir, Ganderbal, Jammu and Kashmir-191201, India

[2] Department of Biotechnology, University of Kashmir, Srinagar, Jammu and Kashmir-190006, India

Abstract: Alzheimer's disease (AD) is an irreversible and progressive neurodegenerative disease, the manifestation of which primarily leads to progressive dementia and ultimately results in death. Currently, 5.5 million people are afflicted with this disease in the USA alone while 50 million are suffering from this disease across the world. The disease is recognised by two pathological markers *viz.* senile plaques and neurofibrillary tangles (NFT). About 50% of the population beyond the age group of over 85 years exhibit symptoms of AD since the majority of cases of AD are sporadic or idiopathic and only 5-10 percent cases are genetic. At present, there is no effective treatment for this disease and most of the drug trials used to control or stop the disease have failed as they are directed to target symptoms and not the cause of the disease. The therapeutic agents that are most commonly used for AD include cholinesterase inhibitors (CI), which enhance cholinergic neurotransmission. Diagnosis of AD still poses a significant challenge regarding the lack of information about the manifestation as well as the status *viz.* progression of this disease. However, there is enough optimism to believe that we will soon be able to develop biomarkers which would help us to accurately detect the progression of the disease. Therefore, the present chapter will provide an extensive overview of the disease and focus on possible ways to develop and formulate effective strategies to control this dreadful disease. The purpose of this citation is to guide researchers and personnel associated with pharmaceutical sciences into a new domain of investigations. This paper depicts the present scenario and projects the future challenges posed by Alzheimer's disease.

Keywords: Alzheimer's disease, Alpha Secretase, Amyloid precursor protein, Beta amyloid, Beta Secretase Biomarkers, cholinesterase, Gamma Secretase, INNOTEST, Neurofibrillary tangle, Neurogranin, Tau.

INTRODUCTION

Alzheimer's disease (AD) is the most common neurological disorder in which the

---

* **Corresponding author Umar Mushtaq:** Department of Biotechnology, School of life Science, Central University of Kashmir, Ganderbal, Jammu and Kashmir-191201, India; Tel: 9797396411; E-mail: umar6403@gmail.com

Atta-ur-Rahman (Ed.)
All rights reserved-© 2020 Bentham Science Publishers

death of neuronal cells leads to dementia and eventually results in fatality [1]. The disease was first described by a German psychiatrist, Alois Alzheimer, in 1906, when he carried out the autopsy of the brain of a patient named Auguste Deter. He found two prominent pathological markers, which were later named as neurofibrillary tangles and neurotic plaques. Neurofibrillary tangles are composed of hyperphosphorylated tau protein and neurotic plaques are composed of beta amyloid peptides [2, 3]. AD as the leading cause of dementia is now reaching an epidemic level in which 5.5 million people in the USA and 50 million people all around the world are suffering from this disease [4]. AD starts in the midbrain with mild cognitive impairment and damages the formation of new memory [5, 6]. The disease then progresses to other parts of the brain which results in problems with speech, mood swings, personality disorders, decision making, and emotions. At later stages of the disease, old memories are also destroyed and the patient becomes completely dependent on care and remains bed-ridden [5, 7, 8]. At present, there is not a single drug, which could cure or stop the progression of this disease [9]. The reason being the pathology of the disease, which starts 20-30 years prior to the appearance of symptoms in the patients and most of the drug trials are used to target symptoms and not the cause of this disease. To stop the progression of the disease at an early stage, researchers in Colombia are targeting the people prone to the early onset of AD due to rare genetic mutations. If researchers are able to develop a drug that could prevent cognitive decline in these individuals, it could revolutionize the drug therapy for this disease. To decrease the symptoms of AD, few drugs are approved by FDA, which target cholinesterase [10, 11] and N-methyl-D-Aspartate (NMDA) glutamate receptor [12]. This is not the only hurdle in the way to find a cure for this disease. AD cannot be diagnosed by a single test. INNOTEST assay, which detects total tau, phosphorylated tau, and beta amyloid 1-42 in cerebrospinal fluid (CSF) was developed 10-15 years ago [13 - 15]. Since then there have been many modifications and advances in the diagnosis of this disease. Novel Ultrasensitive Immunoassay and Mass Spectrometry are holding great promises for early diagnosis of this disease. Efforts are being made for better biomarkers which could detect the exact progression of this disease. One of the recent additions to this group is neurogranin, this protein specifically detects cognitive deterioration in AD [16]. For neurological disorders, including AD, CSF acts as an efficient matrix for biomarkers as compared to the blood. It has an advantage, which is in close proximity to the brain and the brain secretes proteins into it.

PATHOLOGY

In order to develop effective therapeutic drugs to control or stop the progression of this threatening disease, it is imperative to obtain a better understanding of this disease. The disease is caused by abnormal folding and processing of proteins.

AD is recognised by beta amyloid plaques (Aβ), neurofibrillary tangles, dystrophic neuritis, and neuropil threads [2, 3]. Aβ plaques are composites of 39 to 43-amino acid peptides generated by sequentially proteolytic cleavage of 110-130 kDa amyloid precursor protein (APP) by beta-secretase (β-secretase) and gamma-secretase (γ-secretase) via the amyloidogenic pathway. In fact, APP is processed *via* two distinct pathways: the amyloidogenic and the non-amyloidogenic pathway. In the non-amyloidogenic pathway, alpha-secretase (α-secretase) cleaves APP, twelve amino acids from the transmembrane domain at the N-terminal, releasing a large soluble fragment known as soluble amyloid precursor protein alpha (sAPPα). The remaining membrane-bound C-terminal fragment of 83 amino-acids called C-terminal fragment alpha (CTFα) is then processed by γ-secretase to liberate the short p3 fragment and the APP intracellular domain (AICD). The amyloidogenic pathway starts with the cleavage of 16 amino acids by β-secretase upstream from the α-secretase cleavage site, generating a soluble amino terminal APP derivative called soluble amyloid precursor protein beta (sAPPβ) and a membrane inserted C-terminal fragment beta (CTFβ) which is 99 amino acids long. This is followed by further cleavage of CTFβ fragment in a progressive manner by γ-secretase, giving rise to Aβ peptides of different lengths from 38–43 amino acids and an additional ACID fragment [17]. Fig. (**1**) Amyloid plaques which are spherical protein aggregates can be divided into two types: neuritic and diffuse plaques. Neuritic plaques consist of microscopic foci of extracellular filamentous Aβ protein that exhibit a cross-sectional diameter ranging from 10 μm to 120μm. Dystrophic neurites contain enlarged lysosomes, degenerating mitochondria, and paired helical filaments in the vicinity of the plaques along with microglia and astrocytes [18]. The most common isoforms of Aβ are Aβ40 and Aβ42; the shorter form is typically produced by cleavage that occurs in the endoplasmic reticulum, while the longer form is produced by cleavage in the trans-Golgi network [19]. The biological functions of APP and the factor(s) that trigger the APP proteolytic cascade remain unclear. The importance of APP in the pathogenesis of Alzheimer's disease is suggested by several findings. Notably, mutations in APP or presenilin; two proteins that are implicated in familial forms of Alzheimer's disease leading to an increase in the amyloidogenic form of Aβ [3]. γ-secretase is an enzyme complex consisting of at least four subunits: presenilin (PS, PS1, or PS2), presenilin enhancer-2 (PEN-2), nicastrin, and anterior pharynx-defective-1 (APH-1). Presenilins 1 and 2 (PS1 and PS2) are highly homologous polytopic membrane proteins of mammals principally localized to the ER and Golgi apparatus [20]. It is the presenilin subunit that possesses the γ-secretase active site responsible for executing the intra-membrane sequential processing of APP [21]. More than 50 miss sense mutations in PS1 and PS2 have been found in families with early-onset Alzheimer's disease and cellular and transgenic modelling of such mutations shows that they selectively increase the production of Aβ-42 about twofold [22 - 26]. The Aβ so produced is toxic to neurons *"in vitro"* and the load of Aβ *"in vivo"* causes the loss of synapses and neurons in the brain in animal models [27]. Although Aβ induced cell death has been shown to involve a variety of damaging events including oxidative stress; disturbances in intracellular Ca^{2+} homeostasis and activation of inflammatory cytokine cascades involving microglia [28 - 30]. Oxidative stress appears as a common transducer, which is mainly responsible for the pathology of this disease. There are many reports which suggest a link between oxidative stress and neuronal cell death. Activated microglia cells produce increased reactive oxygen species (ROS) radicals and secrete pro-inflammatory cytokines, which could lead to neuronal damage and cell death [31 - 34]. Neuronal damage stimulates inflammatory responses by the up-regulation of chemokine release, cytokine expression and ROS production [34, 35]. Such Aβ - evoked ROS generation appears to be critical for the pathogenesis of AD. Hydrogen peroxide (H_2O_2), in particular, appears to be an important intermediate in Aβ neurotoxicity [36]. Aβ also deregulates the neuronal nicotinic receptor signaling and triggers the production of ROS leading to the sequential phosphorylation of c-Jun N-terminal kinase (JNK), p66Shc, and forkhead proteins [37]. It is the fibrillar β-amyloid and not soluble β-amyloid, which is specifically toxic to cultured neurons *"in vitro"* [2, 38]. Growth of neurons *"in vitro"* in the absence of oligomeric Aβ suggests that they are actually an oligomeric form of beta amyloid, which are most toxic to neuronal cells [39]. The monomeric Aβ forms small aggregates called nucleus onto which more monomeric Aβ molecules are added to form oligomeric Aβ. The oligomeric forms of Aβ form secondary nuclei which finally leads to the formation of the fibrillar form [40].

The second hallmark of AD is the presence of neurofibrillary tangles. Neurofibrillary tangles are a composite of hyper-phosphorylated tau protein. Tau is the major microtubule-associated protein (MAP) in neurons and is found as six molecular isoforms in the human brain [41]. It is a phosphoprotein, which stabilizes the microtubules and facilitates their assembly [42]. When this protein is hyper-phosphorylated in the brain, it leads to the formation of the paired helical filaments (PHFs) that make up the NFTs in AD in the brain [43 - 46]. The pathological development of Tau appears to be caused by sequential events that are triggered by tau phosphorylation. In the early stages of AD, the predominant phosphorylation events are probably those that decrease the ability of tau to bind microtubules rather than those that increase the ability of tau to self-association. This might be caused by an imbalance in the activity of specific protein kinases or phosphatases [47]. In pretangle neurons in Alzheimer's disease, it has been found that tau protein is phosphorylated at Thr231 and Ser262 [48] and phosphorylation of both of these sites significantly decreases interactions of tau with microtubules [49, 50]. Subsequently, tau can be phosphorylated at additional sites, such as

Ser422 [51, 52] and Ser396/404 [53], which increases the tendency of tau to oligomerize and eventually form filamentous aggregates.

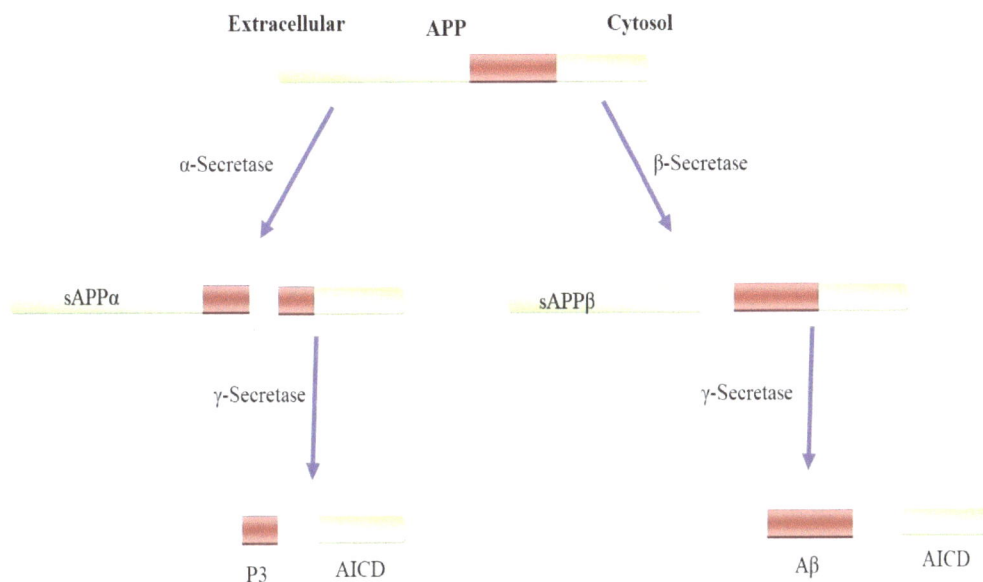

Fig. (1). Processing of Amyloid beta in neuronal cells.

On the other hand, AD is also associated with an increase in the production of ROS. An increase in oxidative stress is regarded as an early sign of Alzheimer's disease pathology, although the source of ROS and the mechanism, through which oxidative stress increases, have not been adequately investigated. The brain constitutes 2-3% of the total body mass and has high ATP demand; therefore, it has a high level of oxygen consumption. Approximately 20% of the body's total basal oxygen is consumed in the brain and subsequently generates a relatively high level of ROS which in some diseases could lead to oxidative stress. Oxidative stress is defined as an imbalance between biochemical processes leading to the production of ROS and those responsible for the removal of ROS. Under physiological conditions, ROS production is a normal consequence of cellular processes that are closely controlled by antioxidants, including glutathione, α-tocopherol (vitamin E), carotenoids, and ascorbic acid, as well as by antioxidant enzymes, such as superoxide dismutase (SOD), catalase and glutathione peroxidases, which detoxify H_2O_2 by converting it to O_2 and H_2O [54]. However, when ROS levels exceed the antioxidant capacity of a cell under disease condition or by age or metabolic demand, a deleterious condition, "oxidative stress" causes molecular damage, promoting neuronal adaptation and leading to a critical failure of biological function [54]. The neurons in the brain are exposed to an environment with considerable ROS compared to other cellular systems of other organs. The early and critical role involves the oxidative damage to proteins including direct oxidation. A number of studies have shown that Aβ toxicity is exhibited through the generation of oxidative stress and therefore, the oxidation of different biomolecules, including peroxidation of membrane lipids [55] and lipoproteins [56], the formation of H_2O_2 [36] and hydroxynonenal (HNE) [57] in neurons, is damaging DNA [58] and inactivating transport enzymes [57]. However, three important processes are prerequisite for Aβ to procure such toxicity: fibrillation, the presence of transition metals and methionine [35], aggregation and fibrillation of Aβ can take place only if the peptide is "aged" and present in a relatively high concentration (micromolar range) [28, 56]. Glycogen synthase kinase 3 (GSK-3β), which has been implicated in hyperphosphorylation of tau shows increased activity under oxidative stress [59]. It has been reported GSK-3β phosphorylates tau at Ser396, Ser404, and Thr231 in human embryonic kidney 293T cells when H_2O_2 increases its activity [60]. Recently, it has been found that PP2A plays a role in oxidative stress. Okadaic acid, which inhibits PP2A activity results in an abnormal increase in mitochondrial ROS production, which results in apoptosis in neuronal cells *via* the activation of JNK and Erk1/2 pathways [61] (Fig. **2**).

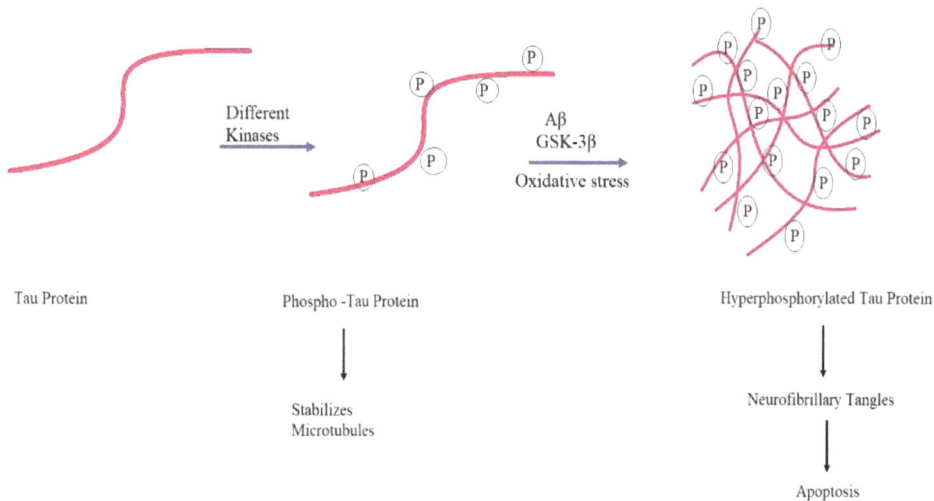

Fig. (2). Aβ and ROS production induces tau protein hyperphosphorylation.

There has been a paradigm shift about the aetiology of AD from the last few years [62, 63]. Many reports suggest that infectious agents, such as Herpesviridae, Spirochetes, Chlamydophila are the causative agents for AD [63, 64]. Increased

Aβ production, hyper-phosphorylation of Tau, increased oxidative stress, and lysosomal abnormalities have been observed in herpes virus infection, which is similar to pathological markers of AD [62, 65]. This evidence strongly supports that infection or reactivation of herpes viruses is a major cause that initiates the early neuropathology of AD.

Symptoms

Scientists are continuously trying to unravel the changes in the brain during the onset and progression of AD. The problem with this disease is that it starts twenty to thirty years before it shows its symptoms. The disease starts from the hippocampus in the midbrain, which initially appears as mild cognitive impairment [5, 6]. The problems which appear at this stage include problems with the memory; speech, and appearance of mood swings in the patient; problems in decision making and attention deficiency are also common. It is important to mention here that all patients with mild cognitive impairment will not progress to AD [6, 66]. Ongoing research at the Wicking Dementia Research and Education Centre is currently investigating whether we can better predict which patients with a mild cognitive impairment will progress to a diagnosis of AD [5, 67]. The next stage which appears in the progression of AD is the moderate form of Alzheimer's disease. In this stage of AD, the patient suffers from increased memory loss, is unable to form new memories, takes a longer time to carry out fairly routine tasks, and displays anxiety, restlessness, agitated behaviour, hallucinations, and paranoia (Fig. **3**). In this stage, the patient has difficulty in recognizing his/her own family members. In the last stage, the disease progresses to the frontal lobe which is more dangerous and devastating. This stage of Alzheimer's disease is known as severe Alzheimer's disease. In this stage, the patient is completely dependent on care and remains bedridden. The major symptoms which appear in the patient are; the patient is unable to communicate, loses weight, suffers from seizures, loses control of bowel and bladder movements, suffers from recurrent infections, has a problem in swallowing, suffers from skin infections and shows increased sleeping pattern [5, 7, 8]. It is pertinent to mention here that patients with AD usually die due to dehydration or pulmonary infection. Because of the problems in swallowing the food, patients take food and liquids into the lungs and develop pneumonia known as aspiration pneumonia.

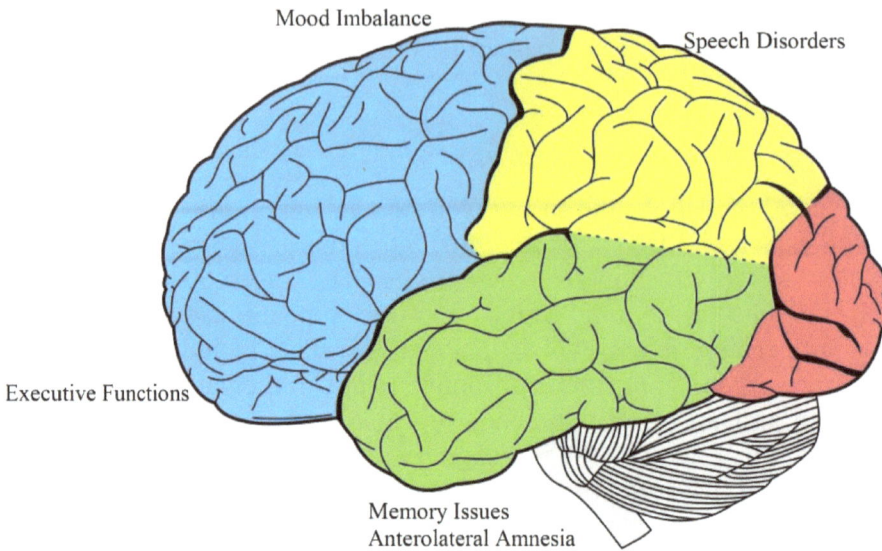

Fig. (3). Diagrammatic representation of the progression of the disease: The disease starts in the midbrain and progresses towards the speech area. Then, the disease progresses towards the frontal lobe and mood swings appear in the patients. Finally, this disease affects executive functions.

Diagnosis

The drugs which are used to target Alzheimer's disease only alleviate the symptoms of the disease and do not treat the disease. To develop an effective drug therapy against this disease, diagnosis at an early stage is a key to success. The most common biomarkers which hold the central position in the diagnosis of the disease include CSF levels of amyloid-β (Ab42), total tau (T-tau), and phosphorylated tau (P-tau) [13 - 15]. Research is still on to find more stringent biomarkers, which could correctly diagnose the proper stage of the disease. Neurogranin, synaptic protein holds immense promise in early diagnosis of this disease. The protein shows higher expression in cortical and hippocampus neurons. The expression drastically decreases in the hippocampus and the frontal lobe in AD, depicting the loss of synaptic functions. Neurogranin levels in CSF could be used as biomarkers for early diagnosis of this disease [68 - 70]. The levels of this protein decrease in AD brain tissue [71, 72] while its concentration drastically increases in CSF of AD patients [73]. There are increasing reports which suggest that neurogranin specifically shows increased expression in AD and in mild cognitive impairment as compared to other types of dementia [74 - 87]. This leads to a growing interest in using neurogranin as a specific biomarker for the diagnosis of AD. It has also been found that neurogranin could be used as a presymptomatic marker for early diagnosis of the disease as the levels of this protein increase way before the symptoms appear [75]. Thus, neurogranin could serve as a predictive and possibly monitoring marker for AD-specific cognitive decline, which is highly needed for treatment intervention studies.

In 1987, the first diagnostic test was developed for the detection of AD. It was immunoblot based assay, monoclonal Alz-50 antibody was used against tau protein [88, 89]. ELISA based test known as INNOTEST, developed in 1995, was used to detect all the six isoforms of tau (total tau) in CSF [14]. Tau protein is hyperphosphorylated at various serine and threonine residues in AD that are used to detect the progression of the disease. These residues include threonine 181, threonine 231, serine 199, serine 235, serine 396, and serine 404, but p-tau 181, 199, and 231 are commonly used for detection [14, 15, 90 - 92].

Another most commonly used diagnostic marker to detect AD is Aβ. ELISA based quantification of Aβ42 was developed, in 1995 [93]. It has been observed that Aβ42 concentration decreases in CSF fluid due to aggregation of this hydrophobic peptide in the brain and decreased release to CSF [13, 93]. Apart from Aβ42, several other forms of Aβ are also released in CSF fluid and the most abundant form is Aβ40 [94, 95]. It has been observed that the Aβ42/Aβ40 ratio diagnoses AD more accurately than Aβ42 levels alone [96 - 98].

Therapy

At present, there is not a single drug available in the market which could cure or stop the AD. Few drugs are available which could only alleviate the symptoms and do not treat the disease. These drugs either target acetylcholinesterase or NMDA receptor. Food and drug administration (FDA) has only approved four cholinesterase inhibitors, which include donepezil, tacrine, rivastigmine, galantamine, and one NMDA antagonist known as memantine [99 - 101].

If we have to stop or cure this disease, we have to target the agents *viz* Aβ42 and tau of two main pathological hallmarks Aβ plaques and NTF which are considered as main causative agents of AD. The pathways involved in the processing of APP and NFT formation are currently the central targets for therapeutic explorations. The toxic Aβ42, which differs from the normal form of Aβ40, containing additional isoleucine and alanine residues, increases its hydrophobic and aggregation propensity [102]. Therapies are devised to target enzymes that are involved in APP processing and the formation of neurotoxic Aβ42. β-secretase and γ-secretase which are involved in the formation of Aβ42 are prime targets for AD therapy.

β-secretase, a key enzyme involved in the production of neurotoxic Aβ42, is an aspartyl protease and like other proteases has a large active site (>1000A°) [103]. Large inhibitors are needed which could fit into its active site and inhibit its function. Usually, 6-10 amino acids fit into the active site of this enzyme [103, 104] and the development of such huge inhibitors creates a problem for crossing the blood-brain barrier. Ideally, molecules need to be hydrophobic and less than 700 kDa in size to cross the blood-brain barrier. Peptidomimetic strategy has been used to develop inhibitors against β-secretase. The other challenge that arises is that these inhibitors get eliminated from the cell by p-glycoproteins after being recognised as foreign molecules. GSK188909 co-administrated with P-glycoprotein inhibitors significantly reduces Aβ42 production [105]. This is not the only hurdle in developing drugs against β-secretase. The inhibitors developed against this enzyme also show selectivity against other protease enzymes, especially against the pepsin family.

The other enzyme involved in APP processing and generation of toxic Aβ42 is γ-secretase. This enzyme is composed of four subunits presenilin (PS, PS1, or PS2), presenilin enhancer-2 (PEN-2), nicastrin, and anterior pharynx-defective-1 (APH-1). Two aspartate residues of PS and PEN-2 subunit are sufficient enough to carry out the cleavage of APP [106 - 108]. Many γ-Secretase inhibitors have been developed which are under clinical trials. Semagacestat and Avagacestat have undergone clinical trials [109, 110]. Semagacestat inhibits the APP cleavage and notch signalling whereas Avagacestat badly affected the cognitive and routine functions of the recipient. Both of these drugs did not provide the desired results [111].

Many such inhibitors are also developed which allosterically modify the functioning of γ-Secretase enzyme and do not completely inhibit its activity. These allosteric inhibitors increase the specificity of the enzyme for Aβ42 and have very little impact on other signalling pathways. One such inhibitor is BMS-708163, which shows much higher specificity (approx. 193 fold) for APP as compared to Notch [111]. Clinical trials of gamma-secretase inhibitors have been ceased because this enzyme is involved in the processing of more than 100 substrates and more importantly notch protein [112 - 114].

Tau, microtubule associated protein exists in different forms in neurons and performs diverse functions. It is found at a conc. of 10-25pg/ml as phospho-tau and 300-400pg/ml as total tau [14, 115, 116]. The protein undergoes numerous post-translational modifications. Many antibodies against different epitopes have been developed. The passive immunotherapy is used to target different forms of tau protein including monomeric [117 - 119], oligomeric [120], or conformationally altered forms [117, 121 - 125]. Many antibodies have also been developed against different phosphor residues of tau protein, which inhibit its aggregation and formation of NFT [126 - 128]. It has been observed that tau protein isoforms are truncated at N-terminus [129]. Thus, oligomeric and fibrillar forms of these proteins mainly contain C-terminal sequences in AD [130]. Recently, it has been reported that antibodies against C-terminus of tau protein are more effective as compared to N-terminal antibodies [130 - 132]. Although the efficiency of antibodies against N-terminus of tau protein in inhibiting the oligomerization of this protein is still controversial, the paradigm has shifted to developing antibodies against the mid-region of tau protein. Many antibodies against different phospho residues in the mid-region domain of tau protein have been developed [126]. T231 phosphorylation in the mid domain of tau protein is regarded as an early event in the progression of AD [127, 128].

Recent Signaling Aspect of AD

Various signaling pathways that are deregulated in AD lead to misfolding of proteins, mitochondrial dysfunction, imbalance in ROS production, deregulation of autophagy, and impairment of Ca^{2+} homeostasis [133 - 137]. An impaired mitochondrial function is an early event in AD. Aβ is involved in disturbing the functioning of multiple components of mitochondria directly or indirectly [138 - 140]. Mitochondria are also a major source of ROS production in AD [140 - 142] and this increased ROS production has been implicated in apoptotic cell death in AD [143 - 145]. Recently, it has been found that mitochondria transfer ameliorates the symptoms of AD in mice by the increase of citrate-synthase and cytochrome c oxidase [146]. Another organelle whose function is deregulated in AD is Endoplasmic reticulum (ER), which is associated with the regulation of Ca^{2+} homeostasis [147], and presenilin PS2, a component of β-secretase, plays a vital role in regulating Ca^{2+} within ER [148, 149]. Mutated PS2 increases the influx of Ca^{2+} and its downregulation leads to decreased transfer of Ca^{2+} [148]. Presenilins are also implicated in the coupling of ER and mitochondria and PS2 increases tethering of these organelles in AD by sequestering MFN2 [150, 151].

Autophagy is a physiological process by which cells get rid of the waste materials. Cells convert complex material into their basic constituents that are reused through different metabolic pathways. Autophagy is a highly regulated process and many proteins which are involved in the formation of autophagosome are deregulated in AD. Autophagy has been implicated in reducing both soluble and aggregated Aβ and tau proteins [152 - 157]. Autophagy-related protein-7 (ATG7) deletion in transgenic mice decreases the secretion of Aβ and inhibits the formation of plaques [158]. PS1, a subunit of β-secretase, is also involved in the formation of autophagosome and clearance of Aβ. Mutations and defects in Ser367 in PS1 reduce the levels of CTF and inhibit the fusion of autophagosome and lysosome [159]. PS1 promotes the interaction of Vamp8 SNARE Syn-taxin

17 on autophagosome [159]. Recently, it has been found that oligomeric Aβ disturbs the normal transport of autophagic vacuoles (AVs) by binding to the dynein motor proteins and block sites for vacuoles [160]. Many other proteins that are involved in autophagy show decreased expression in AD including Nuclear receptor binding factor 2(NRBF2), component of PI3K complex [161], Transient receptor potential Mulcolipin-1 (TRPML1), which inhibit mTOR and activate AMPK signaling pathway [162]. The triggering receptor expressed on myeloid cells 2 (TREM2), which also regulates autophages via mTOR and AMPK pathways [163] and Toll interacting protein (Tollip) that disrupts endosome-lysosome fusion and leads to the accumulation of Aβ in neurons is noticeable [164].

Many kinases are also deregulated in Alzheimer's disease. These proteins are involved in the overproduction of Aβ, hyperphosphorylation of tau, and increased ROS production in Alzheimer's conditions. Kinases that are involved in hyperphosphorylation of tau protein include microtubule affinity regulatory kinases (MARK), mitogen-activated protein kinase (MAPK), protein kinase II, protein kinase C (PKC), cyclin-dependent kinase 5 (cdk5)/p39, and glycogen synthase kinase 3β (GSK3β). Several non-kinase proteins such as apolipoprotein E (ApoE), adaptor protein 14-3-3z are also involved in hyperphosphorylation of tau protein [165 - 167].

Many novel targets have been identified, which opens up new avenues for understanding this disease very closely [168]. NFκβ has been implicated in the pathogenesis of AD and many drugs have been found to inhibit the neuronal dysfunction by targeting the pathways mediated by NFκβ [169]. Recently, melatonin receptor has been reported to regulate Aβ production by regulating the secretases via NFκβ pathways involving Pin1 and GSK3β proteins [170].

Current research is also focusing on discovering non-invasive biomarkers for the disease. CSF, which in close proximity to the brain is a suitable sample to detect the levels of biomarkers and the progression of the disease. But to collect the CSF sample for testing need a complicated procedure. Other sources of samples that could be used for the evaluation of biomarkers for this disease could be saliva, sweat, blood/serum, and urine. Thus there is a need to look for the molecules which show differential levels in these fluids in the disease. One of the potential non-invasive biomarkers could be microRNAs (miRNAs) in the blood/serum. miRNAs are 22-23 long non-coding RNAs that are involved in the regulation of many genes by binding to a 3-untranslated region of the genes. The most common miRNAs that are early deregulated in the disease include miRNA-107, miRNA-26b, miRNA-30e, miRNA-34a, miRNA-485, miRNA200c, miRNA-210, miRNA-146a, miRNA-34c, miRNA-125b [171 - 176]. The possible targets for these miRNAs involve the genes that are associated with stress responses, growth factors, wnt signaling, immune system, cell cycle, Rho GTPases, and cellular senescence in neurons. All these pathways have already been implicated in AD. miRNAs open up new avenues to control and cure the disease.

CONCLUSION

Despite many hiccups in all aspects of the AD, scientists have made great progress in understanding this dreadful disease. With advancing knowledge of genomics, proteomics, and transcriptomics, the search for better biomarkers, advancing detection techniques, and drugs to contain and control the progression of the disease is underway. Current research is looking for drugs that not only alleviate the symptoms of the AD but stop the progression of the disease. Days are not far when we shall see drugs available against this horrifying disease. Inhibitors against gamma secretase do not provide the favourable results whereas beta secretase inhibitors hold promising approaches to inhibit the formation of Aβ. Many drugs against this enzyme have entered in to phase III trials. Passive immunotherapies have also been devised to stop the aggregation of amyloid beta peptides and phosphor tau protein. Various miRNAs show differential expression in AD can be explored as possible biomarkers and targets for drug discovery. For the last few years, there has been a paradigm shift regarding the etiology of the disease to infectious agents. This will open up the scope of the immune system and defense mechanisms involved in the health of the brain. This chapter outlined the progress in understanding the disease and showing how research in this field is making strides in the right direction.

CONSENT FOR PUBLICATION

Not applicable.

CONFLICT OF INTEREST

The authors declare no conflict of interest, financial or otherwise.

ACKNOWLEDGMENTS

This work was supported by the Council of Scientific and Industrial Research Grants Commission No F. 09/251(0060)/2015-EMR-I to Umar Mushtaq. The agency sponsored my fellowship during PhD. Dr. Munazeh Fazal Qureshi and Mr. Amir are highly acknowledged for helping me in editing the language of the chapter.

REFERENCES

[1]   Querfurth HW, LaFerla FM. Alzheimer's disease. N Engl J Med 2010; 362(4): 329-44.
[http://dx.doi.org/10.1056/NEJMra0909142] [PMID: 20107219]

[2]   Yankner BA. Mechanisms of neuronal degeneration in Alzheimer's disease. Neuron 1996; 16(5): 921-32.
[http://dx.doi.org/10.1016/S0896-6273(00)80115-4] [PMID: 8630250]

[3]   Selkoe DJ. Translating cell biology into therapeutic advances in Alzheimer's disease. Nature 1999; 399(6738) (Suppl.): A23-31.
[http://dx.doi.org/10.1038/399a023] [PMID: 10392577]

[4]   Hebert LE, Weuve J, Scherr PA, Evans DA. Alzheimer disease in the United States (2010-2050) estimated using the 2010 census. Neurology 2013; 80(19): 1778-83.
[http://dx.doi.org/10.1212/WNL.0b013e31828726f5] [PMID: 23390181]

[5]   Price BH, Gurvit H, Weintraub S, Geula C, Leimkuhler E, Mesulam M. Neuropsychological patterns and language deficits in 20 consecutive cases of autopsy-confirmed Alzheimer's disease. Arch Neurol 1993; 50(9): 931-7.
[http://dx.doi.org/10.1001/archneur.1993.00540090038008] [PMID: 8363447]

[6]   Arnáiz E, Almkvist O. Neuropsychological features of mild cognitive impairment and preclinical Alzheimer's disease. Acta Neurol Scand Suppl 2003; 179: 34-41.
[http://dx.doi.org/10.1034/j.1600-0404.107.s179.7.x] [PMID: 12603249]

[7]   Mega MS, Cummings JL, Fiorello T, Gornbein J. The spectrum of behavioral changes in Alzheimer's disease. Neurology 1996; 46(1): 130-5.
[http://dx.doi.org/10.1212/WNL.46.1.130] [PMID: 8559361]

[8]   Karttunen K, Karppi P, Hiltunen A, *et al.* Neuropsychiatric symptoms and quality of life in patients with very mild and mild Alzheimer's disease. Int J Geriatr Psychiatry 2011; 26(5): 473-82.
[http://dx.doi.org/10.1002/gps.2550] [PMID: 21445998]

[9]   Parnetti L, Senin U, Mecocci P. Cognitive enhancement therapy for Alzheimer's disease. The way forward. Drugs 1997; 53(5): 752-68.
[http://dx.doi.org/10.2165/00003495-199753050-00003] [PMID: 9129864]

[10]   Brion JP. The neurobiology of Alzheimer's disease. Acta Clin Belg 1996; 51(2): 80-90.
[http://dx.doi.org/10.1080/17843286.1996.11718490] [PMID: 8693872]

[11]   Brodaty H, Ames D, Boundy KL, *et al.* Pharmacological treatment of cognitive deficits in Alzheimer's disease. Med J Aust 2001; 175(6): 324-9.
[http://dx.doi.org/10.5694/j.1326-5377.2001.tb143593.x] [PMID: 11665948]

[12]   Livingston G, Katona C. The place of memantine in the treatment of Alzheimer's disease: a number needed to treat analysis. Int J Geriatr Psychiatry 2004; 19(10): 919-25.
[http://dx.doi.org/10.1002/gps.1166] [PMID: 15449303]

[13]   Andreasen N, Hesse C, Davidsson P, *et al.* Cerebrospinal fluid beta-amyloid(1-42) in Alzheimer disease: differences between early- and late-onset Alzheimer disease and stability during the course of disease. Arch Neurol 1999; 56(6): 673-80.
[http://dx.doi.org/10.1001/archneur.56.6.673] [PMID: 10369305]

[14]   Blennow K, Wallin A, Agren H, Spenger C, Siegfried J, Vanmechelen E. Tau protein in cerebrospinal fluid: a biochemical marker for axonal degeneration in Alzheimer disease? Mol Chem Neuropathol 1995; 26(3): 231-45.
[http://dx.doi.org/10.1007/BF02815140] [PMID: 8748926]

[15]   Vanmechelen E, Vanderstichele H, Davidsson P, *et al.* Quantification of tau phosphorylated at threonine 181 in human cerebrospinal fluid: a sandwich ELISA with a synthetic phosphopeptide for standardization. Neurosci Lett 2000; 285(1): 49-52.
[http://dx.doi.org/10.1016/S0304-3940(00)01036-3] [PMID: 10788705]

[16]   Wellington H, Paterson RW, Portelius E, *et al.* Increased CSF neurogranin concentration is specific to Alzheimer disease. Neurology 2016; 86(9): 829-35.
[http://dx.doi.org/10.1212/WNL.0000000000002423] [PMID: 26826204]

[17]   Xu X. Gamma-secretase catalyzes sequential cleavages of the AbetaPP transmembrane domain. J Alzheimers Dis 2009; 16(2): 211-24.
[http://dx.doi.org/10.3233/JAD-2009-0957] [PMID: 19221413]

[18]   Selkoe DJ. Alzheimer's disease: genes, proteins, and therapy. Physiol Rev 2001; 81(2): 741-66.
[http://dx.doi.org/10.1152/physrev.2001.81.2.741] [PMID: 11274343]

[19]   Hartmann T, Bieger SC, Brühl B, *et al.* Distinct sites of intracellular production for Alzheimer's disease A beta40/42 amyloid peptides. Nat Med 1997; 3(9): 1016-20.
[http://dx.doi.org/10.1038/nm0997-1016] [PMID: 9288729]

[20]   Levitan D, Greenwald I. Facilitation of lin-12-mediated signalling by sel-12, a Caenorhabditis elegans S182 Alzheimer's disease gene. Nature 1995; 377(6547): 351-4.
[http://dx.doi.org/10.1038/377351a0] [PMID: 7566091]

[21]   Wolfe MS. gamma-Secretase in biology and medicine. Semin Cell Dev Biol 2009; 20(2): 219-24.
[http://dx.doi.org/10.1016/j.semcdb.2008.12.011] [PMID: 19162210]

[22]   Scheuner D, Eckman C, Jensen M, *et al.* Secreted amyloid beta-protein similar to that in the senile plaques of Alzheimer's disease is increased in vivo by the presenilin 1 and 2 and APP mutations linked to familial Alzheimer's disease. Nat Med 1996; 2(8): 864-70.
[http://dx.doi.org/10.1038/nm0896-864] [PMID: 8705854]

[23]   Duff K, Eckman C, Zehr C, *et al.* Increased amyloid-beta42(43) in brains of mice expressing mutant presenilin 1. Nature 1996; 383(6602): 710-3.
[http://dx.doi.org/10.1038/383710a0] [PMID: 8878479]

[24]   Borchelt DR, Thinakaran G, Eckman CB, *et al.* Familial Alzheimer's disease-linked presenilin 1 variants elevate Abeta1-42/1-40 ratio in vitro and in vivo. Neuron 1996; 17(5): 1005-13.
[http://dx.doi.org/10.1016/S0896-6273(00)80230-5] [PMID: 8938131]

[25]   Citron M, Westaway D, Xia W, *et al.* Mutant presenilins of Alzheimer's disease increase production of 42-residue amyloid beta-protein in both transfected cells and transgenic mice. Nat Med 1997; 3(1): 67-72.
[http://dx.doi.org/10.1038/nm0197-67] [PMID: 8986743]

[26]   Xia W, Zhang J, Kholodenko D, *et al.* Enhanced production and oligomerization of the 42-residue amyloid beta-protein by Chinese hamster ovary cells stably expressing mutant presenilins. J Biol Chem 1997; 272(12): 7977-82.
[http://dx.doi.org/10.1074/jbc.272.12.7977] [PMID: 9065468]

[27]   Smith WW, Gorospe M, Kusiak JW. Signaling mechanisms underlying Abeta toxicity: potential therapeutic targets for Alzheimer's disease. CNS Neurol Disord Drug Targets 2006; 5(3): 355-61.
[http://dx.doi.org/10.2174/187152706784111515] [PMID: 16787235]

[28]   Iversen LL, Mortishire-Smith RJ, Pollack SJ, Shearman MS. The toxicity in vitro of beta-amyloid protein. Biochem J 1995; 311(Pt 1): 1-16.
[http://dx.doi.org/10.1042/bj3110001] [PMID: 7575439]

[29]   Mattson MP, Barger SW, Cheng B, Lieburg I, Smith-Swintosky VL, Rydel RE. beta-Amyloid precursor protein metabolites and loss of neuronal Ca2+ homeostasis in Alzheimer's disease. Trends Neurosci 1993; 16(10): 409-14.
[http://dx.doi.org/10.1016/0166-2236(93)90009-B] [PMID: 7504356]

[30]   Robinson MJ, Cobb MH. Mitogen-activated protein kinase pathways. Curr Opin Cell Biol 1997; 9(2): 180-6.
[http://dx.doi.org/10.1016/S0955-0674(97)80061-0] [PMID: 9069255]

[31]   Bianca VD, Dusi S, Bianchini E, Dal Prà I, Rossi F. beta-amyloid activates the O-2 forming NADPH oxidase in microglia, monocytes, and neutrophils. A possible inflammatory mechanism of neuronal damage in Alzheimer's disease. J Biol Chem 1999; 274(22): 15493-9.
       [http://dx.doi.org/10.1074/jbc.274.22.15493] [PMID: 10336441]

[32]   Bondy SC, Guo-Ross SX, Truong AT. Promotion of transition metal-induced reactive oxygen species formation by beta-amyloid. Brain Res 1998; 799(1): 91-6.
       [http://dx.doi.org/10.1016/S0006-8993(98)00461-2] [PMID: 9666089]

[33]   Markesbery WR. Oxidative stress hypothesis in Alzheimer's disease. Free Radic Biol Med 1997; 23(1): 134-47.
       [http://dx.doi.org/10.1016/S0891-5849(96)00629-6] [PMID: 9165306]

[34]   Johnstone M, Gearing AJ, Miller KM. A central role for astrocytes in the inflammatory response to beta-amyloid; chemokines, cytokines and reactive oxygen species are produced. J Neuroimmunol 1999; 93(1-2): 182-93.
       [http://dx.doi.org/10.1016/S0165-5728(98)00226-4] [PMID: 10378882]

[35]   Heck S, Lezoualc'h F, Engert S, Behl C. Insulin-like growth factor-1-mediated neuroprotection against oxidative stress is associated with activation of nuclear factor kappaB. J Biol Chem 1999; 274(14): 9828-35.
       [http://dx.doi.org/10.1074/jbc.274.14.9828] [PMID: 10092673]

[36]   Behl C, Davis JB, Lesley R, Schubert D. Hydrogen peroxide mediates amyloid beta protein toxicity. Cell 1994; 77(6): 817-27.
       [http://dx.doi.org/10.1016/0092-8674(94)90131-7] [PMID: 8004671]

[37]   Smith WW, Norton DD, Gorospe M, *et al.* Phosphorylation of p66Shc and forkhead proteins mediates Abeta toxicity. J Cell Biol 2005; 169(2): 331-9.
       [http://dx.doi.org/10.1083/jcb.200410041] [PMID: 15837797]

[38]   Yankner BA, Duffy LK, Kirschner DA. Neurotrophic and neurotoxic effects of amyloid beta protein: reversal by tachykinin neuropeptides. Science 1990; 250(4978): 279-82.
       [http://dx.doi.org/10.1126/science.2218531] [PMID: 2218531]

[39]   Dunkelmann T, Teichmann K, Ziehm T, *et al.* Aβ oligomer eliminating compounds interfere successfully with pEAβ(3-42) induced motor neurodegenerative phenotype in transgenic mice. Neuropeptides 2018; 67: 27-35.
       [http://dx.doi.org/10.1016/j.npep.2017.11.011] [PMID: 29273382]

[40]   Linse S. Mechanism of amyloid protein aggregation and the role of inhibitors. Pure Appl Chem 2019; 91(2): 211-29.
       [http://dx.doi.org/10.1515/pac-2018-1017]

[41]   Goedert M, Spillantini MG, Jakes R, Rutherford D, Crowther RA. Multiple isoforms of human microtubule-associated protein tau: sequences and localization in neurofibrillary tangles of Alzheimer's disease. Neuron 1989; 3(4): 519-26.
       [http://dx.doi.org/10.1016/0896-6273(89)90210-9] [PMID: 2484340]

[42]   Weingarten MD, Lockwood AH, Hwo SY, Kirschner MW. A protein factor essential for microtubule assembly. Proc Natl Acad Sci USA 1975; 72(5): 1858-62.
       [http://dx.doi.org/10.1073/pnas.72.5.1858] [PMID: 1057175]

[43]   Iqbal K, Grundke-Iqbal I, Zaidi T, *et al.* Defective brain microtubule assembly in Alzheimer's disease. Lancet 1986; 2(8504): 421-6.
       [http://dx.doi.org/10.1016/S0140-6736(86)92134-3] [PMID: 2874414]

[44]   Grundke-Iqbal I, Iqbal K, Quinlan M, Tung YC, Zaidi MS, Wisniewski HM. Microtubule-associated protein tau. A component of Alzheimer paired helical filaments. J Biol Chem 1986; 261(13): 6084-9.
       [PMID: 3084478]

[45]   Kosik KS, Joachim CL, Selkoe DJ. Microtubule-associated protein tau (tau) is a major antigenic

component of paired helical filaments in Alzheimer disease. Proc Natl Acad Sci USA 1986; 83(11): 4044-8.
[http://dx.doi.org/10.1073/pnas.83.11.4044] [PMID: 2424016]

[46]   Wood JG, Mirra SS, Pollock NJ, Binder LI. Neurofibrillary tangles of Alzheimer disease share antigenic determinants with the axonal microtubule-associated protein tau (tau). Proc Natl Acad Sci USA 1986; 83(11): 4040-3.
[http://dx.doi.org/10.1073/pnas.83.11.4040] [PMID: 2424015]

[47]   Johnson GV, Stoothoff WH. Tau phosphorylation in neuronal cell function and dysfunction. J Cell Sci 2004; 117(Pt 24): 5721-9.
[http://dx.doi.org/10.1242/jcs.01558] [PMID: 15537830]

[48]   Augustinack JC, Schneider A, Mandelkow EM, Hyman BT. Specific tau phosphorylation sites correlate with severity of neuronal cytopathology in Alzheimer's disease. Acta Neuropathol 2002; 103(1): 26-35.
[http://dx.doi.org/10.1007/s004010100423] [PMID: 11837744]

[49]   Biernat J, Gustke N, Drewes G, Mandelkow EM, Mandelkow E. Phosphorylation of Ser262 strongly reduces binding of tau to microtubules: distinction between PHF-like immunoreactivity and microtubule binding. Neuron 1993; 11(1): 153-63.
[http://dx.doi.org/10.1016/0896-6273(93)90279-Z] [PMID: 8393323]

[50]   Cho JH, Johnson GV. Glycogen synthase kinase 3beta phosphorylates tau at both primed and unprimed sites. Differential impact on microtubule binding. J Biol Chem 2003; 278(1): 187-93.
[http://dx.doi.org/10.1074/jbc.M206236200] [PMID: 12409305]

[51]   Ferrari A, Hoerndli F, Baechi T, Nitsch RM, Götz J. beta-Amyloid induces paired helical filament-like tau filaments in tissue culture. J Biol Chem 2003; 278(41): 40162-8.
[http://dx.doi.org/10.1074/jbc.M308243200] [PMID: 12893817]

[52]   Haase C, Stieler JT, Arendt T, Holzer M. Pseudophosphorylation of tau protein alters its ability for self-aggregation. J Neurochem 2004; 88(6): 1509-20.
[http://dx.doi.org/10.1046/j.1471-4159.2003.02287.x] [PMID: 15009652]

[53]   Abraha A, Ghoshal N, Gamblin TC, *et al.* C-terminal inhibition of tau assembly in vitro and in Alzheimer's disease. J Cell Sci 2000; 113(Pt 21): 3737-45.
[PMID: 11034902]

[54]   Yu BP. Cellular defenses against damage from reactive oxygen species. Physiol Rev 1994; 74(1): 139-62.
[http://dx.doi.org/10.1152/physrev.1994.74.1.139] [PMID: 8295932]

[55]   Varadarajan S, Yatin S, Aksenova M, Butterfield DA. Review: Alzheimer's amyloid beta-peptid--associated free radical oxidative stress and neurotoxicity. J Struct Biol 2000; 130(2-3): 184-208.
[http://dx.doi.org/10.1006/jsbi.2000.4274] [PMID: 10940225]

[56]   Kontush A, Berndt C, Weber W, *et al.* Amyloid-beta is an antioxidant for lipoproteins in cerebrospinal fluid and plasma. Free Radic Biol Med 2001; 30(1): 119-28.
[http://dx.doi.org/10.1016/S0891-5849(00)00458-5] [PMID: 11134902]

[57]   Mark RJ, Lovell MA, Markesbery WR, Uchida K, Mattson MP. A role for 4-hydroxynonenal, an aldehydic product of lipid peroxidation, in disruption of ion homeostasis and neuronal death induced by amyloid beta-peptide. J Neurochem 1997; 68(1): 255-64.
[http://dx.doi.org/10.1046/j.1471-4159.1997.68010255.x] [PMID: 8978733]

[58]   Xu J, Chen S, Ku G, *et al.* Amyloid beta peptide-induced cerebral endothelial cell death involves mitochondrial dysfunction and caspase activation. J Cereb Blood Flow Metab 2001; 21(6): 702-10.
[http://dx.doi.org/10.1097/00004647-200106000-00008] [PMID: 11488539]

[59]   Feng Y, Xia Y, Yu G, *et al.* Cleavage of GSK-3β by calpain counteracts the inhibitory effect of Ser9 phosphorylation on GSK-3β activity induced by H_2O_2. J Neurochem 2013; 126(2): 234-42.

[http://dx.doi.org/10.1111/jnc.12285] [PMID: 23646926]

[60]   Chiara F, Gambalunga A, Sciacovelli M, *et al.* Chemotherapeutic induction of mitochondrial oxidative stress activates GSK-3α/β and Bax, leading to permeability transition pore opening and tumor cell death. Cell Death Dis 2012; 3(3)e444
[http://dx.doi.org/10.1038/cddis.2012.184] [PMID: 23235461]

[61]   Chen L, Na R, Gu M, Richardson A, Ran Q. Lipid peroxidation up-regulates BACE1 expression in vivo: a possible early event of amyloidogenesis in Alzheimer's disease. J Neurochem 2008; 107(1): 197-207.
[http://dx.doi.org/10.1111/j.1471-4159.2008.05603.x] [PMID: 18680556]

[62]   Fülöp T, Itzhaki RF, Balin BJ, Miklossy J, Barron AE. Role of Microbes in the Development of Alzheimer's Disease: State of the Art - An International Symposium Presented at the 2017 IAGG Congress in San Francisco. Front Genet 2018; 9(362): 362.
[http://dx.doi.org/10.3389/fgene.2018.00362] [PMID: 30250480]

[63]   Haas JG, Lathe R. Microbes and Alzheimer's Disease: New Findings Call for a Paradigm Change. Trends Neurosci 2018; 41(9): 570-3.
[http://dx.doi.org/10.1016/j.tins.2018.07.001] [PMID: 30033181]

[64]   Harris SA, Harris EA. Herpes Simplex Virus Type 1 and Other Pathogens are Key Causative Factors in Sporadic Alzheimer's Disease. J Alzheimers Dis 2015; 48(2): 319-53.
[http://dx.doi.org/10.3233/JAD-142853] [PMID: 26401998]

[65]   Kristen H, Sastre I, Muñoz-Galdeano T, Recuero M, Aldudo J, Bullido MJ. The lysosome system is severely impaired in a cellular model of neurodegeneration induced by HSV-1 and oxidative stress. Neurobiol Aging 2018; 68: 5-17.
[http://dx.doi.org/10.1016/j.neurobiolaging.2018.03.025] [PMID: 29689425]

[66]   Robert PH, Berr C, Volteau M, *et al.* Apathy in patients with mild cognitive impairment and the risk of developing dementia of Alzheimer's disease: a one-year follow-up study. Clin Neurol Neurosurg 2006; 108(8): 733-6.
[http://dx.doi.org/10.1016/j.clineuro.2006.02.003] [PMID: 16567037]

[67]   Carlesimo GA, Oscar-Berman M. Memory deficits in Alzheimer's patients: a comprehensive review. Neuropsychol Rev 1992; 3(2): 119-69.
[http://dx.doi.org/10.1007/BF01108841] [PMID: 1300219]

[68]   Seubert P, Vigo-Pelfrey C, Esch F, *et al.* Isolation and quantification of soluble Alzheimer's beta-peptide from biological fluids. Nature 1992; 359(6393): 325-7.
[http://dx.doi.org/10.1038/359325a0] [PMID: 1406936]

[69]   Nakamura T, Shoji M, Harigaya Y, *et al.* Amyloid beta protein levels in cerebrospinal fluid are elevated in early-onset Alzheimer's disease. Ann Neurol 1994; 36(6): 903-11.
[http://dx.doi.org/10.1002/ana.410360616] [PMID: 7998778]

[70]   Iwatsubo T, Odaka A, Suzuki N, Mizusawa H, Nukina N, Ihara Y. Visualization of A beta 42(43) and A beta 40 in senile plaques with end-specific A beta monoclonals: evidence that an initially deposited species is A beta 42(43). Neuron 1994; 13(1): 45-53.
[http://dx.doi.org/10.1016/0896-6273(94)90458-8] [PMID: 8043280]

[71]   Davidsson P, Blennow K. Neurochemical dissection of synaptic pathology in Alzheimer's disease. Int Psychogeriatr 1998; 10(1): 11-23.
[http://dx.doi.org/10.1017/S1041610298005110] [PMID: 9629521]

[72]   Reddy PH, Mani G, Park BS, Jacques J, Murdoch G, Whetsell W, *et al.* Differential loss of synaptic proteins in Alzheimer's disease: implications for synaptic dysfunction. J Alzheimers Dis 2005; 7: 103–17-80.
[http://dx.doi.org/10.3233/JAD-2005-7203]

[73]   Thorsell A, Bjerke M, Gobom J, *et al.* Neurogranin in cerebrospinal fluid as a marker of synaptic

degeneration in Alzheimer's disease. Brain Res 2010; 1362: 13-22.
[http://dx.doi.org/10.1016/j.brainres.2010.09.073] [PMID: 20875798]

[74]   De Vos A, Jacobs D, Struyfs H, *et al.* C-terminal neurogranin is increased in cerebrospinal fluid but unchanged in plasma in Alzheimer's disease. Alzheimers Dement 2015; 11(12): 1461-9.
[http://dx.doi.org/10.1016/j.jalz.2015.05.012] [PMID: 26092348]

[75]   Kester MI, Teunissen CE, Crimmins DL, *et al.* Neurogranin as a cerebrospinal fluid biomarker for synaptic loss in symptomatic Alzheimer disease. JAMA Neurol 2015; 72(11): 1275-80.
[http://dx.doi.org/10.1001/jamaneurol.2015.1867] [PMID: 26366630]

[76]   Kvartsberg H, Duits FH, Ingelsson M, *et al.* Cerebrospinal fluid levels of the synaptic protein neurogranin correlates with cognitive decline in prodromal Alzheimer's disease. Alzheimers Dement 2015; 11(10): 1180-90.
[http://dx.doi.org/10.1016/j.jalz.2014.10.009] [PMID: 25533203]

[77]   Kvartsberg H, Portelius E, Andreasson U, *et al.* Characterization of the postsynaptic protein neurogranin in paired cerebrospinal fluid and plasma samples from Alzheimer's disease patients and healthy controls. Alzheimers Res Ther 2015; 7(1): 40.
[http://dx.doi.org/10.1186/s13195-015-0124-3] [PMID: 26136856]

[78]   Portelius E, Zetterberg H, Skillbäck T, *et al.* Cerebrospinal fluid neurogranin: relation to cognition and neurodegeneration in Alzheimer's disease. Brain 2015; 138(Pt 11): 3373-85.
[http://dx.doi.org/10.1093/brain/awv267] [PMID: 26373605]

[79]   Sanfilippo C, Forlenza O, Zetterberg H, Blennow K. Increased neurogranin concentrations in cerebrospinal fluid of Alzheimer's disease and in mild cognitive impairment due to AD. J Neural Transm (Vienna) 2016; 123(12): 1443-7.
[http://dx.doi.org/10.1007/s00702-016-1597-3] [PMID: 27531278]

[80]   Mattsson N, Insel PS, Palmqvist S, *et al.* Cerebrospinal fluid tau, neurogranin, and neurofilament light in Alzheimer's disease. EMBO Mol Med 2016; 8(10): 1184-96.
[http://dx.doi.org/10.15252/emmm.201606540] [PMID: 27534871]

[81]   Tarawneh R, D'Angelo G, Crimmins D, *et al.* Diagnostic and prognostic utility of the synaptic marker neurogranin in Alzheimer disease. JAMA Neurol 2016; 73(5): 561-71.
[http://dx.doi.org/10.1001/jamaneurol.2016.0086] [PMID: 27018940]

[82]   Remnestål J, Just D, Mitsios N, *et al.* CSF profiling of the human brain enriched proteome reveals associations of neuromodulin and neurogranin to Alzheimer's disease. Proteomics Clin Appl 2016; 10(12): 1242-53.
[http://dx.doi.org/10.1002/prca.201500150] [PMID: 27604409]

[83]   De Vos A, Struyfs H, Jacobs D, *et al.* The cerebrospinal fluid neurogranin/BACE1 ratio is a potential correlate of cognitive decline in Alzheimer's disease. J Alzheimers Dis 2016; 53(4): 1523-38.
[http://dx.doi.org/10.3233/JAD-160227] [PMID: 27392859]

[84]   Wellington H, Paterson RW, Portelius E, *et al.* Increased CSF neurogranin concentration is specific to Alzheimer disease. Neurology 2016; 86(9): 829-35.
[http://dx.doi.org/10.1212/WNL.0000000000002423] [PMID: 26826204]

[85]   Hellwig K, Kvartsberg H, Portelius E, *et al.* Neurogranin and YKL-40: independent markers of synaptic degeneration and neuroinflammation in Alzheimer's disease. Alzheimers Res Ther 2015; 7: 74.
[http://dx.doi.org/10.1186/s13195-015-0161-y] [PMID: 26698298]

[86]   Janelidze S, Hertze J, Zetterberg H, *et al.* Cerebrospinal fluid neurogranin and YKL-40 as biomarkers of Alzheimer's disease. Ann Clin Transl Neurol 2015; 3(1): 12-20.
[http://dx.doi.org/10.1002/acn3.266] [PMID: 26783546]

[87]   Lista S, Toschi N, Baldacci F, *et al.* Cerebrospinal fluid neurogranin as a biomarker of neurodegenerative diseases: a crosssectional study. J Alzheimers Dis 2017; 59(4): 1327-34.

[http://dx.doi.org/10.3233/JAD-170368] [PMID: 28731449]

[88]   Ksiezak-Reding H, Binder LI, Yen SH. Immunochemical and biochemical characterization of tau proteins in normal and Alzheimer's disease brains with Alz 50 and Tau-1. J Biol Chem 1988; 263(17): 7948-53.
[PMID: 3131334]

[89]   Wolozin B, Davies P. Alzheimer-related neuronal protein A68: specificity and distribution. Ann Neurol 1987; 22(4): 521-6.
[http://dx.doi.org/10.1002/ana.410220412] [PMID: 3435070]

[90]   Ishiguro K, Ohno H, Arai H, *et al.* Phosphorylated tau in human cerebrospinal fluid is a diagnostic marker for Alzheimer's disease. Neurosci Lett 1999; 270(2): 91-4.
[http://dx.doi.org/10.1016/S0304-3940(99)00476-0] [PMID: 10462105]

[91]   Kohnken R, Buerger K, Zinkowski R, *et al.* Detection of tau phosphorylated at threonine 231 in cerebrospinal fluid of Alzheimer's disease patients. Neurosci Lett 2000; 287(3): 187-90.
[http://dx.doi.org/10.1016/S0304-3940(00)01178-2] [PMID: 10863026]

[92]   Hu YY, He SS, Wang X, *et al.* Levels of nonphosphorylated and phosphorylated tau in cerebrospinal fluid of Alzheimer's disease patients : an ultrasensitive bienzyme-substrate-recycle enzyme-linked immunosorbent assay. Am J Pathol 2002; 160(4): 1269-78.
[http://dx.doi.org/10.1016/S0002-9440(10)62554-0] [PMID: 11943712]

[93]   Motter R, Vigo-Pelfrey C, Kholodenko D, *et al.* Reduction of beta-amyloid peptide42 in the cerebrospinal fluid of patients with Alzheimer's disease. Ann Neurol 1995; 38(4): 643-8.
[http://dx.doi.org/10.1002/ana.410380413] [PMID: 7574461]

[94]   Portelius E, Tran AJ, Andreasson U, *et al.* Characterization of amyloid beta peptides in cerebrospinal fluid by an automated immunoprecipitation procedure followed by mass spectrometry. J Proteome Res 2007; 6(11): 4433-9.
[http://dx.doi.org/10.1021/pr0703627] [PMID: 17927230]

[95]   Shoji M, Matsubara E, Kanai M, *et al.* Combination assay of CSF tau, A beta 1-40 and A beta 1-42(43) as a biochemical marker of Alzheimer's disease. J Neurol Sci 1998; 158(2): 134-40.
[http://dx.doi.org/10.1016/S0022-510X(98)00122-1] [PMID: 9702683]

[96]   Hansson O, Zetterberg H, Buchhave P, *et al.* Prediction of Alzheimer's disease using the CSF Abeta42/Abeta40 ratio in patients with mild cognitive impairment. Dement Geriatr Cogn Disord 2007; 23(5): 316-20.
[http://dx.doi.org/10.1159/000100926] [PMID: 17374949]

[97]   Lewczuk P, Esselmann H, Otto M, *et al.* Neurochemical diagnosis of Alzheimer's dementia by CSF Abeta42, Abeta42/Abeta40 ratio and total tau. Neurobiol Aging 2004; 25(3): 273-81.
[http://dx.doi.org/10.1016/S0197-4580(03)00086-1] [PMID: 15123331]

[98]   Wiltfang J, Esselmann H, Bibl M, *et al.* Amyloid beta peptide ratio 42/40 but not A beta 42 correlates with phospho-Tau in patients with low- and high-CSF A beta 40 load. J Neurochem 2007; 101(4): 1053-9.
[http://dx.doi.org/10.1111/j.1471-4159.2006.04404.x] [PMID: 17254013]

[99]   Parsons CG, Danysz W, Dekundy A, Pulte I. Memantine and cholinesterase inhibitors: complementary mechanisms in the treatment of Alzheimer's disease. Neurotox Res 2013; 24(3): 358-69.
[http://dx.doi.org/10.1007/s12640-013-9398-z] [PMID: 23657927]

[100] Gulati A, Hornick MG, Briyal S, Lavhale MS. A novel neuroregenerative approach using ET(B) receptor agonist, IRL-1620, to treat CNS disorders. Physiol Res 2018; 67 (Suppl. 1): S95-S113.
[http://dx.doi.org/10.33549/physiolres.933859] [PMID: 29947531]

[101] van Dyck CH. Anti-Amyloid-β Monoclonal Antibodies for Alzheimer's Disease: Pitfalls and Promise. Biol Psychiatry 2018; 83(4): 311-9.
[http://dx.doi.org/10.1016/j.biopsych.2017.08.010] [PMID: 28967385]

[102] Jarrett JT, Berger EP, Lansbury PT Jr. The carboxy terminus of the beta amyloid protein is critical for the seeding of amyloid formation: implications for the pathogenesis of Alzheimer's disease. Biochemistry 1993; 32(18): 4693-7.
[http://dx.doi.org/10.1021/bi00069a001] [PMID: 8490014]

[103] Stachel SJ. Progress toward the development of a viable BACE□1 inhibitor. Drug Dev Res 2009; 70(2): 101-10.
[http://dx.doi.org/10.1002/ddr.20289]

[104] Turner RT III, Koelsch G, Hong L, *et al.* Subsite specificity of memapsin 2 (β-secretase): implications for inhibitor design. Biochemistry 2001; 40(34): 10001-6.
[http://dx.doi.org/10.1021/bi015546s] [PMID: 11513577]

[105] Hussain I, Hawkins J, Harrison D, *et al.* Oral administration of a potent and selective non-peptidic BACE-1 inhibitor decreases β-cleavage of amyloid precursor protein and amyloid-β production in vivo. J Neurochem 2007; 100(3): 802-9.
[http://dx.doi.org/10.1111/j.1471-4159.2006.04260.x] [PMID: 17156133]

[106] Ahn K, Shelton CC, Tian Y, *et al.* Activation and intrinsic γ-secretase activity of presenilin 1. Proc Natl Acad Sci USA 2010; 107(50): 21435-40.
[http://dx.doi.org/10.1073/pnas.1013246107] [PMID: 21115843]

[107] Lessard CB, Wagner SL, Koo EH. And four equals one: presenilin takes the γ-secretase role by itself. Proc Natl Acad Sci USA 2010; 107(50): 21236-7.
[http://dx.doi.org/10.1073/pnas.1016284108] [PMID: 21135249]

[108] Li Y-M, Xu M, Lai M-T, *et al.* Photoactivated γ-secretase inhibitors directed to the active site covalently label presenilin 1. Nature 2000; 405(6787): 689-94.
[http://dx.doi.org/10.1038/35015085] [PMID: 10864326]

[109] Fleisher AS, Raman R, Siemers ER, *et al.* Phase 2 safety trial targeting amyloid β production with a γ-secretase inhibitor in Alzheimer disease. Arch Neurol 2008; 65(8): 1031-8.
[http://dx.doi.org/10.1001/archneur.65.8.1031] [PMID: 18695053]

[110] Doody RS, Raman R, Farlow M, *et al.* A phase 3 trial of semagacestat for treatment of Alzheimer's disease. N Engl J Med 2013; 369(4): 341-50.
[http://dx.doi.org/10.1056/NEJMoa1210951] [PMID: 23883379]

[111] Panelos J, Massi D. Emerging role of Notch signaling in epidermal differentiation and skin cancer. Cancer Biol Ther 2009; 8(21): 1986-93.
[http://dx.doi.org/10.4161/cbt.8.21.9921] [PMID: 19783903]

[112] Okochi M, Steiner H, Fukumori A, *et al.* Presenilins mediate a dual intramembranous γ-secretase cleavage of Notch-1. EMBO J 2002; 21(20): 5408-16.
[http://dx.doi.org/10.1093/emboj/cdf541] [PMID: 12374741]

[113] Fortini ME. γ-secretase-mediated proteolysis in cell-surface-receptor signalling. Nat Rev Mol Cell Biol 2002; 3(9): 673-84.
[http://dx.doi.org/10.1038/nrm910] [PMID: 12209127]

[114] Kopan R, Goate A. Aph-2/Nicastrin: an essential component of γ-secretase and regulator of Notch signaling and Presenilin localization. Neuron 2002; 33(3): 321-4.
[http://dx.doi.org/10.1016/S0896-6273(02)00585-8] [PMID: 11832221]

[115] Blennow K, Zetterberg H. Biomarkers for Alzheimer's disease: current status and prospects for the future. J Intern Med 2018; 284(6): 643-63.
[http://dx.doi.org/10.1111/joim.12816] [PMID: 30051512]

[116] Olsson A, Vanderstichele H, Andreasen N, *et al.* Simultaneous measurement of beta-amyloid(1-42), total tau, and phosphorylated tau (Thr181) in cerebrospinal fluid by the xMAP technology. Clin Chem 2005; 51(2): 336-45.
[http://dx.doi.org/10.1373/clinchem.2004.039347] [PMID: 15563479]

[117] Boutajangout A, Ingadottir J, Davies P, Sigurdsson EM. Passive immunization targeting pathological phospho-tau protein in a mouse model reduces functional decline and clears tau aggregates from the brain. J Neurochem 2011; 118(4): 658-67.
[http://dx.doi.org/10.1111/j.1471-4159.2011.07337.x] [PMID: 21644996]

[118] Yanamandra K, Jiang H, Mahan TE, *et al.* Anti-tau antibody reduces insoluble tau and decreases brain atrophy. Ann Clin Transl Neurol 2015; 2(3): 278-88.
[http://dx.doi.org/10.1002/acn3.176] [PMID: 25815354]

[119] Yanamandra K, Kfoury N, Jiang H, *et al.* Anti-tau antibodies that block tau aggregate seeding in vitro markedly decrease pathology and improve cognition in vivo. Neuron 2013; 80(2): 402-14.
[http://dx.doi.org/10.1016/j.neuron.2013.07.046] [PMID: 24075978]

[120] Castillo-Carranza DL, Sengupta U, Guerrero-Muñoz MJ, *et al.* Passive immunization with Tau oligomer monoclonal antibody reverses tauopathy phenotypes without affecting hyperphosphorylated neurofibrillary tangles. J Neurosci 2014; 34(12): 4260-72.
[http://dx.doi.org/10.1523/JNEUROSCI.3192-13.2014] [PMID: 24647946]

[121] Chai X, Wu S, Murray TK, *et al.* Passive immunization with anti-Tau antibodies in two transgenic models: reduction of Tau pathology and delay of disease progression. J Biol Chem 2011; 286(39): 34457-67.
[http://dx.doi.org/10.1074/jbc.M111.229633] [PMID: 21841002]

[122] d'Abramo C, Acker CM, Jimenez HT, Davies P. Tau passive immunotherapy in mutant P301L mice: antibody affinity versus specificity. PLoS One 2013; 8(4)e62402
[http://dx.doi.org/10.1371/journal.pone.0062402] [PMID: 23638068]

[123] Gu J, Congdon EE, Sigurdsson EM. Two novel Tau antibodies targeting the 396/404 region are primarily taken up by neurons and reduce Tau protein pathology. J Biol Chem 2013; 288(46): 33081-95.
[http://dx.doi.org/10.1074/jbc.M113.494922] [PMID: 24089520]

[124] Kontsekova E, Zilka N, Kovacech B, Novak P, Novak M. First-in-man tau vaccine targeting structural determinants essential for pathological tau-tau interaction reduces tau oligomerisation and neurofibrillary degeneration in an Alzheimer's disease model. Alzheimers Res Ther 2014; 6(4): 44.
[http://dx.doi.org/10.1186/alzrt278] [PMID: 25478017]

[125] Walls KC, Ager RR, Vasilevko V, Cheng D, Medeiros R, LaFerla FM. p-Tau immunotherapy reduces soluble and insoluble tau in aged 3xTg-AD mice. Neurosci Lett 2014; 575: 96-100.
[http://dx.doi.org/10.1016/j.neulet.2014.05.047] [PMID: 24887583]

[126] Braak H, Braak E. Neuropathological stageing of Alzheimer-related changes. Acta Neuropathol 1991; 82(4): 239-59.
[http://dx.doi.org/10.1007/BF00308809] [PMID: 1759558]

[127] Luna-Muñoz J, Chávez-Macías L, García-Sierra F, Mena R. Earliest stages of tau conformational changes are related to the appearance of a sequence of specific phospho-dependent tau epitopes in Alzheimer's disease. J Alzheimers Dis 2007; 12(4): 365-75.
[http://dx.doi.org/10.3233/JAD-2007-12410] [PMID: 18198423]

[128] Luna-Muñoz J, García-Sierra F, Falcón V, Menéndez I, Chávez-Macías L, Mena R. Regional conformational change involving phosphorylation of tau protein at the Thr231, precedes the structural change detected by Alz-50 antibody in Alzheimer's disease. J Alzheimers Dis 2005; 8(1): 29-41.
[http://dx.doi.org/10.3233/JAD-2005-8104] [PMID: 16155347]

[129] Zilka N, Kovacech B, Barath P, Kontsekova E, Novák M. The self-perpetuating tau truncation circle. Biochem Soc Trans 2012; 40(4): 681-6.
[http://dx.doi.org/10.1042/BST20120015] [PMID: 22817716]

[130] Zhou Y, Shi J, Chu D, *et al.* Relevance of Phosphorylation and Truncation of Tau to the Etiopathogenesis of Alzheimer's Disease. Front Aging Neurosci 2018; 10(27): 27.

[http://dx.doi.org/10.3389/fnagi.2018.00027] [PMID: 29472853]

[131] Courade JP, Angers R, Mairet-Coello G, *et al.* Epitope determines efficacy of therapeutic anti-Tau antibodies in a functional assay with human Alzheimer Tau. Acta Neuropathol 2018; 136(5): 729-45.
[http://dx.doi.org/10.1007/s00401-018-1911-2] [PMID: 30238240]

[132] Vandermeeren M, Borgers M, Van Kolen K, *et al.* Anti-Tau Monoclonal Antibodies Derived from Soluble and Filamentous Tau Show Diverse Functional Properties in vitro and in vivo. J Alzheimers Dis 2018; 65(1): 265-81.
[http://dx.doi.org/10.3233/JAD-180404] [PMID: 30040731]

[133] Ransohoff RM. How neuroinflammation contributes to neurodegeneration. Science 2016; 353(6301): 777-83.
[http://dx.doi.org/10.1126/science.aag2590] [PMID: 27540165]

[134] Bondi MW, Edmonds EC, Salmon DP. Alzheimer's Disease: Past, Present, and Future. J Int Neuropsychol Soc 2017; 23(9-10): 818-31.
[http://dx.doi.org/10.1017/S135561771700100X] [PMID: 29198280]

[135] Michel PP, Hirsch EC, Hunot S. Understanding Dopaminergic Cell Death Pathways in Parkinson Disease. Neuron 2016; 90(4): 675-91.
[http://dx.doi.org/10.1016/j.neuron.2016.03.038] [PMID: 27196972]

[136] Prentice H, Modi JP, Wu JY. Mechanisms of Neuronal Protection against Excitotoxicity, Endoplasmic Reticulum Stress, and Mitochondrial Dysfunction in Stroke and Neurodegenerative Diseases. Oxid Med Cell Longev 2015; 2015(10)964518
[http://dx.doi.org/10.1155/2015/964518] [PMID: 26576229]

[137] Wirths O. Altered neurogenesis in mouse models of Alzheimer disease. Neurogenesis (Austin) 2017; 4(1)e1327002
[http://dx.doi.org/10.1080/23262133.2017.1327002] [PMID: 29564360]

[138] Cadonic C, Sabbir MG, Albensi BC. Mechanisms of Mitochondrial Dysfunction in Alzheimer's Disease. Mol Neurobiol 2016; 53(9): 6078-90.
[http://dx.doi.org/10.1007/s12035-015-9515-5] [PMID: 26537901]

[139] Abramov AY, Berezhnov AV, Fedotova EI, Zinchenko VP, Dolgacheva LP. Interaction of misfolded proteins and mitochondria in neurodegenerative disorders. Biochem Soc Trans 2017; 45(4): 1025-33.
[http://dx.doi.org/10.1042/BST20170024] [PMID: 28733489]

[140] Swerdlow RH. Mitochondria and Mitochondrial Cascades in Alzheimer's Disease. J Alzheimers Dis 2018; 62(3): 1403-16.
[http://dx.doi.org/10.3233/JAD-170585] [PMID: 29036828]

[141] Desler C, Lillenes MS, Tønjum T, Rasmussen LJ. The Role of Mitochondrial Dysfunction in the Progression of Alzheimer's Disease. Curr Med Chem 2018; 25(40): 5578-87.
[http://dx.doi.org/10.2174/0929867324666170616110111] [PMID: 28618998]

[142] Angelova PR, Abramov AY. Role of mitochondrial ROS in the brain: from physiology to neurodegeneration. FEBS Lett 2018; 592(5): 692-702.
[http://dx.doi.org/10.1002/1873-3468.12964] [PMID: 29292494]

[143] Butterfield DA, Boyd-Kimball D. Oxidative Stress, Amyloid-β Peptide, and Altered Key Molecular Pathways in the Pathogenesis and Progression of Alzheimer's Disease. J Alzheimers Dis 2018; 62(3): 1345-67.
[http://dx.doi.org/10.3233/JAD-170543] [PMID: 29562527]

[144] Han XJ, Hu YY, Yang ZJ, *et al.* Amyloid β-42 induces neuronal apoptosis by targeting mitochondria. Mol Med Rep 2017; 16(4): 4521-8.
[http://dx.doi.org/10.3892/mmr.2017.7203] [PMID: 28849115]

[145] Del Prete D, Suski JM, Oulès B, *et al.* Localization and Processing of the Amyloid-β Protein Precursor in Mitochondria-Associated Membranes. J Alzheimers Dis 2017; 55(4): 1549-70.

[http://dx.doi.org/10.3233/JAD-160953] [PMID: 27911326]

[146] Nitzan K, Benhamron S, Valitsky M, *et al.* Mitochondrial Transfer Ameliorates Cognitive Deficits, Neuronal Loss, and Gliosis in Alzheimer's Disease Mice. J Alzheimers Dis 2019; 72(2): 587-604.
[http://dx.doi.org/10.3233/JAD-190853] [PMID: 31640104]

[147] Mattson MP. ER calcium and Alzheimer's disease: in a state of flux. Sci Signal 2010; 3(114): pe10.
[http://dx.doi.org/10.1126/scisignal.3114pe10] [PMID: 20332425]

[148] Zampese E, Fasolato C, Kipanyula MJ, Bortolozzi M, Pozzan T, Pizzo P. Presenilin 2 modulates endoplasmic reticulum (ER)-mitochondria interactions and Ca2+ cross-talk. Proc Natl Acad Sci USA 2011; 108(7): 2777-82.
[http://dx.doi.org/10.1073/pnas.1100735108] [PMID: 21285369]

[149] Ca 2+ dysregulation mediated by presenilins in familial Alzheimer's disease: causing or modulating factor. Curr Trends Neurol 2009; 3: 1-14.

[150] Veeresh P, Kaur H, Sarmah D, *et al.* Endoplasmic reticulum-mitochondria crosstalk: from junction to function across neurological disorders. Ann N Y Acad Sci 2019; 1457(1): 41-60.
[http://dx.doi.org/10.1111/nyas.14212] [PMID: 31460675]

[151] Leal NS, Schreiner B, Pinho CM, *et al.* Mitofusin-2 knockdown increases ER-mitochondria contact and decreases amyloid β-peptide production. J Cell Mol Med 2016; 20(9): 1686-95.
[http://dx.doi.org/10.1111/jcmm.12863] [PMID: 27203684]

[152] Boland B, Kumar A, Lee S, *et al.* Autophagy induction and autophagosome clearance in neurons: relationship to autophagic pathology in Alzheimer's disease. J Neurosci 2008; 28(27): 6926-37.
[http://dx.doi.org/10.1523/JNEUROSCI.0800-08.2008] [PMID: 18596167]

[153] Spilman P, Podlutskaya N, Hart MJ, *et al.* Inhibition of mTOR by rapamycin abolishes cognitive deficits and reduces amyloid-beta levels in a mouse model of Alzheimer's disease. PLoS One 2010; 5(4)e9979
[http://dx.doi.org/10.1371/journal.pone.0009979] [PMID: 20376313]

[154] Tian Y, Bustos V, Flajolet M, Greengard P. A small-molecule enhancer of autophagy decreases levels of Abeta and APP-CTF via Atg5-dependent autophagy pathway. FASEB J 2011; 25(6): 1934-42.
[http://dx.doi.org/10.1096/fj.10-175158] [PMID: 21368103]

[155] Vingtdeux V, Chandakkar P, Zhao H, d'Abramo C, Davies P, Marambaud P. Novel synthetic small-molecule activators of AMPK as enhancers of autophagy and amyloid-β peptide degradation. FASEB J 2011; 25(1): 219-31.
[http://dx.doi.org/10.1096/fj.10-167361] [PMID: 20852062]

[156] Caccamo A, Majumder S, Richardson A, Strong R, Oddo S. Molecular interplay between mammalian target of rapamycin (mTOR), amyloid-beta, and Tau: effects on cognitive impairments. J Biol Chem 2010; 285(17): 13107-20.
[http://dx.doi.org/10.1074/jbc.M110.100420] [PMID: 20178983]

[157] Wang Y, Martinez-Vicente M, Krüger U, *et al.* Tau fragmentation, aggregation and clearance: the dual role of lysosomal processing. Hum Mol Genet 2009; 18(21): 4153-70.
[http://dx.doi.org/10.1093/hmg/ddp367] [PMID: 19654187]

[158] Nilsson P, Loganathan K, Sekiguchi M, *et al.* Aβ secretion and plaque formation depend on autophagy. Cell Rep 2013; 5(1): 61-9.
[http://dx.doi.org/10.1016/j.celrep.2013.08.042] [PMID: 24095740]

[159] Bustos V, Pulina MV, Bispo A, *et al.* Phosphorylated Presenilin 1 decreases β-amyloid by facilitating autophagosome-lysosome fusion. Proc Natl Acad Sci USA 2017; 114(27): 7148-53.
[http://dx.doi.org/10.1073/pnas.1705240114] [PMID: 28533369]

[160] Tammineni P, Ye X, Feng T, Aikal D, Cai Q. Impaired retrograde transport of axonal autophagosomes contributes to autophagic stress in Alzheimer's disease neurons. eLife 2017; 6e21776
[http://dx.doi.org/10.7554/eLife.21776] [PMID: 28085665]

[161] Yang C, Cai CZ, Song JX, *et al.* NRBF2 is involved in the autophagic degradation process of APP-CTFs in Alzheimer disease models. Autophagy 2017; 13(12): 2028-40.
[http://dx.doi.org/10.1080/15548627.2017.1379633] [PMID: 28980867]

[162] Zhang L, Fang Y, Cheng X, *et al.* TRPML1 Participates in the Progression of Alzheimer's Disease by Regulating the PPARγ/AMPK/Mtor Signalling Pathway. Cell Physiol Biochem 2017; 43(6): 2446-56.
[http://dx.doi.org/10.1159/000484449] [PMID: 29131026]

[163] Ulland TK, Song WM, Huang SC, *et al.* TREM2 Maintains Microglial Metabolic Fitness in Alzheimer's Disease. Cell 2017; 170(4): 649-663.e13.
[http://dx.doi.org/10.1016/j.cell.2017.07.023] [PMID: 28802038]

[164] Chen K, Yuan R, Geng S, *et al.* Toll-interacting protein deficiency promotes neurodegeneration via impeding autophagy completion in high-fat diet-fed ApoE$^{-/-}$ mouse model. Brain Behav Immun 2017; 59: 200-10.
[http://dx.doi.org/10.1016/j.bbi.2016.10.002] [PMID: 27720815]

[165] Morishima-Kawashima M, Hasegawa M, Takio K, *et al.* Proline-directed and non-proline-directed phosphorylation of PHF-tau. J Biol Chem 1995; 270(2): 823-9.
[http://dx.doi.org/10.1074/jbc.270.2.823] [PMID: 7822317]

[166] Seubert P, Mawal-Dewan M, Barbour R, *et al.* Detection of phosphorylated Ser262 in fetal tau, adult tau, and paired helical filament tau. J Biol Chem 1995; 270(32): 18917-22.
[http://dx.doi.org/10.1074/jbc.270.32.18917] [PMID: 7642549]

[167] Hanger DP, Byers HL, Wray S, *et al.* Novel phosphorylation sites in tau from Alzheimer brain support a role for casein kinase 1 in disease pathogenesis. J Biol Chem 2007; 282(32): 23645-54.
[http://dx.doi.org/10.1074/jbc.M703269200] [PMID: 17562708]

[168] Kumar D, Kumar P. Aβ, Tau, and α-Synuclein aggregation and integrated role of PARK2 in the regulation and clearance of toxic peptides. Neuropeptides 2019; 78(101971)101971
[http://dx.doi.org/10.1016/j.npep.2019.101971] [PMID: 31540705]

[169] Jha NK, Jha SK, Kar R, Nand P, Swati K, Goswami VK. Nuclear factor-kappa β as a therapeutic target for Alzheimer's disease. J Neurochem 2019; 150(2): 113-37.
[http://dx.doi.org/10.1111/jnc.14687] [PMID: 30802950]

[170] Chinchalongporn V, Shukla M, Govitrapong P. Melatonin ameliorates Aβ$_{42}$ -induced alteration of βAPP-processing secretases via the melatonin receptor through the Pin1/GSK3β/NF-κB pathway in SH-SY5Y cells. J Pineal Res 2018; 64(4)e12470
[http://dx.doi.org/10.1111/jpi.12470] [PMID: 29352484]

[171] Wang WX, Rajeev BW, Stromberg AJ, *et al.* The expression of microRNA miR-107 decreases early in Alzheimer's disease and may accelerate disease progression through regulation of beta-site amyloid precursor protein-cleaving enzyme 1. J Neurosci 2008; 28(5): 1213-23.
[http://dx.doi.org/10.1523/JNEUROSCI.5065-07.2008] [PMID: 18234899]

[172] Absalon S, Kochanek DM, Raghavan V, Krichevsky AM. MiR-26b, upregulated in Alzheimer's disease, activates cell cycle entry, tau-phosphorylation, and apoptosis in postmitotic neurons. J Neurosci 2013; 33(37): 14645-59.
[http://dx.doi.org/10.1523/JNEUROSCI.1327-13.2013] [PMID: 24027266]

[173] Sarkar S, Jun S, Rellick S, Quintana DD, Cavendish JZ, Simpkins JW. Expression of microRNA-34a in Alzheimer's disease brain targets genes linked to synaptic plasticity, energy metabolism, and resting state network activity. Brain Res 2016; 1646: 139-51.
[http://dx.doi.org/10.1016/j.brainres.2016.05.026] [PMID: 27235866]

[174] Lau P, Bossers K, Janky R, *et al.* Alteration of the microRNA network during the progression of Alzheimer's disease. EMBO Mol Med 2013; 5(10): 1613-34.
[http://dx.doi.org/10.1002/emmm.201201974] [PMID: 24014289]

[175] Cogswell JP, Ward J, Taylor IA, *et al.* Identification of miRNA changes in Alzheimer's disease brain

and CSF yields putative biomarkers and insights into disease pathways. J Alzheimers Dis 2008; 14(1): 27-41.
[http://dx.doi.org/10.3233/JAD-2008-14103] [PMID: 18525125]

[176] Müller M, Kuiperij HB, Claassen JA, Küsters B, Verbeek MM. MicroRNAs in Alzheimer's disease: differential expression in hippocampus and cell-free cerebrospinal fluid. Neurobiol Aging 2014; 35(1): 152-8.
[http://dx.doi.org/10.1016/j.neurobiolaging.2013.07.005] [PMID: 23962497]

CHAPTER 7

Metals Linked to Alzheimer's Disease

Hani Nasser Abdelhamid[1,*]

[1] Advanced Multifunctional Materials Laboratory, Department of Chemistry, Assiut University, Assiut 71516, Egypt

Abstract: Exposure to metals including copper, zinc, aluminum, and iron ions occurs inevitably. Any disturbance in metal homeostasis develops diseases and abnormalities. Metal ions undergo an electric charge balance *via* gaining or losing electrons from surrounded biomolecules. They bind to amyloid fibrils or tau proteins in the brain in a way that links to the development of neurodegenerative diseases including Alzheimer's disease (AD). For several decades, scientists have been exploring possible links between metals imbalance and AD. However, very little is known about the exact mechanisms governing the links of metals to AD. This book chapter summarizes recent thoughts in the research studies that focus on the links between metals and AD. Most of the current results suggested that metal binding to amyloid binds affects the architecture of the protein fibrils and rate of propagation.

Keywords: Alzheimer's disease, Bioinorganic, Metals, Neurodegenerative diseases.

INTRODUCTION

Dementia, including Alzheimer's disease (AD), is a syndrome in which there is deterioration in memory, learning functions, and performance of normal activities [1, 2]. Alzheimer's disease (AD) is a progressive, and chronic neurodegenerative disorder of the brain [1]. According to WHO, it is considered as the most common (60–70%) cause of dementia in elderly people age over 65 years old [1]. It was estimated that 5.3 million people were living with AD, and the patient's number will rise to 13.8 million by 2050. In 2019, the direct costs of caring for people with AD and other dementias are estimated to be $233.9 billion [3]. Patient of AD at the early stage of dementia suffer from forgetfulness, losing track of the time, and becoming lost in familiar places. These symptoms are progressed to becoming unaware of the time and place, and having difficulty recognizing relatives and friends. Several factors such as mutations in the β-amyloid (Aβ) rel-

* **Corresponding author Hani Nasser Abdelhamid:** Advanced Multifunctional Materials Laboratory, Department of Chemistry, Assiut University, Assiut 71516, Egypt; Tel: +20102995242; E-mails: hany.abdelhamid@aun.edu.eg and chemist.hani@yahoo.com

Atta-ur-Rahman (Ed.)
All rights reserved-© 2020 Bentham Science Publishers

ated genes amyloid protein precursor (APP), presenilin-1 (PSEN1), and presenilin-2 (PSEN2) (<1% of total AD cases), diabetes, hypertension, sleep disorder, and psychological stress may increase the risk of AD [4].

Despite decades of research, scientists still don't fully understand what causes the neurodegenerative disease, although metals are expected to play a significant role in dementia [5]. Metals are ubiquities and can be found in our daily nutrition [6]. Modern fields such as metallobiology are punctuated by diseases that have specific abnormalities in metals or metal transport proteins, such as acrodermatitis enteropathica (failure of zinc absorption across the intestine) [7], amyotrophic lateral sclerosis (ALS, where Cu/Zn superoxide dismutase 1 plays a role) [8], and Wilson's disease (mutations in the copper transport gene, adenosine triphosphatase ATPase, *ATP7B*) [9]. The links between metals imbalance and neurodegenerative disorders including AD are unknown. Several endogenous proteins are involved in metal transportation on body [10]. The interactions of metal ions with amyloid precursor protein (APP), and Aβ enhance reactive oxygen species (ROS) formation, leading to Aβ over-production, aggregation, and enhanced neurotoxicity [11].

This book chapter summarizes the links between metals and Alzheimer's disease [12 - 14]. Different metals such as copper, iron, zinc, magnesium, calcium [15], aluminum [15], magnesium [16], and selenium [17], showed a link to AD [18]. These metals can be essential biometals (*e.g.*, iron, zinc, copper, manganese, magnesium, and calcium), or neurotoxic metals (*e.g.*, aluminum, lead, cadmium, and mercury) [11, 19].

KEY PROTEINS IN THE BRAIN

Aβ and tau are two key proteins in the brain. Misfolding and aggregation of Aβ or tau proteins leading to the formation of plaques, and neurofibrillary tangles (NFTs), respectively.

The predominant Aβ peptide isoforms comprise 40–42 amino acids chains (sequence: AEFRHDSGYEVHHQKLVFFAEDVGSNKGAIIGLMVGGVVIA) [20, 21]. Amyloid is misfolding, and aggregation of amyloid monomers with ordered and stabled structures [22]. Amyloids are typically hard, and waxy deposits consisting of protein. Normally, these folded proteins are cleared from the brain. In contrast, those who suffer from Alzheimer's disease have an unusual amount of sticky amyloid fibrils in their brains. Due to ageing, these fibrils accumulate instead of decomposition, and increasingly hamper functions of the brain.

Tau proteins (τ-proteins) are abundant proteins in neurons of the central nervous system. The defect in tau proteins prevents stabilization of microtubules properly causing neurodegenerative diseases such as Alzheimer's disease.

LINKS BETWEEN METALS AND AD

The exposure of human to metals occurs daily. Metals are present naturally in the body and can be provided *via* external sources. They can be classified to biological and non-biological metals (Fig. **1**). The links between metals and AD for both metals are summarized in Fig. (**1**), and discussed in the following sections.

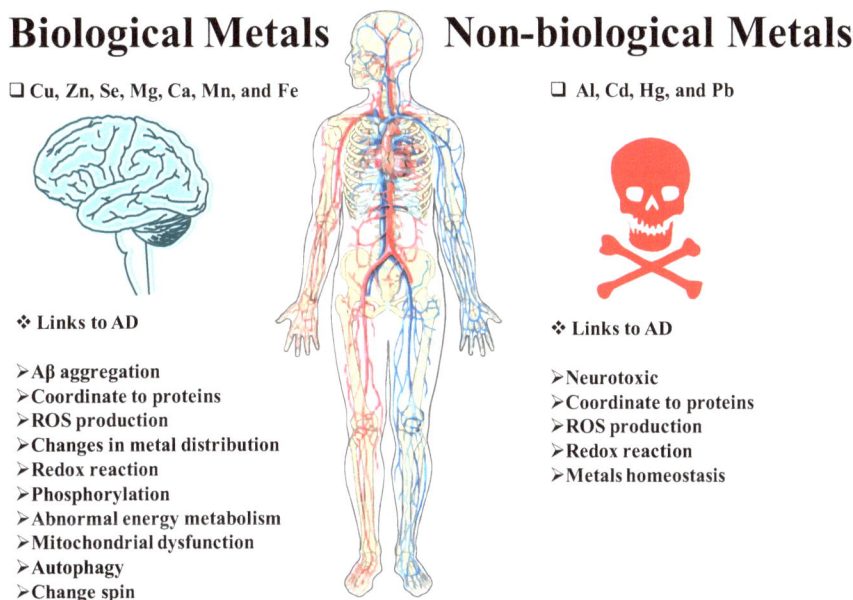

Biological Metals   **Non-biological Metals**

❑ **Cu, Zn, Se, Mg, Ca, Mn, and Fe**   ❑ **Al, Cd, Hg, and Pb**

❖ **Links to AD**

➢ Aβ aggregation
➢ Coordinate to proteins
➢ ROS production
➢ Changes in metal distribution
➢ Redox reaction
➢ Phosphorylation
➢ Abnormal energy metabolism
➢ Mitochondrial dysfunction
➢ Autophagy
➢ Change spin

❖ **Links to AD**

➢ Neurotoxic
➢ Coordinate to proteins
➢ ROS production
➢ Redox reaction
➢ Metals homeostasis

Fig. (1). Overview of human exposure to metals causing AD and their links.

BIOMETALS LINK TO ALZHEIMER'S DISEASE

Calcium and Magnesium

Calcium (Ca), and magnesium (Mg) ions serve as messengers and display role in controlling cellular function, neuronal growth, and signal transmission [23].

Excessive Ca^{2+} in the endoplasmic reticulum (ER) is released *via* type 2 ryanodine receptors (RyanR2) in AD spines due to increase in expression and function of RyanR2 [15]. Store-operated Ca^{2+} entry (nSOC) pathway is disrupted in AD spines due to down regulation of stromal interaction molecule 2 (STIM2) proteins. Because of these Ca^{2+} signaling abnormalities, imbalance in activities of Ca^{2+}-calmodulin dependent kinase II (CaMKII), and Ca^{2+}-dependent phosphatase calcineurin (CaN) takes place at the synapse. As a result, synapses are weakened, and eliminated in AD brains *via* long-term depression (LTD) mechanism, causing memory loss. The increase of Ca levels increases the expression of APP and ApoE facilitating the formation of Aβ aggregation through a mechanism involving the stabilization of γ-secretase [24], and reciprocally, Aβ aggregation alters membrane Ca permeability that further worsens AD [25].

Magnesium (Mg) is an abundant divalent cation in human cells. Mg ions are an antagonist of Ca ions. It shows antioxidant, and neuroprotective properties [16]. The insufficient level of Mg causes AD [26]. The low level of Mg is due to several reasons including high intake of a neurotoxic metal, such as aluminum (Al), which inhibits the activity of Mg-requiring enzymes [27].

Copper

Copper is involved in cellular respiration, free radical defense, and neurotransmitter synthesis [28, 29]. It can be found in transporter such as copper transporter 1 (CTR1), and the copper transporting P-type ATPases, such as ATP7A and ATP7B. These are the major transporters involved in the cellular regulation of monovalent copper [30, 31]. Copper ions are a redox active element. It can be reduced from Cu^{2+} to Cu^{+} *via* binding of the N-terminal domain of APP [32]. Divalent metal transporter 1 (DMT1) may participate in delivering Cu^{2+} into cells for the synthesis of copper-based enzymes [33]. Copper is present in a cluster of intracellular proteins, recognized as molecule chaperones such as antioxidant protein-1, cytochrome oxidase enzyme complex, and superoxide dismutase (SOD) [34]. Copper ions are involved in many diseases including ALS [8], and Wilson's disease [9].

Copper levels in all AD brain regions are lower by 52.8–70.2% corresponding control values [35]. Copper deficiency in the brain has an adverse effect on the development and maintenance of myelin and can induce degeneration of the nervous system. Copper can be removed from the brain after reduction from Cu^{2+} to Cu^{+} [36]. Excessive levels of copper generate free radicals, and reactive oxygen species (ROS) [37]. In the presence of Cu^{2+}, $O_2{}^{-}$ generates from glycation of $Aβ_{1-40}$ with sugars converted into hydroxyl radical (HO·) [38]. Excessive level causes also tau phosphorylation *via* the activation of cyclin-dependent kinase 5 (CDK5), and glycogen synthase kinase 3β (GSK3β) pathways [39]. Furthermore, it binds directly to Aβ with a high affinity, and enhances oligomer formation [40]. Copper ions bind to amyloid in several specific ways, and affect their structure, and rate of propagation. Copper induces spin change of heme–Aβ complex from low spin to the detrimental high spin form [41]. It shows also binding to tau proteins, facilitating the formation of NFTs [42].

The toxicity of the Cu-Aβ complex could be reversed by copper chelators, such as clioquinol [43], chitosan, chitooligosaccharide, and their derivatives [44], and 8-hydroxyquinoline analogs [45].

Iron

Iron is an essential transition metal for several neuronal functions in the brain, such as oxygen transport, mitochondrial respiration, and myelin synthesis. Iron is a cofactor for a large number of metalloproteins and importers/exporters, including ferritin (iron storage), transferrin (iron mobilization), DMT1, lactoferrin (Lf), melanotransferrin (MTf), and ferroportin (Fpn), involved in metabolism, and signal transduction. A computational study showed that the iron metabolic network consists of 151 chemical species, 107 reactions and transport steps [46]. The intracellular metabolism of iron may take place through iron-regulatory proteins (IRP) that bind iron-responsive elements (IREs) in regulated messenger RNAs (mRNAs) [47]. Iron levels in the brain are increased with aging. Several studies reported that imbalance in iron may have a link to several neurodegenerative disorders including AD [48]. Iron displays high affinity toward Aβ causing modulation of their redox potential. This causes the aggregation of Aβ. The affinity of Aβ for Fe^{3+}, and Fe^{2+} is 8 orders of magnitude stronger than that of transferrin [49]. Thus, Aβ changes the iron homeostasis and compete with other iron-containing proteins.

Excessive levels of iron cause an over-abundant production of ROS through Fenton reaction. Aβ reduced ROS at the beginning of neurodegeneration [50]. Iron may be loaded around senile plaques, and elevates the production of Aβ *via* increasing the expression of APP [51]. Iron modulates APP transcription through IREs present in the 5′-untranslated region of APP mRNA [52]. High concentration of iron is associated with the dissociation of IRPs from their IRE binding sites and increasing APP translation [53]. Excessive iron causes ferroptosis, which is programmed cell death based on the accumulation of lipid peroxides and regulated necrosis [54]. A report showed that there is a correlation between iron accumulations, and the amount of Aβ plaques, and tau pathology in the same block [55].

Iron promotes aggregation, induces phosphorylation [56], and creates iron-enriched regions in tau proteins [57] causing neurodegenerative diseases *via* impairing APP-mediated iron export [58]. It enhances the activation of CDK5, and GSK3β pathways [56].

Excessive level of iron may be suppressed using iron chelator [59], such as curcumin [60], lithium treatment [61], and lactoferrin (Lf) [62]. Iron chelators attenuate their oxidizing capability, thereby decreasing oxidative stress by reducing free radical levels [60]. Lipoprotein receptor-related protein (LRP) is a cell surface receptor involved in Aβ clearance *via* an endocytic process. Lactoferrin (Lf) can bind to LRP and substantially enhance the clearance of soluble Aβ rather than the production of Aβ [62].

Zinc

Zinc is the second most abundant metal in the body after iron. It can be found in 70% of proteins present in the brain [63]. It exists in 2000 transcription factors and more than 300 enzymes [63]. It can be present as free ionic form (Zn^{2+}) within synaptic vesicles at glutamatergic nerve terminals [64]. It plays vital roles as a component of several enzymes, proteins, zinc transporters (ZnTs), such as zinc-regulated transporter-like, metallothioneins (MTs), and neurotransmitter for cell signaling [65]. Thus, disruption of zinc homeostasis is involved in several diseases such as acrodermatitis enteropathica (failure of zinc absorption across the intestine) [7].

Low level of synaptic zinc affects the proper development of the brain, causes depression-like symptoms, dysfunctions in learning and memory [66], and suppresses cell signaling and synaptic plasticity [67]. The low level of zinc increases the overload of other metals, such as copper, nickel and possibly toxic metals, *via* upregulating zinc transporters [68]. Excessive zinc can be sequestered by the mitochondria and triggers the generation of ROS [69]. ROS promote zinc release from metallothioneins [70]. Literature shows inconsistent observations regarding the link of zinc levels with AD. Several studies showed high levels of zinc [71, 72], while other reports reported no difference or even reduced zinc levels in AD patients [36, 73, 74].

Zinc binds strongly to histidine residues His-13 and His-14 of Aβ compared to other metals such as iron and copper [75]. The high affinity for Aβ causes inhibition of APP ferroxidase activity, thereby increasing the levels of iron and ROS [76]. It causes proteolytic cleavage of Aβ [77]. It can also bind to tau proteins, and modulates phosphorylation by activation of GSK3β, extracellular regulated protein kinase 1/2 (ERK1/2), and c-Jun N-terminal kinase (JNK) [78, 79]. It promotes tau hyperphosphorylation indirectly *via* the inactivation of tau phosphatases protein phosphates 2A (PP2A) [80]. This induces the loss of zinc bioavailability in the synaptic cleft, thus contributing to changes in synaptic plasticity [81]. Zinc inhibits the activity of γ-secretase, which is involved in the generation of Aβ from APP [82]. Chiral penicillamine-capped selenium nanoparticles (l-/d-Pen@Se NPs) could be used to reduce Zn^{2+}-induced intracellular $Aβ_{40}$ fibrillation [83].

Manganese

Manganese (Mn) is a trace element present in human body. It plays important physiological roles for growth and intracellular homeostasis [84]. It is a component of Mn superoxide dismutase (Mn-SOD), which is an antioxidant enzyme that plays an important role in maintaining mitochondria vitality. It forms transporter 1 (MagT1), and transient receptor potential melastatins 6 and 7 (TRPM6/TRPM7), as well as cyclin M (CNNM) transporter. Manganese plays a role as a cofactor for enzymes such as SOD and glutamine synthetase. Amyotrophic lateral sclerosis (ALS) patient showed an elevated level of manganese in spinal cord sections [85].

Several reports showed evidence for a link between Mn overload and neurodegenerative diseases [86]. Manganese displays weak affinity Aβ revealing low effect on plaque formation [87]. A cell surface-localized efflux exporter SLC30A10 was identified to transport both zinc and Mn [88]. Mutation in SLC30A10 showed accumulation of Mn in the brain [88]. Manganese level is higher in the brain of AD patients compared to healthy brains, while the highest level of Mn was detected in the parietal cortex [89]. Mn mediates cytotoxicity *via* over-production of ROS, abnormal energy metabolism, accumulation of intracellular toxic metabolites, mitochondrial dysfunction, the depletion of cellular antioxidant defense, and autophagy [90, 91].

Selenium

Selenium is an essential element for healthy human. It shows neurological function [17]. It exists in selenocysteine, the 21st amino acid. A review article on the role of Se in AD summarized that current studies showed no evidence for a role of Se in the treatment of AD [92]. However, a recent study reported that chiral penicillamine-capped selenium nanoparticles (l-/d-Pen@Se NPs) showed inhibition of chiral Aβ [83]. The role of Se in AD is still like most other metals which are not fully understood [93].

Non-biological Metals

Heavy metals such as aluminum, cadmium, lead, and mercury are neurotoxic elements, and cause neurofibrillary tangles (aggregates of hyperphosphorylated tau protein), and aggregation Aβ [94].

Aluminum

Aluminum is ubiquitous element, and is widely used in daily life. High levels of Al affect three important genes; APP, presenilin-1, and presenilin-2 [95]. In 1965, researchers found that rabbits injected with an extremely high dose of aluminum cause tangles to tau in brains. This leads to speculation that the released aluminum from food containers/cookware, and even the water pipelines could cause dementia. Other studies have reported that Al reduced the activity of some key enzymes related to Aβ catabolism by activating the amyloidogenic pathway [96]. It affects the expression of iron-binding proteins expression with IRE/IRP sequences in their mRNA, causing an increase in iron concentration [97, 98]. Aluminum decreases the enzymatic activity of catalase, SOD, and glutathione peroxidase [99, 100], and glutamine synthetase and inhibition of glutaminase activity in astrocytes [101], and cholinergic systems [102], causing neuronal damage *via* oxidative stress. Aluminum decreases the level of neurotransmitters, such as serotonin, dopamine, glutamate and aspartate [103, 104].

The human body cannot absorb easily the available form of aluminum in food and drink. The maximum absorbed amount is less than 1% of the total amount present in food and drink. Kidneys take the major part of aluminum away. A good coating of the cookware may reduce aluminum exposure, and metal release [105]. Deferoxamine can form chelating bonds to aluminum and reduces its toxicity [106].

Cadmium

Cadmium (Cd) is a highly neurotoxic element [107 - 112]. It is water-soluble element, and can be transported from soil to plants and concentrated in the food chain [113]. Tobacco products increase the risk of cadmium-related morbidities for smokers [114]. Cadmium species are accumulated in the kidney and liver with extremely long half-life of 20–40 years [115 - 117]. A link between cadmium metal and AD is reported [118]. Cadmium shows the aggregation of Aβ plaques [119], and increases self-aggregation of tau [120]. It increases Aβ levels in the brain *via* downregulating the expression of α-secretase (ADAM10) and neutral endopeptidase [121]. Studies showed that levels of blood cadmium in U.S. adults were positively associated with elevated AD mortality 7–13 years later [118]. Cadmium causes activation of various signaling pathways involved in mitochondrial apoptotic signaling pathway [122], mitogen-activated protein kinase (MAPK) [123], and neuronal apoptosis [124]. Thus, it induces inflammation, apoptosis, and oxidative stress. Furthermore, it causes a disturbance in Ca^{2+} homeostasis [125].

Other Metals

Mercury (hydrargyrium = Hg) is highly neurotoxic element with high industrial interests [126, 127]. It crosses easily the blood-brain barrier, the lipid bilayers of cells, and mitochondria with a half-life of 30 to 60 days. Even low level of mercury, it decreases glutathione levels (GSH), and increases oxidative stress [128]. Hg^{2+} ions show high affinity to selenium, and selenoproteins [129]. ALS patients showed an elevated level of mercury in spinal cord sections [85]. Eighty percent (80%) of the reported studies reported that mercury showed significant memory deficits [130].

Lead (Pb) is a well-known neurotoxic element [131]. Lead may be released from scratching cookware aluminum [105]. It can be absorbed into the bloodstream with a half-life of 30 days. It affects neurotransmitter release, disrupting the function of neuro systems [132]. It induces AD-like pathology and disturbs cholesterol metabolism [133]. Lead (Pb) represses the IRE-mediated enhancement of APP translation [53]. It also causes chronic glial activation, together with features overproduction of proinflammatory proteins such as inducible nitric oxide synthase, IL-1β and tumor necrosis factor-α (TNF-α) [134]. Lead exposure results extracellular signal-regulated kinase and protein kinase B (AKT) [135], and abnormal microgliosis by triggering TLR4-MyD88-NF-κB signaling [136].

Treatments

Various methods including inhibiting Aβ aggregation, destabilizing Aβ fibrils, and ROS scavenging have been proposed to treat AD [137]. Several drugs, including tacrine (9-amino-1,2,3,4-tetrahydroacridine)-based scaffolds [138], non-steroidal anti-inflammatory drugs (NSAIDs) [110, 139 - 141], MoS_2 nanoparticles [142], nanotechnology [143], antibiotics [144], and rhenium complexes [145], were reported for AD treatment Fig. (**2**). These drugs can be used as cholinesterase inhibitors [138], and targeting insulin signaling using anti-diabetic, antioxidant, and nutrition supplements [146].

It is also unknown whether reducing metals in the brain *via* drugs or reducing agents would have any effect. Balance level of metals is very important for a healthy brain. Low level of metals can be treated with supplements. While, high level of metals can be removed or inactivated in the brain using a drug or ligand. For instance, calcium can be deactivated by targeting synaptic calcium signaling pathways [15]. Several drugs have shown positive results in human trials on either reducing plaques and/or cognitive decline Fig. (**2**). However, very few drugs have been approved due to significant side effects including headaches, renal failure, or life threatening low calcium. Anti-oxidative supplements such as fruits, and vegetables may be highly effective in attenuating the harmful effects of oxidative stress [147]. Reagents such as vitamin C chelate to metals causing suppression of Aβ fibrillogenesis [148]. The chelator reagents should have high affinity and good selectivity for the target metal over the others. The concentration of metal supplementation should be lower than that of important metalloproteins, in order to avoid deleterious side effects.

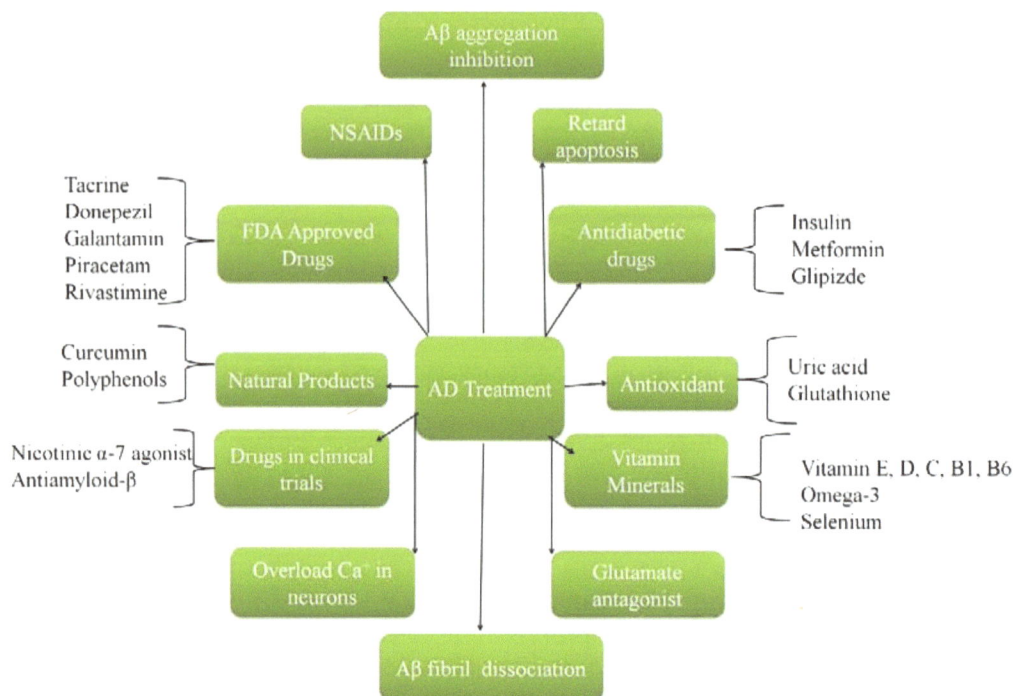

Fig. (2). Various methods, approaches and drug for treatment, prevention and target of Alzheimer's disease.

Prospective

In the case of supplementation of low metal levels, the metal forms and the metal pathways to reach the area of deficit in the target tissue are required. The concentration of the metals is also a key parameter. The presence of other species may reduce cell uptakes. For instance, zinc absorption is significantly impacted by the presence of dietary phytates. The species formed between metal and proteins such as Cu–Aβ catalyze the production of ROS [149].

Advanced technique for characterization AD is highly required [150]. New techniques may offer deeper insights into the structural, interactions, and connective role of metals in the brain [151]. The advanced techniques may solve the present contradicts in literature. For example, some studies have reported a deficiency of total copper brain levels in the AD brain [152]. Meta-analysis showed that the combined levels of plasma, and serum copper were higher in AD patients [153], while there is no difference in cerebral fluid compared to healthy subjects [154]. The rational explanation for this heterogeneity is that a significant quantity of copper precipitates with senile plaques in AD-affected regions, leading to copper deficiency in other regions. There is a common agreement that copper interacts with both Aβ and tau, and exacerbates their pathological consequences [155]. The measurements for bulk concentrations of metals are in most cases inaccurate and show "tip of the iceberg".

Effect of metals on other metal is important. Zinc deficiency has been reported to increase the overload of other metals, such as copper and nickel [68]. Although excessive zinc enhances Aβ toxicity, zinc changes the conformation of Aβ and prevents copper from interacting with Aβ, which ameliorates the oxidative stress burden [156, 157]. It showed strong copper detoxification [158]. Vašák *et al.* found that there is a metal swap between Zn (metallothionein 3, Zn7-MT-3), and Cu (Cu(II)–Aβ complex) [159]. Excessive Mn uptake causes iron deficiency in the Golgi apparatus, consistent with the fact that Mn and iron compete with the same binding sites and transport mechanisms [160]. A number of studies reported that the iron exporter Fpn could also function as a cellular exporter of Mn in a pH-dependent manner [161]. Under physiological conditions, Mg maintains intracellular Ca levels and protects neurons from excitatory responses induced Ca overload [162]. However, disturbances in Mg and Ca homeostasis alter a cascade of events which lead to a variety of diseases including diabetes, cancer, and neurodegeneration [163]. APP is a transmembrane protein which was found to catalytically oxidize Fe^{2+} to Fe^{3+} *via* ferroxidase activity and then interact with Fpn to promote iron export [164]; however, this process is inhibited by extracellular zinc, which originates from zinc-Aβ complexes [76].

Metals levels in the brain may be affected by other diseases. It was reported that copper non-bound ceruloplasmin levels in Type 2 diabetes (T2D) subjects with AD were higher compared to control samples [165]. The results indicated that free Cu^{2+} ions cause catalytic, and glucose dysregulation in oxidative stress reactions leading to tissue damage in AD and diabetes [165]. It is well known that the brain consumes about 18–30% of total body glucose. High level of glucose leads to the copper deficiency that is linked to AD [166]. Insulin resistance increased aggregation, tau phosphorylation, reduced degradation of Aβ, and mitochondrial dysfunction due to the malfunction of glucose, and lipid metabolism [146].

Understanding the function of metals in the brain is very difficult since the brain is very complicated making AD intricate disease. The interactions between the metals (hetero-bimetallic) are also complicated, and depend on several parameters such as metal ratios, metal: protein ratios, Aβ peptides ($Aβ_{1–40}$, $Aβ_{4–40}$, $Aβ_{11–40}$, *etc.*), *etc.* Other challenge, such as data heterogeneity, and proteome level should be addressed [167].

Based on current studies, it is likely that disruption of metal homeostasis in intracellular biometal homeostasis may be a hallmark of AD. Toxic metal exposure plays a contributory role in the pathology, and possibly the etiology of AD. However, a little is known about the metal role in AD. It does not appear to be one metal species that plays a major role in Alzheimer's diseases. Several points are required to understand the links between metals and AD. Animal models investigations are highly required with a particular focus on metal-proteins, metal-metal, and metal-enzymes interactions [168]. Advanced techniques may answer the question as to why specific interactions between the metal and brain proteins are more effective compared to the other interactions modes. The relation between AD and other diseases is required to understand the mechanism of drug interactions. The relationship to living conditions may also plays a role in this regard [169].

CONSENT FOR PUBLICATION

Not applicable.

CONFLICT OF INTEREST

The authors confirm that this chapter contents have no conflict of interest.

ACKNOWLEDGEMENTS

The author would thank the Ministry of Higher Education and Scientific Research (MHESR), and Institutional Review Board (IRB) of the Faculty of Science at Assiut University, Egypt for the support.

REFERENCES

[1]   Https://www.who.int/en/news-room/fact-sheets/detail/dementia

[2]   Nasica-Labouze J, Nguyen PH, Sterpone F, *et al.* Amyloid β Protein and Alzheimer's Disease: When Computer Simulations Complement Experimental Studies. Chem Rev 2015; 115(9): 3518-63.
[http://dx.doi.org/10.1021/cr500638n] [PMID: 25789869]

[3]   Association Alzheimer's. 2019 Alzheimer's Disease Facts and Figures. Alzheimers Dement 2019; 15(3): 321-87.
[http://dx.doi.org/10.1016/j.jalz.2019.01.010]

[4]   Ashby EL, Miners JS, Kehoe PG, Love S. Effects of Hypertension and Anti-Hypertensive Treatment on Amyloid-β (Aβ) Plaque Load and Aβ-Synthesizing and Aβ-Degrading Enzymes in Frontal Cortex. J Alzheimers Dis 2016; 50(4): 1191-203.
[http://dx.doi.org/10.3233/JAD-150831] [PMID: 26836178]

[5]   Savelieff MG, Nam G, Kang J, Lee HJ, Lee M, Lim MH. Development of Multifunctional Molecules as Potential Therapeutic Candidates for Alzheimer's Disease, Parkinson's Disease, and Amyotrophic Lateral Sclerosis in the Last Decade. Chem Rev 2019; 119(2): 1221-322.
[http://dx.doi.org/10.1021/acs.chemrev.8b00138] [PMID: 30095897]

[6]   Boyer J, Liu RH. Apple phytochemicals and their health benefits. Nutr J 2004; 3(1): 5.
[http://dx.doi.org/10.1186/1475-2891-3-5] [PMID: 15140261]

[7]　Kasana S, Din J, Maret W. Genetic causes and gene–nutrient interactions in mammalian zinc deficiencies: acrodermatitis enteropathica and transient neonatal zinc deficiency as examples. J Trace Elem Med Biol 2015; 29: 47-62.
[http://dx.doi.org/10.1016/j.jtemb.2014.10.003] [PMID: 25468189]

[8]　Lovejoy DB, Guillemin GJ. The potential for transition metal-mediated neurodegeneration in amyotrophic lateral sclerosis. Front Aging Neurosci 2014; 6: 173.
[http://dx.doi.org/10.3389/fnagi.2014.00173] [PMID: 25100994]

[9]　Bandmann O, Weiss KH, Kaler SG. Wilson's disease and other neurological copper disorders. Lancet Neurol 2015; 14(1): 103-13.
[http://dx.doi.org/10.1016/S1474-4422(14)70190-5] [PMID: 25496901]

[10]　Li Y, Jiao Q, Xu H, *et al.* Biometal Dyshomeostasis and Toxic Metal Accumulations in the Development of Alzheimer's Disease. Front Mol Neurosci 2017; 10: 339.
[http://dx.doi.org/10.3389/fnmol.2017.00339] [PMID: 29114205]

[11]　Wang X, Wang X, Guo Z. Metal-involved theranostics: An emerging strategy for fighting Alzheimer's disease. Coord Chem Rev 2018; 362: 72-84.
[http://dx.doi.org/10.1016/j.ccr.2018.03.010]

[12]　Barnham KJ, Bush AI. Biological metals and metal-targeting compounds in major neurodegenerative diseases. Chem Soc Rev 2014; 43(19): 6727-49.
[http://dx.doi.org/10.1039/C4CS00138A] [PMID: 25099276]

[13]　Cicero CE, Mostile G, Vasta R, *et al.* Metals and neurodegenerative diseases. A systematic review. Environ Res 2017; 159: 82-94.
[http://dx.doi.org/10.1016/j.envres.2017.07.048] [PMID: 28777965]

[14]　Huat TJ, Camats-Perna J, Newcombe EA, Valmas N, Kitazawa M, Medeiros R. Metal Toxicity Links to Alzheimer's Disease and Neuroinflammation. J Mol Biol 2019; 431(9): 1843-68.
[http://dx.doi.org/10.1016/j.jmb.2019.01.018] [PMID: 30664867]

[15]　Popugaeva E, Pchitskaya E, Bezprozvanny I. Dysregulation of neuronal calcium homeostasis in Alzheimer's disease - A therapeutic opportunity? Biochem Biophys Res Commun 2017; 483(4): 998-1004.
[http://dx.doi.org/10.1016/j.bbrc.2016.09.053] [PMID: 27641664]

[16]　Veronese N, Zurlo A, Solmi M, *et al.* Magnesium Status in Alzheimer's Disease: A Systematic Review. Am J Alzheimers Dis Other Demen 2016; 31(3): 208-13.
[http://dx.doi.org/10.1177/1533317515602674] [PMID: 26351088]

[17]　Cardoso BR, Roberts BR, Bush AI, Hare DJ. Selenium, selenoproteins and neurodegenerative diseases. Metallomics 2015; 7(8): 1213-28.
[http://dx.doi.org/10.1039/C5MT00075K] [PMID: 25996565]

[18]　Wang P, Wang Z-Y. Metal ions influx is a double edged sword for the pathogenesis of Alzheimer's disease. Ageing Res Rev 2017; 35: 265-90.
[http://dx.doi.org/10.1016/j.arr.2016.10.003] [PMID: 27829171]

[19]　Chauhan V, Chauhan A. Oxidative stress in Alzheimer's disease. Pathophysiology 2006; 13(3): 195-208.
[http://dx.doi.org/10.1016/j.pathophys.2006.05.004] [PMID: 16781128]

[20]　Hamley IW. The amyloid beta peptide: a chemist's perspective. Role in Alzheimer's and fibrillization. Chem Rev 2012; 112(10): 5147-92.
[http://dx.doi.org/10.1021/cr3000994] [PMID: 22813427]

[21]　Butterfield DA, Swomley AM, Sultana R. Amyloid β-peptide (1-42)-induced oxidative stress in Alzheimer disease: importance in disease pathogenesis and progression. Antioxid Redox Signal 2013; 19(8): 823-35.
[http://dx.doi.org/10.1089/ars.2012.5027] [PMID: 23249141]

[22] Bhattacharya S, Xu L, Thompson D. Revisiting the earliest signatures of amyloidogenesis: Roadmaps emerging from computational modeling and experiment. Wiley Interdiscip Rev Comput Mol Sci 2018; 8(4)e1359
[http://dx.doi.org/10.1002/wcms.1359]

[23] Komuro H, Kumada T. Ca2+ transients control CNS neuronal migration. Cell Calcium 2005; 37(5): 387-93.
[http://dx.doi.org/10.1016/j.ceca.2005.01.006] [PMID: 15820385]

[24] Ho M, Hoke DE, Chua YJ, *et al.* Effect of Metal Chelators on γ-Secretase Indicates That Calcium and Magnesium Ions Facilitate Cleavage of Alzheimer Amyloid Precursor Substrate. Int J Alzheimers Dis 2010; 2011950932
[PMID: 21253550]

[25] Kelly BL, Ferreira A. β-Amyloid-induced dynamin 1 degradation is mediated by N-methyl-D-aspartate receptors in hippocampal neurons. J Biol Chem 2006; 281(38): 28079-89.
[http://dx.doi.org/10.1074/jbc.M605081200] [PMID: 16864575]

[26] Chui D, Chen Z, Yu J, *et al.* 2011.Magnesium in Alzheimer's disease.

[27] Glick JL. Dementias: the role of magnesium deficiency and an hypothesis concerning the pathogenesis of Alzheimer's disease. Med Hypotheses 1990; 31(3): 211-25.
[http://dx.doi.org/10.1016/0306-9877(90)90095-V] [PMID: 2092675]

[28] Scheiber IF, Dringen R. Astrocyte functions in the copper homeostasis of the brain. Neurochem Int 2013; 62(5): 556-65.
[http://dx.doi.org/10.1016/j.neuint.2012.08.017] [PMID: 22982300]

[29] Desai V, Kaler SG. Role of copper in human neurological disorders. Am J Clin Nutr 2008; 88(3): 855S-8S.
[http://dx.doi.org/10.1093/ajcn/88.3.855S] [PMID: 18779308]

[30] Kuo Y-M, Gybina AA, Pyatskowit JW, Gitschier J, Prohaska JR. Copper transport protein (Ctr1) levels in mice are tissue specific and dependent on copper status. J Nutr 2006; 136(1): 21-6.
[http://dx.doi.org/10.1093/jn/136.1.21] [PMID: 16365053]

[31] Yu CH, Dolgova NV, Dmitriev OY. Dynamics of the metal binding domains and regulation of the human copper transporters ATP7B and ATP7A. IUBMB Life 2017; 69(4): 226-35.
[http://dx.doi.org/10.1002/iub.1611] [PMID: 28271598]

[32] Multhaup G, Schlicksupp A, Hesse L, *et al.* The Amyloid Precursor Protein of Alzheimer's Disease in the Reduction of Copper(II) to Copper(I). Science (80-) 1996; 271(5254): 1406-9.

[33] Zheng W, Monnot AD. Regulation of brain iron and copper homeostasis by brain barrier systems: implication in neurodegenerative diseases. Pharmacol Ther 2012; 133(2): 177-88.
[http://dx.doi.org/10.1016/j.pharmthera.2011.10.006] [PMID: 22115751]

[34] Harris ED. Copper homeostasis: the role of cellular transporters. Nutr Rev 2001; 59(9): 281-5.
[http://dx.doi.org/10.1111/j.1753-4887.2001.tb07017.x] [PMID: 11570430]

[35] Xu J, Church SJ, Patassini S, *et al.* Evidence for widespread, severe brain copper deficiency in Alzheimer's dementia. Metallomics 2017; 9(8): 1106-19.
[http://dx.doi.org/10.1039/C7MT00074J] [PMID: 28654115]

[36] Wang Z-X, Tan L, Wang H-F, *et al.* Serum Iron, Zinc, and Copper Levels in Patients with Alzheimer's Disease: A Replication Study and Meta-Analyses. J Alzheimers Dis 2015; 47(3): 565-81.
[http://dx.doi.org/10.3233/JAD-143108] [PMID: 26401693]

[37] Gybina AA, Tkac I, Prohaska JR. Copper deficiency alters the neurochemical profile of developing rat brain. Nutr Neurosci 2009; 12(3): 114-22.
[http://dx.doi.org/10.1179/147683009X423265] [PMID: 19356314]

[38] Fica-Contreras SM, Shuster SO, Durfee ND, *et al.* Glycation of Lys-16 and Arg-5 in amyloid-β and

the presence of Cu^{2+} play a major role in the oxidative stress mechanism of Alzheimer's disease. J Biol Inorg Chem 2017; 22(8): 1211-22.
[http://dx.doi.org/10.1007/s00775-017-1497-5] [PMID: 29038915]

[39]   Crouch PJ, Hung LW, Adlard PA, *et al.* Increasing Cu bioavailability inhibits Abeta oligomers and tau phosphorylation. Proc Natl Acad Sci USA 2009; 106(2): 381-6.
[http://dx.doi.org/10.1073/pnas.0809057106] [PMID: 19122148]

[40]   Jin L, Wu W-H, Li Q-Y, Zhao Y-F, Li Y-M. Copper inducing Aβ42 rather than Aβ40 nanoscale oligomer formation is the key process for Aβ neurotoxicity. Nanoscale 2011; 3(11): 4746-51.
[http://dx.doi.org/10.1039/c1nr11029b] [PMID: 21952557]

[41]   Mukherjee S, Ghosh C, Seal M, Dey SG. Copper induced spin state change of heme-Aβ associated with Alzheimer's disease. Dalton Trans 2017; 46(39): 13171-5.
[http://dx.doi.org/10.1039/C7DT01700F] [PMID: 28682389]

[42]   Zhou L-X, Du J-T, Zeng Z-Y, *et al.* Copper (II) modulates *in vitro* aggregation of a tau peptide. Peptides 2007; 28(11): 2229-34.
[http://dx.doi.org/10.1016/j.peptides.2007.08.022] [PMID: 17919778]

[43]   Matlack KES, Tardiff DF, Narayan P, *et al.* Clioquinol promotes the degradation of metal-dependent amyloid-β (Aβ) oligomers to restore endocytosis and ameliorate Aβ toxicity. Proc Natl Acad Sci USA 2014; 111(11): 4013-8.
[http://dx.doi.org/10.1073/pnas.1402228111] [PMID: 24591589]

[44]   Ouyang Q-Q, Zhao S, Li S-D, Song C. Application of Chitosan, Chitooligosaccharide, and Their Derivatives in the Treatment of Alzheimer's Disease. Mar Drugs 2017; 15(11): 322.
[http://dx.doi.org/10.3390/md15110322] [PMID: 29112116]

[45]   Adlard PA, Cherny RA, Finkelstein DI, *et al.* Rapid restoration of cognition in Alzheimer's transgenic mice with 8-hydroxy quinoline analogs is associated with decreased interstitial Abeta. Neuron 2008; 59(1): 43-55.
[http://dx.doi.org/10.1016/j.neuron.2008.06.018] [PMID: 18614028]

[46]   Hower V, Mendes P, Torti FM, *et al.* A general map of iron metabolism and tissue-specific subnetworks. Mol Biosyst 2009; 5(5): 422-43.
[http://dx.doi.org/10.1039/b816714c] [PMID: 19381358]

[47]   Hentze MW, Muckenthaler MU, Galy B, Camaschella C. Two to tango: regulation of Mammalian iron metabolism. Cell 2010; 142(1): 24-38.
[http://dx.doi.org/10.1016/j.cell.2010.06.028] [PMID: 20603012]

[48]   Jiang H, Wang J, Rogers J, Xie J. Brain Iron Metabolism Dysfunction in Parkinson's Disease. Mol Neurobiol 2017; 54(4): 3078-101.
[http://dx.doi.org/10.1007/s12035-016-9879-1] [PMID: 27039308]

[49]   Jiang D, Li X, Williams R, *et al.* Ternary complexes of iron, amyloid-β, and nitrilotriacetic acid: binding affinities, redox properties, and relevance to iron-induced oxidative stress in Alzheimer's disease. Biochemistry 2009; 48(33): 7939-47.
[http://dx.doi.org/10.1021/bi900907a] [PMID: 19601593]

[50]   Jolivet-Gougeon A, Bonnaure-Mallet M. Treponema, Iron and Neurodegeneration. Curr Alzheimer Res 2018; 15(8): 716-22.
[http://dx.doi.org/10.2174/1567205013666161122093404] [PMID: 27875949]

[51]   Tamagno E, Guglielmotto M, Aragno M, *et al.* Oxidative stress activates a positive feedback between the γ- and β-secretase cleavages of the β-amyloid precursor protein. J Neurochem 2007.071115163316002
[http://dx.doi.org/10.1111/j.1471-4159.2007.05072.x] [PMID: 18005001]

[52]   Rogers JT, Randall JD, Cahill CM, *et al.* An iron-responsive element type II in the 5′-untranslated region of the Alzheimer's amyloid precursor protein transcript. J Biol Chem 2002; 277(47): 45518-28.

[http://dx.doi.org/10.1074/jbc.M207435200] [PMID: 12198135]

[53]   Rogers JT, Venkataramani V, Washburn C, *et al.* A role for amyloid precursor protein translation to restore iron homeostasis and ameliorate lead (Pb) neurotoxicity. J Neurochem 2016; 138(3): 479-94.
[http://dx.doi.org/10.1111/jnc.13671] [PMID: 27206843]

[54]   Lane DJR, Ayton S, Bush AI. Iron and Alzheimer's Disease: An Update on Emerging Mechanisms. J Alzheimers Dis 2018; 64(s1): S379-95.
[http://dx.doi.org/10.3233/JAD-179944] [PMID: 29865061]

[55]   van Duijn S, Bulk M, van Duinen SG, *et al.* Cortical Iron Reflects Severity of Alzheimer's Disease. J Alzheimers Dis 2017; 60(4): 1533-45.
[http://dx.doi.org/10.3233/JAD-161143] [PMID: 29081415]

[56]   Lovell MA, Xiong S, Xie C, Davies P, Markesbery WR. Induction of hyperphosphorylated tau in primary rat cortical neuron cultures mediated by oxidative stress and glycogen synthase kinase-3. J Alzheimers Dis 2004; 6(6): 659-71.
[http://dx.doi.org/10.3233/JAD-2004-6610] [PMID: 15665406]

[57]   Sayre LM, Perry G, Harris PLR, Liu Y, Schubert KA, Smith MA. In situ oxidative catalysis by neurofibrillary tangles and senile plaques in Alzheimer's disease: a central role for bound transition metals. J Neurochem 2000; 74(1): 270-9.
[http://dx.doi.org/10.1046/j.1471-4159.2000.0740270.x] [PMID: 10617129]

[58]   Lei P, Ayton S, Finkelstein DI, *et al.* Tau deficiency induces parkinsonism with dementia by impairing APP-mediated iron export. Nat Med 2012; 18(2): 291-5.
[http://dx.doi.org/10.1038/nm.2613] [PMID: 22286308]

[59]   Lei P, Ayton S, Appukuttan AT, *et al.* Clioquinol rescues Parkinsonism and dementia phenotypes of the tau knockout mouse. Neurobiol Dis 2015; 81: 168-75.
[http://dx.doi.org/10.1016/j.nbd.2015.03.015] [PMID: 25796563]

[60]   Yang F, Lim GP, Begum AN, *et al.* Curcumin inhibits formation of amyloid β oligomers and fibrils, binds plaques, and reduces amyloid *in vivo*. J Biol Chem 2005; 280(7): 5892-901.
[http://dx.doi.org/10.1074/jbc.M404751200] [PMID: 15590663]

[61]   Lei P, Ayton S, Appukuttan AT, *et al.* Lithium suppression of tau induces brain iron accumulation and neurodegeneration. Mol Psychiatry 2017; 22(3): 396-406.
[http://dx.doi.org/10.1038/mp.2016.96] [PMID: 27400857]

[62]   Qiu Z, Strickland DK, Hyman BT, Rebeck GW. Alpha2-macroglobulin enhances the clearance of endogenous soluble beta-amyloid peptide *via* low-density lipoprotein receptor-related protein in cortical neurons. J Neurochem 1999; 73(4): 1393-8.
[http://dx.doi.org/10.1046/j.1471-4159.1999.0731393.x] [PMID: 10501182]

[63]   Takeda A. Movement of zinc and its functional significance in the brain. Brain Res Brain Res Rev 2000; 34(3): 137-48.
[http://dx.doi.org/10.1016/S0165-0173(00)00044-8] [PMID: 11113504]

[64]   Sensi SL, Paoletti P, Koh J-Y, Aizenman E, Bush AI, Hershfinkel M. The neurophysiology and pathology of brain zinc. J Neurosci 2011; 31(45): 16076-85.
[http://dx.doi.org/10.1523/JNEUROSCI.3454-11.2011] [PMID: 22072659]

[65]   Barr CA, Burdette SC. The zinc paradigm for metalloneurochemistry. Essays Biochem 2017; 61(2): 225-35.
[http://dx.doi.org/10.1042/EBC20160073] [PMID: 28487399]

[66]   Swardfager W, Herrmann N, McIntyre RS, *et al.* Potential roles of zinc in the pathophysiology and treatment of major depressive disorder. Neurosci Biobehav Rev 2013; 37(5): 911-29.
[http://dx.doi.org/10.1016/j.neubiorev.2013.03.018] [PMID: 23567517]

[67]   Khan MZ. A possible significant role of zinc and GPR39 zinc sensing receptor in Alzheimer disease and epilepsy. Biomed Pharmacother 2016; 79: 263-72.

[http://dx.doi.org/10.1016/j.biopha.2016.02.026] [PMID: 27044837]

[68]   Antala S, Dempski RE. The human ZIP4 transporter has two distinct binding affinities and mediates transport of multiple transition metals. Biochemistry 2012; 51(5): 963-73.
[http://dx.doi.org/10.1021/bi201553p] [PMID: 22242765]

[69]   Frazzini V, Rockabrand E, Mocchegiani E, Sensi SL. Oxidative stress and brain aging: is zinc the link? Biogerontology 2006; 7(5-6): 307-14.
[http://dx.doi.org/10.1007/s10522-006-9045-7] [PMID: 17028932]

[70]   Jomova K, Vondrakova D, Lawson M, Valko M. Metals, oxidative stress and neurodegenerative disorders. Mol Cell Biochem 2010; 345(1-2): 91-104.
[http://dx.doi.org/10.1007/s11010-010-0563-x] [PMID: 20730621]

[71]   Religa D, Strozyk D, Cherny RA, *et al.* Elevated cortical zinc in Alzheimer disease. Neurology 2006; 67(1): 69-75.
[http://dx.doi.org/10.1212/01.wnl.0000223644.08653.b5] [PMID: 16832080]

[72]   Hozumi I, Hasegawa T, Honda A, *et al.* Patterns of levels of biological metals in CSF differ among neurodegenerative diseases. J Neurol Sci 2011; 303(1-2): 95-9.
[http://dx.doi.org/10.1016/j.jns.2011.01.003] [PMID: 21292280]

[73]   Panayi AE, Spyrou NM, Iversen BS, White MA, Part P. Determination of cadmium and zinc in Alzheimer's brain tissue using inductively coupled plasma mass spectrometry. J Neurol Sci 2002; 195(1): 1-10.
[http://dx.doi.org/10.1016/S0022-510X(01)00672-4] [PMID: 11867068]

[74]   Ventriglia M, Brewer GJ, Simonelli I, *et al.* Zinc in Alzheimer's Disease: A Meta-Analysis of Serum, Plasma, and Cerebrospinal Fluid Studies. J Alzheimers Dis 2015; 46(1): 75-87.
[http://dx.doi.org/10.3233/JAD-141296] [PMID: 25697706]

[75]   Yoshiike Y, Tanemura K, Murayama O, *et al.* New insights on how metals disrupt amyloid β-aggregation and their effects on amyloid-β cytotoxicity. J Biol Chem 2001; 276(34): 32293-9.
[http://dx.doi.org/10.1074/jbc.M010706200] [PMID: 11423547]

[76]   Duce JA, Tsatsanis A, Cater MA, *et al.* Iron-export ferroxidase activity of β-amyloid precursor protein is inhibited by zinc in Alzheimer's disease. Cell 2010; 142(6): 857-67.
[http://dx.doi.org/10.1016/j.cell.2010.08.014] [PMID: 20817278]

[77]   Crouch PJ, Tew DJ, Du T, *et al.* Restored degradation of the Alzheimer's amyloid-β peptide by targeting amyloid formation. J Neurochem 2009; 108(5): 1198-207.
[http://dx.doi.org/10.1111/j.1471-4159.2009.05870.x] [PMID: 19141082]

[78]   Pei J-J, An W-L, Zhou X-W, *et al.* P70 S6 kinase mediates tau phosphorylation and synthesis. FEBS Lett 2006; 580(1): 107-14.
[http://dx.doi.org/10.1016/j.febslet.2005.11.059] [PMID: 16364302]

[79]   Lei P, Ayton S, Bush AI, Adlard PA. GSK-3 in Neurodegenerative Diseases. Int J Alzheimers Dis 2011; 2011189246
[http://dx.doi.org/10.4061/2011/189246] [PMID: 21629738]

[80]   Sun X-Y, Wei Y-P, Xiong Y, *et al.* Synaptic released zinc promotes tau hyperphosphorylation by inhibition of protein phosphatase 2A (PP2A). J Biol Chem 2012; 287(14): 11174-82.
[http://dx.doi.org/10.1074/jbc.M111.309070] [PMID: 22334661]

[81]   Deshpande A, Kawai H, Metherate R, Glabe CG, Busciglio J. A role for synaptic zinc in activity-dependent Abeta oligomer formation and accumulation at excitatory synapses. J Neurosci 2009; 29(13): 4004-15.
[http://dx.doi.org/10.1523/JNEUROSCI.5980-08.2009] [PMID: 19339596]

[82]   Hoke DE, Tan J-L, Ilaya NT, *et al. In vitro* gamma-secretase cleavage of the Alzheimer's amyloid precursor protein correlates to a subset of presenilin complexes and is inhibited by zinc. FEBS J 2005; 272(21): 5544-57.

[http://dx.doi.org/10.1111/j.1742-4658.2005.04950.x] [PMID: 16262694]

[83]   Sun D, Zhang W, Yu Q, *et al.* Chiral penicillamine-modified selenium nanoparticles enantioselectively inhibit metal-induced amyloid β aggregation for treating Alzheimer's disease. J Colloid Interface Sci 2017; 505: 1001-10.
[http://dx.doi.org/10.1016/j.jcis.2017.06.083] [PMID: 28693096]

[84]   Prakash A, Dhaliwal GK, Kumar P, Majeed ABA. Brain biometals and Alzheimer's disease - boon or bane? Int J Neurosci 2017; 127(2): 99-108.
[http://dx.doi.org/10.3109/00207454.2016.1174118] [PMID: 27044501]

[85]   Roos PM. Metals and Motor Neuron Disease.Biometals in Neurodegenerative Diseases. Elsevier 2017; pp. 175-93.
[http://dx.doi.org/10.1016/B978-0-12-804562-6.00010-5]

[86]   Park RM. Neurobehavioral deficits and parkinsonism in occupations with manganese exposure: a review of methodological issues in the epidemiological literature. Saf Health Work 2013; 4(3): 123-35.
[http://dx.doi.org/10.1016/j.shaw.2013.07.003] [PMID: 24106642]

[87]   Wallin C, Kulkarni YS, Abelein A, *et al.* Characterization of Mn(II) ion binding to the amyloid-β peptide in Alzheimer's disease. J Trace Elem Med Biol 2016; 38: 183-93.
[http://dx.doi.org/10.1016/j.jtemb.2016.03.009] [PMID: 27085215]

[88]   Tuschl K, Clayton PT, Gospe SM Jr, *et al.* Syndrome of Hepatic Cirrhosis, Dystonia, Polycythemia, and Hypermanganesemia Caused by Mutations in SLC30A10, a Manganese Transporter in Man. Am J Hum Genet 2016; 99(2): 521.
[http://dx.doi.org/10.1016/j.ajhg.2016.07.015] [PMID: 27486784]

[89]   Tong Y, Yang H, Tian X, *et al.* High manganese, a risk for Alzheimer's disease: high manganese induces amyloid-β related cognitive impairment. J Alzheimers Dis 2014; 42(3): 865-78.
[http://dx.doi.org/10.3233/JAD-140534] [PMID: 24961945]

[90]   Guilarte TR. Manganese neurotoxicity: new perspectives from behavioral, neuroimaging, and neuropathological studies in humans and non-human primates. Front Aging Neurosci 2013; 5: 23.
[http://dx.doi.org/10.3389/fnagi.2013.00023] [PMID: 23805100]

[91]   Martinez-Finley EJ, Gavin CE, Aschner M, Gunter TE. Manganese neurotoxicity and the role of reactive oxygen species. Free Radic Biol Med 2013; 62: 65-75.
[http://dx.doi.org/10.1016/j.freeradbiomed.2013.01.032] [PMID: 23395780]

[92]   Loef M, Schrauzer GN, Walach H. Selenium and Alzheimer's disease: a systematic review. J Alzheimers Dis 2011; 26(1): 81-104.
[http://dx.doi.org/10.3233/JAD-2011-110414] [PMID: 21593562]

[93]   Solovyev N, Drobyshev E, Bjørklund G, Dubrovskii Y, Lysiuk R, Rayman MP. Selenium, selenoprotein P, and Alzheimer's disease: is there a link? Free Radic Biol Med 2018; 127: 124-33.
[http://dx.doi.org/10.1016/j.freeradbiomed.2018.02.030] [PMID: 29481840]

[94]   Lee HJ, Park MK, Seo YR. Pathogenic Mechanisms of Heavy Metal Induced-Alzheimer's Disease. Toxicol Environ Health Sci 2018; 10(1): 1-10.
[http://dx.doi.org/10.1007/s13530-018-0340-x]

[95]   Sato N, Hori O, Yamaguchi A, *et al.* A novel presenilin-2 splice variant in human Alzheimer's disease brain tissue. J Neurochem 1999; 72(6): 2498-505.
[http://dx.doi.org/10.1046/j.1471-4159.1999.0722498.x] [PMID: 10349860]

[96]   Liang RF, Li WQ, Wang H, Wang JX, Niu Q. Impact of sub-chronic aluminium-maltolate exposure on catabolism of amyloid precursor protein in rats. Biomed Environ Sci 2013; 26(6): 445-52.
[PMID: 23816578]

[97]   Cho H-H, Cahill CM, Vanderburg CR, *et al.* Selective translational control of the Alzheimer amyloid precursor protein transcript by iron regulatory protein-1. J Biol Chem 2010; 285(41): 31217-32.

[http://dx.doi.org/10.1074/jbc.M110.149161] [PMID: 20558735]

[98] Yamanaka K, Minato N, Iwai K. Stabilization of iron regulatory protein 2, IRP2, by aluminum. FEBS Lett 1999; 462(1-2): 216-20.
[http://dx.doi.org/10.1016/S0014-5793(99)01533-1] [PMID: 10580122]

[99] Fattoretti P, Bertoni-Freddari C, Balietti M, *et al.* The effect of chronic aluminum(III) administration on the nervous system of aged rats: clues to understand its suggested role in Alzheimer's disease. J Alzheimers Dis 2003; 5(6): 437-44.
[http://dx.doi.org/10.3233/JAD-2003-5603] [PMID: 14757933]

[100] Sánchez-Iglesias S, Méndez-Alvarez E, Iglesias-González J, *et al.* Brain oxidative stress and selective behaviour of aluminium in specific areas of rat brain: potential effects in a 6-OHDA-induced model of Parkinson's disease. J Neurochem 2009; 109(3): 879-88.
[http://dx.doi.org/10.1111/j.1471-4159.2009.06019.x] [PMID: 19425176]

[101] Zielke HR, Jackson MJ, Tildon JT, Max SR. A glutamatergic mechanism for aluminum toxicity in astrocytes. Mol Chem Neuropathol 1993; 19(3): 219-33.
[http://dx.doi.org/10.1007/BF03160001] [PMID: 8104402]

[102] Yellamma K, Saraswathamma S, Kumari BN. Cholinergic system under aluminium toxicity in rat brain. Toxicol Int 2010; 17(2): 106-12.
[http://dx.doi.org/10.4103/0971-6580.72682] [PMID: 21170257]

[103] Abu-Taweel GM, Ajarem JS, Ahmad M. Neurobehavioral toxic effects of perinatal oral exposure to aluminum on the developmental motor reflexes, learning, memory and brain neurotransmitters of mice offspring. Pharmacol Biochem Behav 2012; 101(1): 49-56.
[http://dx.doi.org/10.1016/j.pbb.2011.11.003] [PMID: 22115621]

[104] Liu YQ, Xin TR, Liang JJ, Wang WM, Zhang YY. Memory performance, brain excitatory amino acid and acetylcholinesterase activity of chronically aluminum exposed mice in response to soy isoflavones treatment. Phytother Res 2010; 24(10): 1451-6.
[http://dx.doi.org/10.1002/ptr.3120] [PMID: 20878693]

[105] Weidenhamer JD, Fitzpatrick MP, Biro AM, *et al.* Metal exposures from aluminum cookware: An unrecognized public health risk in developing countries. Sci Total Environ 2017; 579: 805-13.
[http://dx.doi.org/10.1016/j.scitotenv.2016.11.023] [PMID: 27866735]

[106] Yokel RA. Aluminum chelation: chemistry, clinical, and experimental studies and the search for alternatives to desferrioxamine. J Toxicol Environ Health 1994; 41(2): 131-74.
[http://dx.doi.org/10.1080/15287399409531834] [PMID: 8301696]

[107] Abdelhamid HN, El-Bery HM, Metwally AA, Elshazly M, Hathout RM. Synthesis of CdS-modified chitosan quantum dots for the drug delivery of Sesamol. Carbohydr Polym 2019; 214: 90-9.
[http://dx.doi.org/10.1016/j.carbpol.2019.03.024] [PMID: 30926012]

[108] Abdelhamid HN, Wu H-F. Nanoparticles Advanced Drug Delivery for Cancer Cells.Nanoparticulate Drug Delivery Systems. 2019; pp. 121-44.
[http://dx.doi.org/10.1201/9781351137263-4]

[109] Abdelhamid HN, Wu H-F. Synthesis and multifunctional applications of quantum nanobeads for label-free and selective metal chemosensing RSC Adv 5(62)2015;

[110] Abdelhamid HN, Wu H-F. Monitoring metallofulfenamic–bovine serum albumin interactions: a novel method for metallodrug analysis. RSC Advances 2014; 4(96): 53768-76.
[http://dx.doi.org/10.1039/C4RA07638A]

[111] Chen Z-Y, Abdelhamid HN, Wu H-F. Effect of surface capping of quantum dots (CdTe) on proteomics. Rapid Commun Mass Spectrom 2016; 30(12): 1403-12.
[http://dx.doi.org/10.1002/rcm.7575] [PMID: 27197033]

[112] Wu H-F, Gopal J, Abdelhamid HN, Hasan N. Quantum dot applications endowing novelty to analytical proteomics. Proteomics 2012; 12(19-20): 2949-61.

[http://dx.doi.org/10.1002/pmic.201200295] [PMID: 22930415]

[113] Qadir S, Jamshieed S, Rasool S, Ashraf M, Akram NA, Ahmad P. Modulation of Plant Growth and Metabolism in Cadmium-Enriched Environments. 2014; pp. 51-88.
[http://dx.doi.org/10.1007/978-3-319-03777-6_4]

[114] Richter P, Faroon O, Pappas RS. Cadmium and Cadmium/Zinc Ratios and Tobacco-Related Morbidities. Int J Environ Res Public Health 2017; 14(10): 1154.
[http://dx.doi.org/10.3390/ijerph14101154] [PMID: 28961214]

[115] Satarug S. Dietary Cadmium Intake and Its Effects on Kidneys. Toxics 2018; 6(1): 15.
[http://dx.doi.org/10.3390/toxics6010015] [PMID: 29534455]

[116] Fransson MN, Barregard L, Sallsten G, Akerstrom M, Johanson G. Physiologically-based toxicokinetic model for cadmium using Markov-chain Monte Carlo analysis of concentrations in blood, urine, and kidney cortex from living kidney donors. Toxicol Sci 2014; 141(2): 365-76.
[http://dx.doi.org/10.1093/toxsci/kfu129] [PMID: 25015660]

[117] Suwazono Y, Kido T, Nakagawa H, *et al.* Biological half-life of cadmium in the urine of inhabitants after cessation of cadmium exposure. Biomarkers 2009; 14(2): 77-81.
[http://dx.doi.org/10.1080/13547500902730698] [PMID: 19330585]

[118] Peng Q, Bakulski KM, Nan B, Park SK. Cadmium and Alzheimer's disease mortality in U.S. adults: Updated evidence with a urinary biomarker and extended follow-up time. Environ Res 2017; 157: 44-51.
[http://dx.doi.org/10.1016/j.envres.2017.05.011] [PMID: 28511080]

[119] Syme CD, Viles JH. Solution 1H NMR investigation of Zn2+ and Cd2+ binding to amyloid-beta peptide (Abeta) of Alzheimer's disease. Biochim Biophys Acta 2006; 1764(2): 246-56.
[http://dx.doi.org/10.1016/j.bbapap.2005.09.012] [PMID: 16266835]

[120] Del Pino J, Zeballos G, Anadón MJ, *et al.* Cadmium-induced cell death of basal forebrain cholinergic neurons mediated by muscarinic M1 receptor blockade, increase in GSK-3β enzyme, β-amyloid and tau protein levels. Arch Toxicol 2016; 90(5): 1081-92.
[http://dx.doi.org/10.1007/s00204-015-1540-7] [PMID: 26026611]

[121] Endres K, Fahrenholz F. The Role of the anti-amyloidogenic secretase ADAM10 in shedding the APP-like proteins. Curr Alzheimer Res 2012; 9(2): 157-64.
[http://dx.doi.org/10.2174/156720512799361664] [PMID: 21605036]

[122] Yuan Y, Zhang Y, Zhao S, *et al.* Cadmium-induced apoptosis in neuronal cells is mediated by Fas/FasL-mediated mitochondrial apoptotic signaling pathway. Sci Rep 2018; 8(1): 8837.
[http://dx.doi.org/10.1038/s41598-018-27106-9] [PMID: 29891925]

[123] Chen L, Liu L, Huang S. Cadmium activates the mitogen-activated protein kinase (MAPK) pathway *via* induction of reactive oxygen species and inhibition of protein phosphatases 2A and 5. Free Radic Biol Med 2008; 45(7): 1035-44.
[http://dx.doi.org/10.1016/j.freeradbiomed.2008.07.011] [PMID: 18703135]

[124] Xu B, Chen S, Luo Y, *et al.* Calcium signaling is involved in cadmium-induced neuronal apoptosis *via* induction of reactive oxygen species and activation of MAPK/mTOR network. PLoS One 2011; 6(4)e19052
[http://dx.doi.org/10.1371/journal.pone.0019052] [PMID: 21544200]

[125] Jiang JH, Ge G, Gao K, *et al.* Calcium Signaling Involvement in Cadmium-Induced Astrocyte Cytotoxicity and Cell Death Through Activation of MAPK and PI3K/Akt Signaling Pathways. Neurochem Res 2015; 40(9): 1929-44.
[http://dx.doi.org/10.1007/s11064-015-1686-y] [PMID: 26248512]

[126] Abdelhamid HN, Wu HF. Thymine chitosan nanomagnets for specific preconcentration of mercury (II) prior to analysis using SELDI-MS Microchim Acta 2017.

[127] Abdelhamid HN, Wu H-F. Ultrasensitive, rapid, and selective detection of mercury using graphene

assisted laser desorption/ionization mass spectrometry. J Am Soc Mass Spectrom 2014; 25(5): 861-8.
[http://dx.doi.org/10.1007/s13361-014-0825-z] [PMID: 24590364]

[128]   Olivieri G, Brack C, Müller-Spahn F, *et al.* Mercury induces cell cytotoxicity and oxidative stress and increases β-amyloid secretion and tau phosphorylation in SHSY5Y neuroblastoma cells. J Neurochem 2000; 74(1): 231-6.
[http://dx.doi.org/10.1046/j.1471-4159.2000.0740231.x] [PMID: 10617124]

[129]   Ganther HE. Interactions of vitamin e and selenium with mercury and silver Ann N Y Acad Sci 1980; 355(1 Micronutrient): 212-226.

[130]   Mutter J, Curth A, Naumann J, Deth R, Walach H. Does inorganic mercury play a role in Alzheimer's disease? A systematic review and an integrated molecular mechanism. J Alzheimers Dis 2010; 22(2): 357-74.
[http://dx.doi.org/10.3233/JAD-2010-100705] [PMID: 20847438]

[131]   Needleman H. Low level lead exposure: history and discovery. Ann Epidemiol 2009; 19(4): 235-8.
[http://dx.doi.org/10.1016/j.annepidem.2009.01.022] [PMID: 19344860]

[132]   Stansfield KH, Ruby KN, Soares BD, McGlothan JL, Liu X, Guilarte TR. Early-life lead exposure recapitulates the selective loss of parvalbumin-positive GABAergic interneurons and subcortical dopamine system hyperactivity present in schizophrenia. Transl Psychiatry 2015; 5(3): e522-2.
[http://dx.doi.org/10.1038/tp.2014.147] [PMID: 25756805]

[133]   Zhou C-C, Gao Z-Y, Wang J, *et al.* Lead exposure induces Alzheimers's disease (AD)-like pathology and disturbes cholesterol metabolism in the young rat brain. Toxicol Lett 2018; 296: 173-83.
[http://dx.doi.org/10.1016/j.toxlet.2018.06.1065] [PMID: 29908845]

[134]   Liu M-C, Liu X-Q, Wang W, *et al.* Involvement of microglia activation in the lead induced long-term potentiation impairment. PLoS One 2012; 7(8)e43924
[http://dx.doi.org/10.1371/journal.pone.0043924] [PMID: 22952811]

[135]   Liu J-T, Chen B-Y, Zhang J-Q, Kuang F, Chen L-W. Lead exposure induced microgliosis and astrogliosis in hippocampus of young mice potentially by triggering TLR4-MyD88-NFκB signaling cascades. Toxicol Lett 2015; 239(2): 97-107.
[http://dx.doi.org/10.1016/j.toxlet.2015.09.015] [PMID: 26386401]

[136]   Kumawat KL, Kaushik DK, Goswami P, Basu A. Acute exposure to lead acetate activates microglia and induces subsequent bystander neuronal death *via* caspase-3 activation. Neurotoxicology 2014; 41: 143-53.
[http://dx.doi.org/10.1016/j.neuro.2014.02.002] [PMID: 24530660]

[137]   Pardeshi R, Bolshette N, Gadhave K, *et al.* Insulin signaling: An opportunistic target to minify the risk of Alzheimer's disease. Psychoneuroendocrinology 2017; 83: 159-71.
[http://dx.doi.org/10.1016/j.psyneuen.2017.05.004] [PMID: 28624654]

[138]   Sameem B, Saeedi M, Mahdavi M, Shafiee A. A review on tacrine-based scaffolds as multi-target drugs (MTDLs) for Alzheimer's disease. Eur J Med Chem 2017; 128: 332-45.
[http://dx.doi.org/10.1016/j.ejmech.2016.10.060] [PMID: 27876467]

[139]   Abdelhamid HN, Wu H-F. Soft Ionization of Metallo-Mefenamic Using Electrospray Ionization Mass Spectrometry. Mass Spectrom Lett 2015; 6(2): 43-7.
[http://dx.doi.org/10.5478/MSL.2015.6.2.43]

[140]   Abdelhamid HN, Wu HF. Furoic and mefenamic acids as new matrices for matrix assisted laser desorption/ionization-(MALDI)-mass spectrometry. Talanta 2013; 115: 442-50.
[http://dx.doi.org/10.1016/j.talanta.2013.05.050] [PMID: 24054616]

[141]   Abdelhamid HN, Wu H-F. A method to detect metal-drug complexes and their interactions with pathogenic bacteria *via* graphene nanosheet assist laser desorption/ionization mass spectrometry and biosensors. Anal Chim Acta 2012; 751: 94-104.
[http://dx.doi.org/10.1016/j.aca.2012.09.012] [PMID: 23084057]

[142] Han Q, Cai S, Yang L, *et al.* Molybdenum Disulfide Nanoparticles as Multifunctional Inhibitors against Alzheimer's Disease. ACS Appl Mater Interfaces 2017; 9(25): 21116-23.
[http://dx.doi.org/10.1021/acsami.7b03816] [PMID: 28613069]

[143] Leszek J, Md Ashraf G, Tse WH, *et al.* Nanotechnology for Alzheimer Disease. Curr Alzheimer Res 2017; 14(11): 1182-9.
[http://dx.doi.org/10.2174/1567205014666170203125008] [PMID: 28164767]

[144] Helmuth L. NEUROSCIENCE: An Antibiotic to Treat Alzheimer's? Science (80-) 2000; 290(5495): 1273a-4..

[145] Chan CY, Noor A, McLean CA, Donnelly PS, Barnard PJ. Rhenium(i) complexes of N-heterocyclic carbene ligands that bind to amyloid plaques of Alzheimer's disease. Chem Commun (Camb) 2017; 53(15): 2311-4.
[http://dx.doi.org/10.1039/C6CC10066J] [PMID: 28111677]

[146] Pardeshi R, Bolshette N, Gadhave K, *et al.* Insulin signaling: An opportunistic target to minify the risk of Alzheimer's disease. Psychoneuroendocrinology 2017; 83: 159-71.
[http://dx.doi.org/10.1016/j.psyneuen.2017.05.004] [PMID: 28624654]

[147] Thapa A, Carroll NJ. Dietary Modulation of Oxidative Stress in Alzheimer's Disease. Int J Mol Sci 2017; 18(7): 1583.
[http://dx.doi.org/10.3390/ijms18071583] [PMID: 28753984]

[148] Monacelli F, Acquarone E, Giannotti C, Borghi R, Nencioni A. Vitamin C, Aging and Alzheimer's Disease. Nutrients 2017; 9(7): 670.
[http://dx.doi.org/10.3390/nu9070670] [PMID: 28654021]

[149] Cheignon C, Jones M, Atrián-Blasco E, *et al.* Identification of key structural features of the elusive Cu-Aβ complex that generates ROS in Alzheimer's disease. Chem Sci (Camb) 2017; 8(7): 5107-18.
[http://dx.doi.org/10.1039/C7SC00809K] [PMID: 28970897]

[150] Abdelhamid HN, Wu H-F. Biological Mass Spectrometry for Diagnosis of Alzheimer's Disease.Frontiers in Clinical Drug Research - Alzheimer Disorders. BENTHAM SCIENCE PUBLISHERS 2017; pp. 110-26.
[http://dx.doi.org/10.2174/9781681083391117060007]

[151] Everett J, Collingwood JF, Tjendana-Tjhin V, *et al.* Nanoscale synchrotron X-ray speciation of iron and calcium compounds in amyloid plaque cores from Alzheimer's disease subjects. Nanoscale 2018; 10(25): 11782-96.
[http://dx.doi.org/10.1039/C7NR06794A] [PMID: 29688240]

[152] Deibel MA, Ehmann WD, Markesbery WR. Copper, iron, and zinc imbalances in severely degenerated brain regions in Alzheimer's disease: possible relation to oxidative stress. J Neurol Sci 1996; 143(1-2): 137-42.
[http://dx.doi.org/10.1016/S0022-510X(96)00203-1] [PMID: 8981312]

[153] Klevay LM. Alzheimer's disease as copper deficiency. Med Hypotheses 2008; 70(4): 802-7.
[http://dx.doi.org/10.1016/j.mehy.2007.04.051] [PMID: 17928161]

[154] Vaz FNC, Fermino BL, Haskel MVL, *et al.* The Relationship Between Copper, Iron, and Selenium Levels and Alzheimer Disease. Biol Trace Elem Res 2018; 181(2): 185-91.
[http://dx.doi.org/10.1007/s12011-017-1042-y] [PMID: 28500578]

[155] Kitazawa M, Cheng D, Laferla FM. Chronic copper exposure exacerbates both amyloid and tau pathology and selectively dysregulates cdk5 in a mouse model of AD. J Neurochem 2009; 108(6): 1550-60.
[http://dx.doi.org/10.1111/j.1471-4159.2009.05901.x] [PMID: 19183260]

[156] Cuajungco MP, Goldstein LE, Nunomura A, *et al.* Evidence that the β-amyloid plaques of Alzheimer's disease represent the redox-silencing and entombment of abeta by zinc. J Biol Chem 2000; 275(26): 19439-42.

[http://dx.doi.org/10.1074/jbc.C000165200] [PMID: 10801774]

[157]   Pedersen JT, Hureau C, Hemmingsen L, *et al.* Rapid exchange of metal between Zn(7)-metallothionein-3 and amyloid-β peptide promotes amyloid-related structural changes. Biochemistry 2012; 51(8): 1697-706.
[http://dx.doi.org/10.1021/bi201774z] [PMID: 22283439]

[158]   Atrián-Blasco E, Conte-Daban A, Hureau C. Mutual interference of Cu and Zn ions in Alzheimer's disease: perspectives at the molecular level. Dalton Trans 2017; 46(38): 12750-9.
[http://dx.doi.org/10.1039/C7DT01344B] [PMID: 28937157]

[159]   Meloni G, Sonois V, Delaine T, *et al.* Metal swap between Zn7-metallothionein-3 and amyloid-β-Cu protects against amyloid-β toxicity. Nat Chem Biol 2008; 4(6): 366-72.
[http://dx.doi.org/10.1038/nchembio.89] [PMID: 18454142]

[160]   Carmona A, Devès G, Roudeau S, Cloetens P, Bohic S, Ortega R. Manganese accumulates within golgi apparatus in dopaminergic cells as revealed by synchrotron X-ray fluorescence nanoimaging. ACS Chem Neurosci 2010; 1(3): 194-203.
[http://dx.doi.org/10.1021/cn900021z] [PMID: 22778823]

[161]   Seo YA, Wessling-Resnick M. Ferroportin deficiency impairs manganese metabolism in flatiron mice. FASEB J 2015; 29(7): 2726-33.
[http://dx.doi.org/10.1096/fj.14-262592] [PMID: 25782988]

[162]   Levitsky DO, Takahashi M. Interplay of Ca2+ and Mg2+ in Sodium-Calcium Exchanger and in Other Ca2+-Binding Proteins: Magnesium, Watchdog That Blocks Each Turn if Able. 2013; p. 65-78.

[163]   Volpe SL. Magnesium in disease prevention and overall health. Adv Nutr 2013; 4(3): 378S-83S.
[http://dx.doi.org/10.3945/an.112.003483] [PMID: 23674807]

[164]   McCarthy RC, Park Y-H, Kosman DJ. sAPP modulates iron efflux from brain microvascular endothelial cells by stabilizing the ferrous iron exporter ferroportin. EMBO Rep 2014; 15(7): 809-15.
[http://dx.doi.org/10.15252/embr.201338064] [PMID: 24867889]

[165]   Squitti R, Mendez AJ, Simonelli I, Ricordi C. Diabetes and Alzheimer's Disease: Can Elevated Free Copper Predict the Risk of the Disease? J Alzheimers Dis 2017; 56(3): 1055-64.
[http://dx.doi.org/10.3233/JAD-161033] [PMID: 27983558]

[166]   Xu J, Begley P, Church SJ, *et al.* Elevation of brain glucose and polyol-pathway intermediates with accompanying brain-copper deficiency in patients with Alzheimer's disease: metabolic basis for dementia. Sci Rep 2016; 6(1): 27524.
[http://dx.doi.org/10.1038/srep27524] [PMID: 27276998]

[167]   Kepp KP. Ten Challenges of the Amyloid Hypothesis of Alzheimer's Disease. J Alzheimers Dis 2017; 55(2): 447-57.
[http://dx.doi.org/10.3233/JAD-160550] [PMID: 27662304]

[168]   Mercer SW, Wang J, Burke R. *In Vivo* Modeling of the Pathogenic Effect of Copper Transporter Mutations That Cause Menkes and Wilson Diseases, Motor Neuropathy, and Susceptibility to Alzheimer's Disease. J Biol Chem 2017; 292(10): 4113-22.
[http://dx.doi.org/10.1074/jbc.M116.756163] [PMID: 28119449]

[169]   Chen H, Kwong JC, Copes R, *et al.* Living near major roads and the incidence of dementia, Parkinson's disease, and multiple sclerosis: a population-based cohort study. Lancet 2017; 389(10070): 718-26.
[http://dx.doi.org/10.1016/S0140-6736(16)32399-6] [PMID: 28063597]

SUBJECT INDEX

Atta-ur-Rahman (Ed.)
All rights reserved-© 2020 Bentham Science Publishers

S

www.ingramcontent.com/pod-product-compliance
Lightning Source LLC
Chambersburg PA
CBHW050823220326
41598CB00006B/299